Lyapunov Functionals and Stability of Stochastic Functional Differential Equations

Leonid Shaikhet

Lyapunov Functionals and Stability of Stochastic Functional Differential Equations

 Springer

Leonid Shaikhet
Department of Higher Mathematics
Donetsk State University of Management
Donetsk, Ukraine

ISBN 978-3-319-00100-5   ISBN 978-3-319-00101-2 (eBook)
DOI 10.1007/978-3-319-00101-2
Springer Cham Heidelberg New York Dordrecht London

Library of Congress Control Number: 2013935832

Printed on acid-free paper

Springer is part of Springer Science+Business Media (www.springer.com)

Preface

This book deals with stability of stochastic functional differential equations and continues and complements the previous book of the author, "Lyapunov Functionals and Stability of Stochastic Difference Equations" [278].

Functional differential equations (also called hereditary systems [9], or systems with aftereffect [140], or equations with memory [11], or equations with deviating arguments [66, 167], or equations with delays [10, 24–29, 54, 93, 97, 226], or equations with time lag [65, 72, 103], or retarded differential equations [52, 78], or differential difference equations [22]) are infinite-dimensional ones [88, 181], contrary to ordinary differential equations, and describe the processes whose behavior depends not only on their present state but also on their past history [104–106, 131–133]. Systems of such type are widely used to model processes in physics, mechanics, automatic regulation, economy, finance, biology, ecology, sociology, medicine, etc. (see, e.g., [15, 18, 20, 24–26, 32, 36–41, 57, 58, 71, 91, 92, 98, 160, 161, 165, 177, 186, 189, 192, 199, 212, 213, 218, 219, 224, 245, 253, 257, 286, 287, 293, 303]).

The first mathematical models with functional differential equations have been studied during the 18th century (L. Euler, J. Bernoulli, M.J. Condorcet, J. Lagrange, P. Laplace). In the beginning of 20th century the development of the delay systems study started from the works by Vito Volterra [302], linked to viscoelasticity and ecology. This pioneer work was continued by Tsypkin [298–300], Myshkis [220, 221], Kac and Krasovskii [120, 156–158], Elsgoltz [66], Razumikhin [239–242], and many other (see, for example, [5, 9, 22, 43, 59, 62, 67, 100, 103–106, 111, 126, 132, 133, 140, 194, 206, 308, 313]).

An important direction in the study of hereditary systems is their stability [3, 12, 55, 56, 88, 96, 103, 125, 134, 135, 181, 182, 191, 195–197, 204–207, 211, 215–217, 231, 236, 258–260, 301]. As it was proposed by Krasovskii in the 1950s [156–159], a stability condition for differential equation with delays can be obtained using an appropriate Lyapunov functional. The construction of different Lyapunov functionals for one differential equation with delay allows one to get different stability conditions for the solution of this equation. The method of Lyapunov–Krasovskii functionals is very popular and developing until now [78, 79]. However, the construction of each Lyapunov functional required a unique work from its author. In

1975, Shaikhet [261] introduced a parametric family of Lyapunov functionals, so that an infinite number of Lyapunov functionals were used simultaneously. This approach allowed one to get different stability conditions for the considered equation using only one Lyapunov functional.

During twenty last years, a general method of constructing Lyapunov functionals was proposed and developed by Kolmanovskii and Shaikhet for stochastic functional differential equations, for stochastic difference equations with discrete time and continuous time, and for partial differential equations [48, 130, 131, 136–139, 145–149, 265, 269–272, 278, 280, 281]. This method was successfully applied to stability research of some mathematical models in mechanics, biology, ecology, etc. (see, for instance, [27, 36–39, 267, 273, 276]). Nevertheless, it should be noted that the stability theory for stochastic hereditary system dynamically develops and has yet a number of unsolved problems [277, 279].

Usually the books devoted to the stability theory for functional differential equations do not concern numerical determination of stability domains or the behavior of the solutions. In spite of that, this book, along with the modern theoretical results, includes also many numerical investigations showing both the stability domains and structure of the solutions. It offers a certain amount of analytical mathematics, practical numerical procedures, and actual implementations of these approaches.

In this book, consisting of twelve chapters, a general method of construction of Lyapunov functionals for stochastic functional differential equations is expounded.

Introductory Chap. 1 presents general classification and some peculiarities of functional differential equations, some properties of their solutions, the method of steps and the characteristic equation for retarded differential equations, the dependence of solution stability on small delay in equation, and the Routh–Hurwitz stability conditions for systems without delay. This section covers some theoretical backgrounds of the differential equations used in the book with concentration on mathematical rigor.

In Chap. 2 short introduction to stochastic functional differential equations is presented, in particular, the definition of the Wiener process and its numerical simulation, the Itô integral, and the Itô formula. Different definitions of stability for stochastic functional differential equations are also considered, basic Lyapunov-type stability theorems, and description of the procedure of constructing Lyapunov functionals for stability investigation. In this section some useful statements, some useful inequalities, and some unsolved problems are also included.

In Chap. 3 the procedure of constructing Lyapunov functionals is used to obtain conditions for stability of scalar stochastic linear delay differential equations with constant and variable coefficients and with constant and variable delays. It is shown that different ways of constructing Lyapunov functionals for a given equation allow us to get different conditions for asymptotic mean-square stability of the zero solution of this equation.

In Chap. 4 the procedure of constructing Lyapunov functionals is demonstrated for stability investigation of stochastic linear systems of two equations with constant and distributed delays and with constant and variable coefficients.

In Chap. 5 the stability of the zero solution and positive equilibrium points for nonlinear systems is studied. In particular, differential equations with nonlinearities in deterministic and stochastic parts and with fractional nonlinearity are considered. It is shown that investigation of stability in probability for nonlinear systems with the level of nonlinearity higher than one can be reduced to investigation of the asymptotic mean-square stability of the linear part of the considered nonlinear system.

In Chap. 6 the general method of construction of Lyapunov functionals is used to get the asymptotic mean-square stability conditions for stochastic linear differential equations with constant delay, with distributed delay, and with variable bounded and unbounded delays. Sufficient stability conditions are formulated in terms of the existence of positive definite solutions of some matrix Riccati equations. Using the procedure of constructing Lyapunov functionals, it is shown that for one stochastic linear differential equation, several different matrix Riccati equations can be obtained that allow one to get different stability conditions.

In Chap. 7 sufficient conditions for asymptotic mean-square stability of the solutions of stochastic differential equations with delay and Markovian switching are obtained. Taking into account that it is difficult enough in each case to get analytical stability conditions, a numerical procedure for investigation of stability of stochastic systems with Markovian switching is considered. This procedure can be used in the cases where analytical conditions of stability are absent. Some examples of using the proposed numerical procedure are considered. Results of the calculations are presented by a lot of figures.

Chapter 8 is devoted to the classical problem of stabilization of the controlled inverted pendulum. The problem of stabilization for the mathematical model of the controlled inverted pendulum during many years is very popular among the researchers. Unlike the classical way of stabilization in which the stabilized control is a linear combination of the states and velocities of the pendulum, here another way of stabilization is proposed. It is supposed that only the trajectory of the pendulum can be observed and stabilized control depends on the whole trajectory of the pendulum. Via the general method of construction of Lyapunov functionals, sufficient conditions for stabilization by stochastic perturbations are obtained, and nonzero steady-state solutions are investigated.

In Chap. 9 the well-known Nicholson blowflies equation with stochastic perturbations is considered. Sufficient conditions for stability in probability of the trivial and positive equilibrium points of this nonlinear differential equation with delay are obtained.

In Chap. 10 the mathematical model of the type of predator–prey with aftereffect and stochastic perturbations is considered. Sufficient conditions for stability in probability of the positive equilibrium point of the considered nonlinear system are obtained.

Chapter 11 deals with a mathematical model of the spread of infectious diseases, the so-called SIR epidemic model. Sufficient conditions for stability in probability of two equilibrium points of the SIR epidemic model with distributed delays and stochastic perturbations are obtained.

In Chap. 12 mathematical models are considered that describe human behaviors related to some addictions: consumption of alcohol and obesity. The existence of positive equilibrium points for these models are shown, and sufficient conditions for stability in probability of these equilibrium points are obtained.

The bibliography at the end of the book does not pretend to be complete and includes some of the author's publications [261–279], his publications jointly with coauthors [27, 36–39, 48, 64, 77, 136–148, 155, 197, 232, 246, 280, 281], and the literature used by the author during preparation of this book.

The book is addressed both to experts in stability theory and to a wider audience of professionals and students in pure and computational mathematics, physics, engineering, biology, and so on.

The book is mostly based on the results obtained by the author independently or jointly with coauthors, in particular, with the friend and colleague V. Kolmanovskii, with whom the author is glad and happy to collaborate for more than 30 years.

Taking into account that the possibilities for further improvement and development are endless, the author will appreciate receiving useful remarks, comments, and suggestions.

Donetsk, Ukraine   Leonid Shaikhet

Contents

Chapter 1
Short Introduction to Stability Theory of Deterministic Functional Differential Equations

1.1 Some Peculiarities of Functional Differential Equations

In comparison with ordinary differential equations, functional differential equations have some peculiarities. Below we consider some of them for deterministic functional differential equations.

1.1.1 Description of Functional Differential Equations

This section covers some theoretical backgrounds of the equations used in the book with concentration on mathematical rigor. General classification of the equations and some properties of their solutions will be discussed.

Let us consider equations with an unknown function depending on a continuous argument t, which may be treated as time. The equations can be scalar or vector equations and have the same dimension as the unknown function. It is assumed that all variables under consideration are real.

A *functional equation* is an equation involving an unknown function for different argument values. The equations

$$2x(3t) + 3x(2t) = 1, \qquad x\big(x(t)\big) = x\big(t^2\big) + 1$$

are examples of functional equations. The differences between the argument values of an unknown function and t in a functional equation are called *argument deviations*. If all argument deviations are constant, as in the example

$$x(t) = t^2 x(t + 1) - x^2(t - 2),$$

then the functional equation is called a *difference equation* [278].

Above we have given some examples of functional equations with *discrete* (or *concentrated*) *argument deviations*. By increasing in the equation "the number of

L. Shaikhet, *Lyapunov Functionals and Stability of Stochastic Functional Differential Equations*, DOI 10.1007/978-3-319-00101-2_1,
© Springer International Publishing Switzerland 2013

summands" and simultaneously decreasing the differences between neighboring argument values, one naturally arrives at functional equations with *continuous* (or
distributed) *argument deviations*

$$x(t) = x(0) + \int_0^t x(s)\,ds$$

and *mixed* (both continuous and discrete) *argument deviations* [35, 66, 167, 233,
234, 269–272]

$$x(t) = x(t-1) + \int_{t-1}^t x(s)\,ds.$$

These equations are called *integral* and *integral functional equations* (in particular,
integral difference equations).

Combining the notions of differential and functional equations, we obtain the
notion of *functional differential equation* [5, 44, 104–106, 122, 126, 131–133]
or, equivalently, *differential equation with deviating argument* [66, 167]. Thus,
this is an equation connecting the unknown function and some of its derivatives
for, in general, different argument values. Here also the argument values can be
discrete, continuous, or mixed. Correspondingly, one introduces the notions of
*differential–difference equation, differential equation of neutral type, integral or
Volterra integro–differential equation* [4, 13, 55, 57, 63, 64, 74–78, 89, 90, 104,
139, 150, 151, 171, 172, 185, 208, 222, 229, 230, 237, 238, 292], etc.

The *order* of a functional differential equation is the order of the highest derivative of the unknown function entering in the equation. So, a functional equation may
be regarded as a functional differential equation of order zero. Hence the notion of
functional differential equation generalizes all equations of mathematical analysis
for function of a continuous argument. A similar assertion holds for function depending on several arguments.

1.1.2 Reducing to Ordinary Differential Equations

Sometimes Volterra-type integro–differential equations can be reduced to equivalent ordinary differential equations. We explain this phenomenon by the following
equation with exponentially fading memory [281]:

$$\dot{x}(t) = f\left(t, \int_{t_0}^t x(\theta)e^{-k(t-\theta)}\,d\theta\right), \quad t \geq t_0.$$

Put

$$y(t) = \int_{t_0}^t x(\theta)e^{-k(t-\theta)}\,d\theta.$$

Differentiating this equality, we obtain that the initial equation is equivalent to the
system of ordinary differential equations

$$\dot{x}(t) = f\big(t, y(t)\big), \qquad \dot{y}(t) = x(t) - k y(t),$$

with the additional initial condition $y(t_0) = 0$.

Consider another example of Volterra-type equation that can be reduced to ordinary differential equations. Consider the equation

$$\dot{x}(t) = -2x(t) + \int_0^t \sin(t - s) x(s)\, ds.$$

Differentiating this equation two times,

$$\ddot{x}(t) = -2\dot{x}(t) + \int_0^t \cos(t - s) x(s)\, ds,$$

$$\dddot{x}(t) = -2\ddot{x}(t) + x(t) - \int_0^t \sin(t - s) x(s)\, ds,$$

and adding the last equation to the initial equation, we obtain

$$\dddot{x}(t) + 2\ddot{x}(t) + \dot{x}(t) + x(t) = 0.$$

This equivalence looks like a surprise because integro–differential equations describe processes whose rate is determined by all previous states, while ordinary
differential equations describe processes determined by the current states only.

1.2 Method of Steps for Retarded Functional Differential Equations

For discrete delays, very often to solve the Cauchy problem on a finite interval, one
can apply a step method. Let us explain this method using the scalar equation with
one discrete delay

$$\dot{x}(t) = f\big(t, x(t), x(t - h)\big), \quad t \geq t_0, \ h = \text{const} > 0,$$
$$x_{t_0}(s) = \phi(s), \quad s \in [-h, 0]. \tag{1.1}$$

Let $f : [t_0, \infty) \times \mathbf{R}^2 \to \mathbf{R}$ be a continuous function satisfying the Lipschitz condition with respect to the second argument. The initial function ϕ for (1.1) must be
assigned on the interval $[t_0 - h, t_0]$.

If $t \in [t_0, t_0 + h]$ (it is the first step), then $t - h \in [t_0 - h, t_0]$. Therefore, the delay
differential equation (1.1) converts into the ordinary differential equation

$$\dot{x}(t) = f\big(t, x(t), \phi(t - t_0 - h)\big), \quad t_0 \leq t \leq t_0 + h,$$
$$x(t_0) = \phi(0).$$

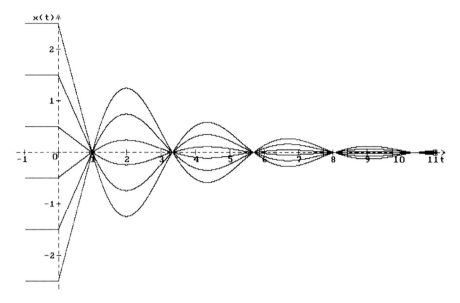

Fig. 1.1 Solutions of (1.2) for $a = -1$, $h = 1$, and different initial values of x_0

Solving this equation for the initial value $x(t_0) = \phi(0)$, we get the solution on the interval $[t_0, t_0 + h]$. If now $t_0 + h \le t \le t_0 + 2h$ (it is the second step), then $t - h \in [t_0, t_0 + h]$, and so $x(t - h)$ is known from the first step. Hence, (1.1) for $t_0 + h \le t \le t_0 + 2h$ once again converts into the ordinary differential equation, which with the known initial value $x(t_0 + h)$ defines the solution $x(t)$ on this interval. After that we consider the interval $[t_0 + 2h, t_0 + 3h]$, etc. In this way the solution can be obtained for arbitrarily large t (theoretically, for the whole semiaxis $[t_0, \infty)$).

For the simplest case where the right-hand side of (1.1) does not contain $x(t)$, at each step the solution is reduced to the integration of a given function.

Example 1.1 Consider the scalar differential equation with discrete delay

$$\dot{x}(t) = ax(t - h), \quad x(s) \equiv x_0 = \text{const}, \quad -h \le s \le 0. \tag{1.2}$$

Using the method of steps, it is easy to get the solution of this equation in the form

$$x(t) = x_0 \left(1 + \sum_{l=0}^{k} a^{l+1} \frac{(t - lh)^{l+1}}{(l + 1)!} \right), \quad k = \left[\frac{t}{h} \right]. \tag{1.3}$$

The solutions of (1.2) obtained via (1.3) are shown in Fig. 1.1 for $a = -1$, $h = 1$, and different initial values x_0. One can see that the solutions for different x_0 intersect each other at the point $t = 1$ and many other points. It is known that this situation is impossible for ordinary differential equations.

1.3 Characteristic Equation for Differential Equation with Discrete Delays

Consider the nth-order differential equation with discrete delays

$$x^{(n)}(t) + \sum_{i=0}^{n-1}\sum_{j=1}^{m(i)} a_{ij}x^{(i)}(t - h_{ij}) = 0, \quad h_{ij} \geq 0. \tag{1.4}$$

With this equation, the following function is associated:

$$\Delta(z) = z^n + \sum_{i=0}^{n-1}\sum_{j=1}^{m(i)} a_{ij}z^i e^{-zh_{ij}}, \tag{1.5}$$

which is called the characteristic quasipolynomial of (1.4).

It is known [80] that the trivial solution of (1.4) is asymptotically stable if and only if all zeros $z = \alpha + i\beta$ of the characteristic quasipolynomial (1.5) (or all roots of the characteristic equation $\Delta(z) = 0$) satisfy the condition $\alpha < 0$. The bound of the region of asymptotic stability can be defined by the equality $\Delta(i\beta) = 0$, $i^2 = -1$, $\beta \in \mathbf{R}$.

Example 1.2 The characteristic quasipolynomial for (1.2) is $\Delta(z) = z - ae^{-hz}$. Thus, the characteristic equation $\Delta(i\beta) = 0$ gives the system of two equations

$$a\cos\beta h = 0, \qquad \beta + a\sin\beta h = 0$$

with the solution $\beta h = \frac{1}{2}\pi$ or $h = -\frac{1}{2a}\pi$. This means that the trivial solution of (1.2) is asymptotically stable if and only if $h < \frac{1}{2|a|}\pi$, $a < 0$.

Put $a = -1$. Then, for $h = 1$, the trivial solution of (1.2) is asymptotically stable, and all solutions go to zero (Fig. 1.1). If $h = \frac{1}{2}\pi$, then the trivial solution of (1.2) is stable but not asymptotically stable, and all solutions are bounded (Fig. 1.2). If $h > \frac{1}{2}\pi$, then the trivial solution of (1.2) is unstable, and all solutions go to infinity (Fig. 1.3, $h = 2$).

Note also that as $h \to 0$, (1.2) goes to an ordinary differential equation with asymptotically (for $a < 0$) stable trivial solution and different solutions without intersections. In Fig. 1.4 the solutions of (1.2) are shown for $h = 0.5$.

Example 1.3 Consider the scalar differential equation

$$\dot{x}(t) + ax(t - h_1) + bx(t \cdot h_2) = 0, \quad 0 \leq h_1 < h_2. \tag{1.6}$$

The characteristic quasipolynomial of this equation is

$$\Delta(z) = z + ae^{-h_1 z} + be^{-h_2 z}.$$

Suppose that $a + b \leq 0$. Then $\Delta(0) = a + b \leq 0$ and $\Delta(\infty) = \infty$. Therefore, there exists $\alpha \geq 0$ such that $\Delta(\alpha) = 0$. This means that by the condition $a + b \leq 0$

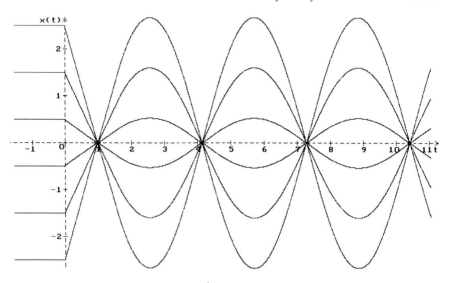

Fig. 1.2 Solutions of (1.2) for $a = -1$, $h = \frac{1}{2}\pi$, and different initial values of x_0

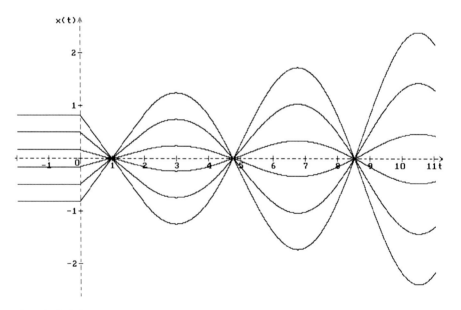

Fig. 1.3 Solutions of (1.2) for $a = -1$, $h = 2$, and different initial values of x_0

the trivial solution of (1.6) cannot be asymptotically stable. So, we will suppose that $a + b > 0$.

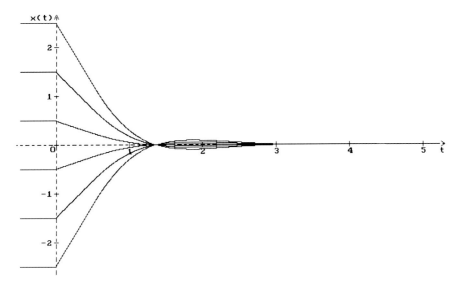

Fig. 1.4 Solutions of (1.2) for $a = -1$, $h = 0.5$, and different initial values of x_0

The equality $\Delta(i\beta) = 0$ gives the system of two equations

$$a \cos \beta h_1 + b \cos \beta h_2 = 0,$$
$$a \sin \beta h_1 + b \sin \beta h_2 = \beta. \tag{1.7}$$

From (1.7) it follows that the bounds of the stability region for (1.6) are formed in the (a, b)-plane by the straight line $a + b = 0$ and the parametric curve

$$a = -\frac{\beta \cos \beta h_2}{\sin \beta (h_2 - h_1)}, \qquad b = \frac{\beta \cos \beta h_1}{\sin \beta (h_2 - h_1)}. \tag{1.8}$$

Example 1.4 Consider the first-order scalar linear differential equation with delay

$$\dot{x}(t) + ax(t) + bx(t - h) = 0, \tag{1.9}$$

which is the special case of (1.6) with $h_1 = 0$, $h_2 = h$. Thus, (1.8) takes the form

$$a = -\frac{\beta \cos \beta h}{\sin \beta h}, \qquad b = \frac{\beta}{\sin \beta h}, \qquad \beta h \in [0, \pi).$$

The bound of the asymptotic stability region for the trivial solution of (1.9) is defined by the conditions

$$a + b = 0, \quad bh < 1,$$
$$a + b \cos\left(h \sqrt{b^2 - a^2}\right) = 0, \quad bh \geq 1. \tag{1.10}$$

The bound of stability region given by conditions (1.10) is shown in Fig. 1.5 for $h = 1.3$.

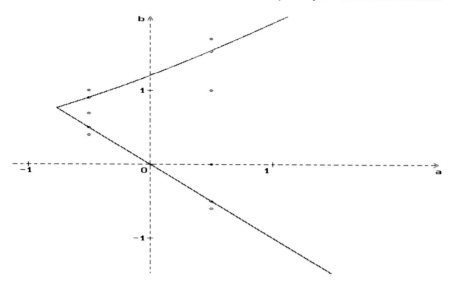

Fig. 1.5 Stability region for (1.9) given by conditions (1.10) for $h = 1.3$

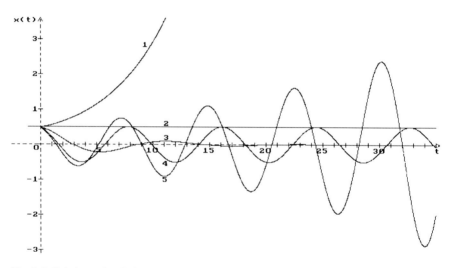

Fig. 1.6 Solutions of (1.9) for $h = 1.3$, $x(s) = 0.5$, $-h \leq s \leq 0$, $a = -0.5$, and different values of b: (*1*) $b = 0.4$, (*2*) $b = 0.5$, (*3*) $b = 0.7$, (*4*) $b = 0.909$, (*5*) $b = 1$

In Fig. 1.6 the solutions of (1.9) are shown for $x(s) = 0.5$, $-1.3 \leq s \leq 0$, $a = -0.5$ and different values of b: (1) $b = 0.4$ ($x(t) \to \infty$); (2) $b = 0.5$ ($x(t) = \text{const}$); (3) $b = 0.7$ ($x(t) \to 0$); (4) $b = 0.909$ ($x(t)$ is a periodical solution); (5) $b = 1$ ($x(t) \to \pm\infty$).

In Fig. 1.7 the solutions of (1.9) are shown for $x(s) = 0.5$, $-1.3 \leq s \leq 0$, $a = 0.5$ and different values of b: (1) $b = -0.6$ ($x(t) \to \infty$); (2) $b = -0.5$ ($x(t) = \text{const}$);

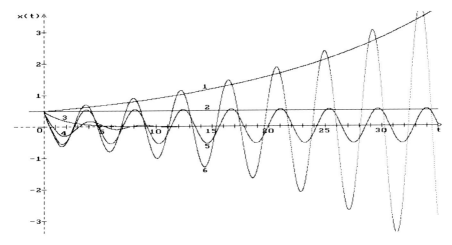

Fig. 1.7 Solutions of (1.9) for $h = 1.3$, $x(s) = 0.5$, $-h \le s \le 0$, $a = 0.5$, and different values of b: (*1*) $b = -0.6$, (*2*) $b = -0.5$, (*3*) $b = 0$, (*4*) $b = 1$, (*5*) $b = 1.537$, (*6*) $b = 1.7$

(3) $b = 0$ $(x(t) \to 0)$; (4) $b = 1$ $(x(t) \to 0)$; (5) $b = 1.537$ $(x(t)$ is a periodical solution); (6) $b = 1.7$ $(x(t) \to \pm\infty)$.

The points that were used in Figs. 1.6 and 1.7 are shown also in Fig. 1.5. The solutions of (1.9) were obtained via its difference analogue in the form

$$x_{i+1} = (1 - a\Delta)x_i - b\Delta x_{i-m},$$

where Δ is the step of discretization, and $x_i = x(i\Delta)$, $m = h/\Delta$, $\Delta = 0.01$.

Example 1.5 Consider the first-order scalar linear differential equation of neutral type

$$\dot{x}(t) + ax(t) + bx(t - h) + c\dot{x}(t - h) = 0, \quad |c| < 1, \qquad (1.11)$$

which is a generalization of (1.9).

The corresponding characteristic quasipolynomial is $\Delta(z) = z + a + be^{-hz} + cze^{-hz}$. The equality $\Delta(i\beta) = 0$ gives the system of two equations

$$a + b\cos\beta h + c\beta\sin\beta h = 0,$$
$$\beta - b\sin\beta h + c\beta\cos\beta h = 0,$$

with the solution

$$a = -\frac{\beta(c + \cos\beta h)}{\sin\beta h}, \qquad b = \frac{\beta(1 + c\cos\beta h)}{\sin\beta h}, \quad \beta h \in [0, \pi).$$

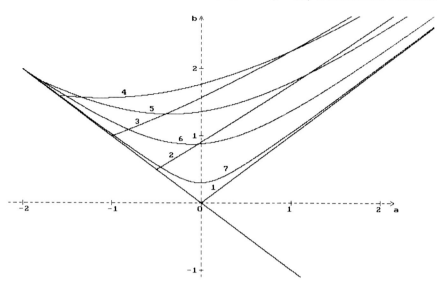

Fig. 1.8 Stability regions for (1.11) given by (1.12) are shown in space of parameters (a, b) for $h = 1$ and different values of c: (*1*) $c = -0.999$, (*2*) $c = -0.5$, (*3*) $c = 0$, (*4*) $c = 0.6$, (*5*) $c = 0.85$, (*6*) $c = 0.95$, (*7*) $c = 0.995$

The bound of the region of asymptotic stability of the trivial solution of (1.11) is defined by the conditions

$$a + b = 0, \quad bh < 1 + c,$$

$$a + bc + (ac + b)\cos\left(h\sqrt{\frac{b^2 - a^2}{1 - c^2}}\right) = 0, \quad b > |a|. \tag{1.12}$$

In Fig. 1.8 the stability regions given by (1.12) for (1.11) are shown in the space of the parameters (a, b) for $h = 1$ and different values of c: (1) $c = -0.999$, (2) $c = -0.5$, (3) $c = 0$, (4) $c = 0.6$, (5) $c = 0.85$, (6) $c = 0.95$, (7) $c = 0.995$. In Fig. 1.8 one can see that if $a > |b|$, then the trivial solution of (1.11) is asymptotically stable for all $h > 0$ and $|c| < 1$.

In Fig. 1.9 the stability regions for (1.11) are shown in the space of the parameters (c, b) for $a = 0.4$ and different values of h: (1) $h = 2$, (2) $h = 1$, (3) $h = 0.7$. In Figs. 1.10 and 1.11 the similar stability regions are shown for $a = 0$ and $a = -0.4$, respectively.

In Fig. 1.12 the solutions of (1.11) are shown for $c = 0.1$, $x(s) = 0.5$, $s \leq 0$, $a = -0.5$, $h = 1.3$ and different values of b: (1) $b = 0.4$ ($x(t)$ goes to infinity), (2) $b = 0.5$ ($x(t) = $ const), (3) $b = 0.7$ ($x(t)$ goes to zero), (4) $b = 1.004$ ($x(t)$ is a bounded periodical solution), (5) $b = 1.1$ ($|x(t)|$ goes to infinity). In Fig. 1.13 the similar solutions are shown for the initial condition $x(s) = \cos(s)$, $s \leq 0$, and the same values of other parameters.

Fig. 1.9 Stability regions for (1.11) given by (1.12) are shown in space of parameters (c, b) for $a = 0.4$ and different values of h: (*1*) $h = 2$, (*2*) $h = 1$, (*3*) $h = 0.7$

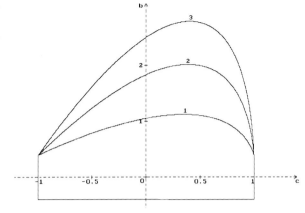

Fig. 1.10 Stability regions for (1.11) given by (1.12) are shown in space of parameters (c, b) for $a = 0$ and different values of h: (*1*) $h = 2$, (*2*) $h = 1$, (*3*) $h = 0.7$

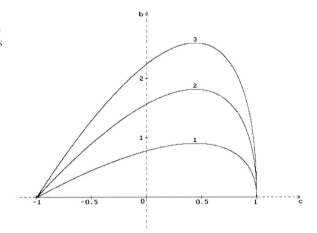

Fig. 1.11 Stability regions for (1.11) given by (1.12) are shown in space of parameters (c, b) for $a = -0.4$ and different values of h: (*1*) $h = 2$, (*2*) $h = 1$, (*3*) $h = 0.7$

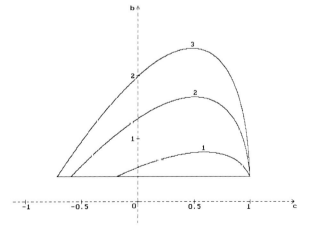

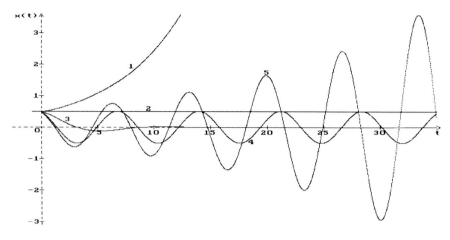

Fig. 1.12 Solutions of (1.11) are shown for $c = 0.1$, $h = 1.3$, $x(s) = 0.5$, $s \in [-h, 0]$, $a = -0.5$, and different values of b: (1) $b = 0.4$, (2) $b = 0.5$, (3) $b = 0.7$, (4) $b = 1.004$, (5) $b = 1.1$

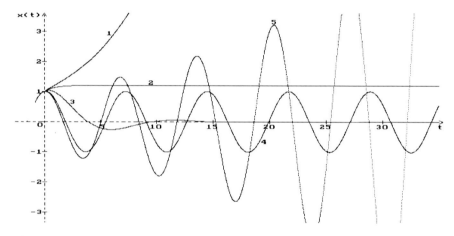

Fig. 1.13 Solutions of (1.11) are shown for $c = 0.1$, $h = 1.3$, $x(s) = \cos(s)$, $s \in [-h, 0]$, $a = -0.5$, and different values of b: (1) $b = 0.4$, (2) $b = 0.5$, (3) $b = 0.7$, (4) $b = 1.004$, (5) $b = 1.1$

In Fig. 1.14 the solutions of (1.11) are shown for $c = -0.1$, $x(s) = 0.5$, $s \leq 0$, $a = -0.5$, $h = 1.3$ and different values of b: (1) $b = 0.4$ ($x(t)$ goes to infinity), (2) $b = 0.5$ ($x(t) = $ const), (3) $b = 0.7$ ($x(t)$ goes to zero), (4) $b = 0.804$ ($x(t)$ is a bounded periodical solution), (5) $b = 0.9$ ($|x(t)|$ goes to infinity).

The solutions of (1.11) were obtained via its difference analogue in the form

$$x_{i+1} = (1 - a\Delta)x_i + (c - b\Delta)x_{i-m} - cx_{i+1-m},$$

where Δ is the step of discretization, and $x_i = x(i\Delta)$, $m = h/\Delta$, $\Delta = 0.01$.

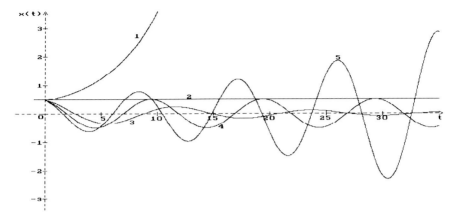

Fig. 1.14 Solutions of (1.11) are shown for $c = -0.1$, $h = 1.3$, $x(s) = 0.5$, $s \in [-h, 0]$, $a = -0.5$, and different values of b: (*1*) $b = 0.4$, (*2*) $b = 0.5$, (*3*) $b = 0.7$, (*4*) $b = 0.804$, (*5*) $b = 0.9$

Example 1.6 Let us consider the scalar second-order differential equation

$$\ddot{x}(t) + a\dot{x}(t - h_1) + bx(t - h_2) = 0. \tag{1.13}$$

The characteristic quasipolynomial of (1.13) is $\Delta(z) = z^2 + aze^{-h_1 z} + be^{-h_2 z}$. The equality $\Delta(i\beta) = 0$ gives the following system of two equations:

$$\beta a \cos \beta h_1 - b \sin \beta h_2 = 0,$$
$$\beta a \sin \beta h_1 + b \cos \beta h_2 = \beta^2, \tag{1.14}$$

with the solution

$$a = \frac{\beta \sin \beta h_2}{\cos \beta (h_1 - h_2)}, \qquad b = \frac{\beta^2 \cos \beta h_1}{\cos \beta (h_1 - h_2)}.$$

In Fig. 1.15 the stability regions for (1.13) are shown for $h_1 = 1$ and different values of h_2: (1) $h_2 = 0.2$, (2) $h_2 = 0.6$, (3) $h_2 = 1$, (4) $h_2 = 1.4$, (5) $h_2 = 1.8$. In Fig. 1.16 the stability regions for (1.13) are shown for $h_2 = 1$ and different values of h_1: (1) $h_1 = 0.8$, (2) $h_1 = 0.9$, (3) $h_1 = 1$, (4) $h_1 = 1.1$, (5) $h_1 = 1.2$.

Let us consider some particular cases of (1.13).

(a) $h_1 = h_2 = h$. In this case the solution of (1.14) has the form $a = \beta \sin \beta h$, $b - \beta^2 \cos \beta h$, $0 \le \beta h \le \frac{\pi}{2}$. In Fig. 1.17 the stability regions are shown for different values of h: (1) $h = 0.5$, (2) $h = 0.75$, (3) $h = 1$, (4) $h = 1.25$, (5) $h = 1.5$.

(b) $h_1 = 0$, $h_2 = h$. In this case we obtain $a = \beta \tan \beta h$, $b = \frac{\beta^2}{\cos \beta h}$, $0 \le \beta h < \frac{\pi}{2}$. In Fig. 1.18 the stability regions are shown for different values of h: (1) $h = 1$, (2) $h = 2$, (3) $h = 3$, (4) $h = 4$, (5) $h = 5$.

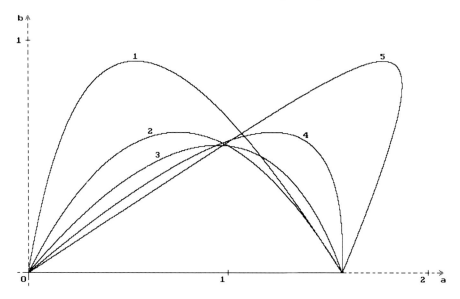

Fig. 1.15 Stability regions for (1.13) for $h_1 = 1$ and different values of h_2: (*1*) $h_2 = 0.2$, (*2*) $h_2 = 0.6$, (*3*) $h_2 = 1$, (*4*) $h_2 = 1.4$, (*5*) $h_2 = 1.8$

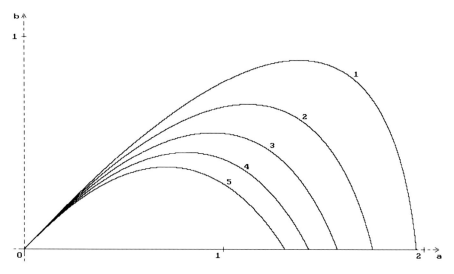

Fig. 1.16 Stability regions for (1.13) for $h_2 = 1$ and different values of h_1: (*1*) $h_1 = 0.8$, (*2*) $h_1 = 0.9$, (*3*) $h_1 = 1$, (*4*) $h_1 = 1.1$, (*5*) $h_1 = 1.2$

(c) $h_1 = h$, $h_2 = 0$. In this case from the first equation of (1.14) it follows that $\cos \beta h = 0$. Therefore,

$$\beta = \frac{\pi}{2h}(2l + 1), \quad l = 0, 1, \ldots,$$

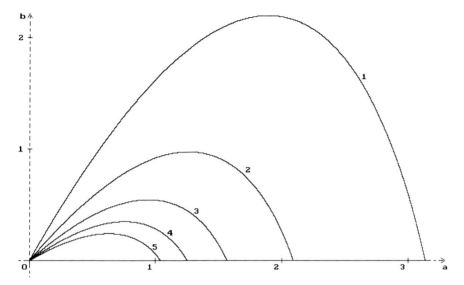

Fig. 1.17 Stability regions for (1.13) in the case $h_1 = h_2 = h$ for different values of h: (*1*) $h = 0.5$, (*2*) $h = 0.75$, (*3*) $h = 1$, (*4*) $h = 1.25$, (*5*) $h = 1.5$

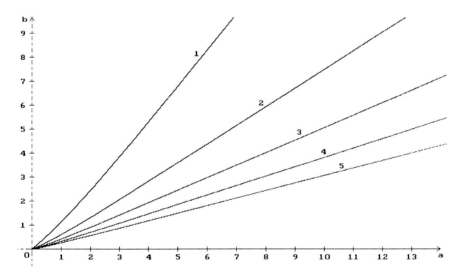

Fig. 1.18 Stability regions for (1.13) in the case $h_1 = 0$, $h_2 = h$ for different values of h: (*1*) $h = 1$, (*2*) $h = 2$, (*3*) $h = 3$, (*4*) $h = 4$, (*5*) $h = 5$

and from system (1.14) it follows that the stability region consists of a sequence of triangles, formed by parts of the a- and the b-axes and the line segments

$$b = (-1)^{l+1} \frac{2l+1}{2h} \pi a + \left(\frac{2l+1}{2h} \pi \right)^2, \quad l = 0, 1, \dots.$$

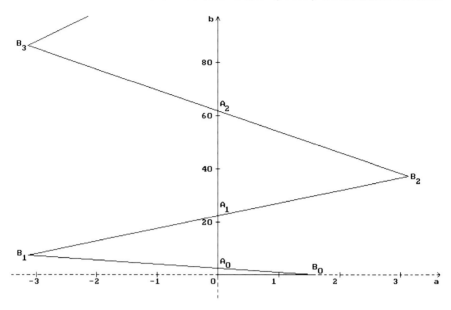

Fig. 1.19 Stability regions for (1.13) in the case $h_1 = 1$, $h_2 = 0$

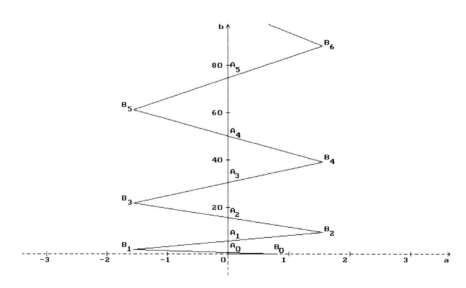

Fig. 1.20 Stability regions for (1.13) in the case $h_1 = 2$, $h_2 = 0$

The stability regions (the triangles $O B_0 A_0$, $A_0 B_1 A_1$, $A_1 B_2 A_2$, $A_2 B_3 A_3$, ...)
for this case are shown in Fig. 1.19 ($h = 1$) and Fig. 1.20 ($h = 2$). The points A_l

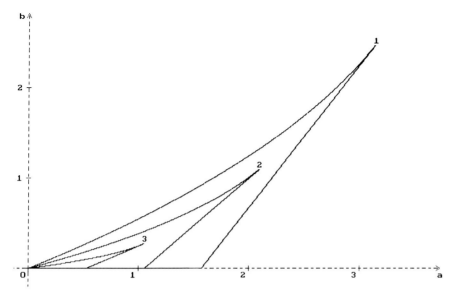

Fig. 1.21 Stability regions for (1.13) in the case $h_1 = h$, $h_2 = 2h$ for different values of h: (1) $h = 1$, (2) $h = 1.5$, (3) $h = 3$

and B_l have the coordinates respectively

$$A_l = \left(0, \left(\frac{2l+1}{2h}\pi\right)^2\right), \quad l = 0, 1, \ldots,$$

$$B_0 = \left(\frac{\pi}{2h}, 0\right), \qquad B_l = \left((-1)^l\frac{\pi}{h}, \frac{(2l-1)(2l+1)}{4h^2}\pi^2\right), \quad l = 1, 2, \ldots.$$

(d) $h_1 = h$, $h_2 = 2h$. In this case from the first equation of system (1.14) it follows that

$$\beta a \cos \beta h - b \sin 2\beta h = (\beta a - 2b \sin \beta h)\cos \beta h = 0.$$

If $\cos \beta h = 0$, i.e., $\beta h = \frac{\pi}{2}$, then from the second equation of system (1.14) it follows that a part of the stability region bound is defined by the straight line

$$b = \frac{\pi}{2h}\left(a - \frac{\pi}{2h}\right). \tag{1.15}$$

The stability regions in this case are shown in Fig. 1.21 for different values of h: (1) $h = 1$, (2) $h = 1.5$, (3) $h = 3$. One can see that the part of the stability region bound is defined by the straight line (1.15).

Example 1.7 Consider the scalar second-order differential equation

$$\ddot{x}(t) = ax(t) + b_1 x(t - h_1) + b_2 x(t - h_2), \quad h_1 > h_2. \tag{1.16}$$

The characteristic quasipolynomial of (1.16) is

$$\Delta(z) = z^2 - a - b_1 e^{-h_1 z} - b_2 e^{-h_2 z}, \quad z = \alpha + i\beta.$$

Suppose that $a + b_1 + b_2 \geq 0$. Then $\Delta(0) \leq 0$ and $\Delta(\infty) = \infty$. Therefore, there exists $\alpha \geq 0$ such that $\Delta(\alpha) = 0$. This means that by the condition $a + b_1 + b_2 \geq 0$ the trivial solution of (1.16) cannot be asymptotically stable. So, the inequality $a + b_1 + b_2 < 0$ is the necessary condition for asymptotic stability of the trivial solution of (1.16).

The equality $\Delta(i\beta) = 0$ gives the following system of two equations:

$$\begin{aligned}
b_1 \cos \beta h_1 + b_2 \cos \beta h_2 &= -(a + \beta^2), \\
b_1 \sin \beta h_1 + b_2 \sin \beta h_2 &= 0,
\end{aligned} \tag{1.17}$$

with the solution

$$b_1 = \frac{(a + \beta^2) \sin \beta h_2}{\sin \beta (h_1 - h_2)}, \qquad b_2 = -\frac{(a + \beta^2) \sin \beta h_1}{\sin \beta (h_1 - h_2)}.$$

Example 1.8 Consider the scalar differential equation

$$\ddot{x}(t) = -ax(t) + bx(t - h), \tag{1.18}$$

where a, b, h are nonnegative constants. The characteristic quasipolynomial of (1.18) is

$$\Delta(z) = z^2 + a - be^{-zh}, \quad z = \alpha + \beta i.$$

By the condition $b \geq a$ we have $\Delta(0) \leq 0$, $\Delta(\infty) = \infty$. So, there exists $\alpha \geq 0$ such that $\Delta(\alpha) = 0$. Therefore, the inequality $b < a$ is the necessary condition for asymptotic stability of the trivial solution of (1.18).

The characteristic equation $\Delta(i\beta) = 0$ can be represented in the form of the system of two equations

$$b \cos \beta h = a - \beta^2, \qquad b \sin \beta h = 0$$

with the following solutions: (1) $b = a$, (2) $b = 0$, $a \geq 0$, (3) $a + b = \frac{\pi^2}{h^2}$. So, the necessary and sufficient condition for asymptotic stability of the trivial solution of (1.18) takes the form

$$b < a < \frac{\pi^2}{h^2} - b. \tag{1.19}$$

In Fig. 1.22 the stability regions given by condition (1.19) are shown for different values of h: (1) $h = 2$, (2) $h = 1.5$, (3) $h = 1.2$, (4) $h = 1$, (5) $h = 0.8$.

Put $x(s) = e^s$, $s \in [-h, 0]$, $a = 1$, and consider the behavior of the solution of (1.18) for different values of b and h.

In Fig. 1.23 the solutions of (1.18) are shown for $h = 0$ and the values of b: (1) $b = 0.3$, (2) $b = 0.9$, (3) $b = 1.01$. For $h = 0$, condition (1.19) is the necessary

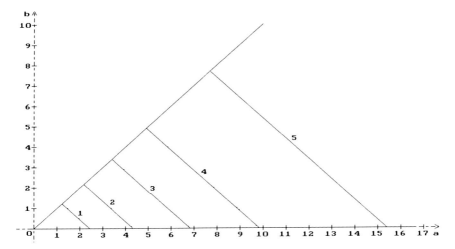

Fig. 1.22 Stability regions for (1.18) given by condition (1.19) for different values of h: (*1*) $h = 2$, (*2*) $h = 1.5$, (*3*) $h = 1.2$, (*4*) $h = 1$, (*5*) $h = 0.8$

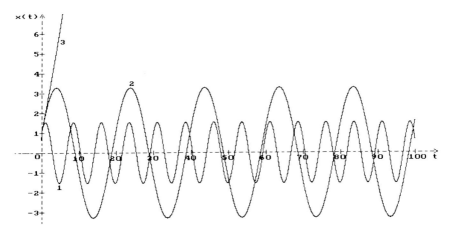

Fig. 1.23 Solutions of (1.18) for $a = 1$, $h = 0$ and different values of b: (*1*) $b = 0.3$, (*2*) $b = 0.9$, (*3*) $b = 1.01$

stability condition only, so, the solution is periodical in the cases (1), (2) and goes to infinity in the case (3).

In Fig. 1.24 the solutions of (1.18) are shown for $h = 0.5$ and the same values of a and b. Condition (1.19) holds in the cases (1), (2) and does not hold in the case (3). So, the solution goes to zero in the cases (1), (2) and goes to infinity in the case (3). The similar picture one can see in Fig. 1.25 for $h = 2$.

In Fig. 1.26 the solutions of (1.18) are shown for $h = 2.755359$. In this case condition (1.19) does not hold since $a = \frac{\pi^2}{h^2} - b = 1$ for $b = 0.3$ and $a = 1 > \frac{\pi^2}{h^2} - b = 0.4$ for $b = 0.9$. So, the solution of (1.18) is periodical for $b = 0.3$ and goes

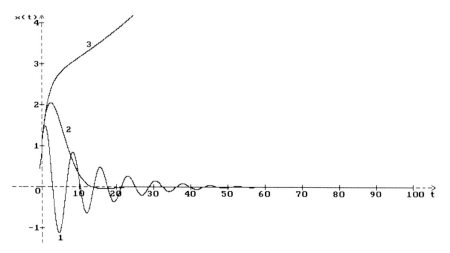

Fig. 1.24 Solutions of (1.18) for $a = 1$, $h = 0.5$ and different values of b: (*1*) $b = 0.3$, (*2*) $b = 0.9$, (*3*) $b = 1.01$

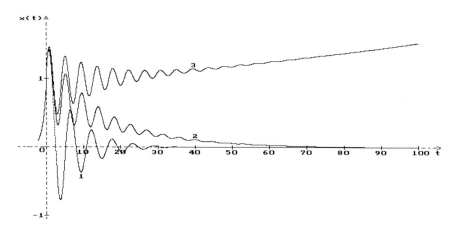

Fig. 1.25 Solutions of (1.18) for $a = 1$, $h = 2$ and different values of b: (*1*) $b = 0.3$, (*2*) $b = 0.9$, (*3*) $b = 1.01$

to infinity for $b = 0.9$. In Fig. 1.27 the solutions of (1.18) are shown for $h = 0.8$, $b = 0.6$ and different conditions on a: (1) $a = b$, (2) $a = 1.3b$, (3) $a = \frac{\pi^2}{h^2} - b$.

In Figs. 1.22–1.27 one can see that the solution of (1.18) goes to constant on the bound $a = b$ of stability region, is a periodical on the bound $a + b = \frac{\pi^2}{h^2}$, goes to zero between these bounds, and goes to infinity out of these bounds. The solutions of (1.18) were obtained via its difference analogue in the form

$$x_{i+1} = \left(2 - a\Delta^2\right)x_i - x_{i-1} + b\Delta^2 x_{i-m},$$

where Δ is the step of discretization, and $x_i = x(i\Delta)$, $m = h/\Delta$, $\Delta = 0.01$.

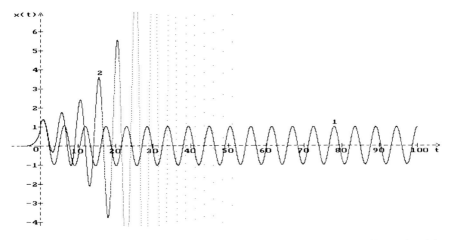

Fig. 1.26 Solutions of (1.18) for $a = 1$, $h = 2.755359$, and different values of b: (*1*) $b = 0.3$, (*2*) $b = 0.9$

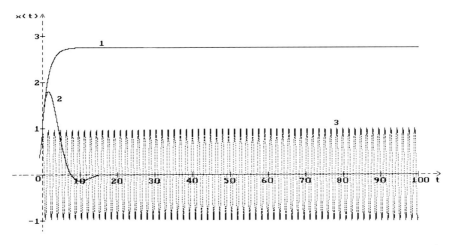

Fig. 1.27 Solutions of (1.18) for $h = 0.8$, $b = 0.6$, and different values of a: (*1*) $a = b$, (*2*) $a = 1.3b$, (*3*) $a = \frac{\pi^2}{h^2} - b$

1.4 The Influence of Small Delays on Stability

The effect of time delay is very essential in different real processes [46, 109, 225]. Many actual phenomena involve small delays, which are often neglected in the process of mathematical modeling. Sometimes this leads to false conclusions. We shall show that sometimes even small delay can change essentially the properties of the solution.

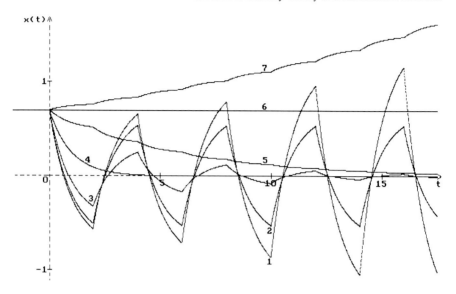

Fig. 1.28 Solutions of (1.20) for $a = 1$, $h = 2$, $x_0(s) = 0.7$, $s \in [-h, 0]$, and different values of b: (1) $b = -1.1$, (2) $b = -1$, (3) $b = -0.7$, (4) $b = 0$, (5) $b = 0.7$, (6) $b = 1$, (7) $b = 1.1$

Example 1.9 Consider the first-order differential equation of neutral time

$$\dot{x}(t) + ax(t) - b\big[\dot{x}(t - h) + ax(t - h)\big] = 0, \quad a > 0. \tag{1.20}$$

For $h = 0$, we have the equation $(1 - b)[\dot{x}(t) + ax(t)] = 0$ with the solution $x(t) = x(0)\exp(-at)$, which is asymptotically stable for $a > 0$ and arbitrary b. If $h > 0$, then the appropriate characteristic equation $(z + a)(1 - be^{-hz}) = 0$ has all roots

$$z_k = \frac{1}{h}\left(\ln|b| + i2k\pi\right), \quad i^2 = -1, \ k = 0, \pm 1, \pm 2, \dots,$$

with real parts $\mathrm{Re}z_k = \frac{1}{h}\ln|b|$. So, if $|b| > 1$, then the trivial solution of (1.20) is asymptotically stable for $h = 0$ and unstable for each $h > 0$. If $|b| \leq 1$, we have another picture.

In Fig. 1.28 the solutions of (1.20) are shown for the initial function $x_0(s) = 0.7$, $s \in [-h, 0]$, $h = 2$, $a = 1$, and for different values of the parameter b: (1) $b = -1.1$, (2) $b = -1$, (3) $b = -0.7$, (4) $b = 0$, (5) $b = 0.7$, (6) $b = 1$, (7) $b = 1.1$. We can see that in the cases (1) and (7) the solution of (1.20) goes to infinity, in the case (2) it is a periodic solution, in the cases (3), (4), and (5) the solution goes to zero, and in the case (6) it is a constant. In Fig. 1.29 a similar picture is shown for the initial function $x_0(s) = \cos s$, $s \in [-h, 0]$, and the same values of other parameters. We can see again that in the cases (1) and (7) the solution of (1.20) goes to infinity, in the cases (2) and (6) it is a periodic solution, and in the cases (3), (4), and (5) it goes to zero. In Fig. 1.30 the same situation with the solution of (1.20) is shown for small

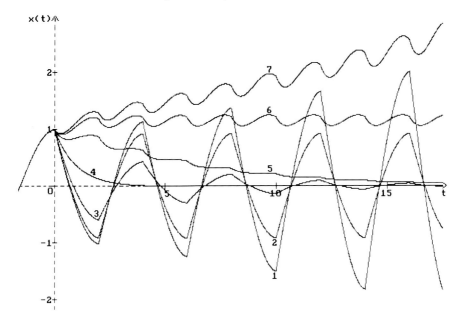

Fig. 1.29 Solutions of (1.20) for $a = 1$, $h = 2$, $x_0(s) = \cos(s)$, $s \in [-h, 0]$, and different values of b: (1) $b = -1.1$, (2) $b = -1$, (3) $b = -0.7$, (4) $b = 0$, (5) $b = 0.7$, (6) $b = 1$, (7) $b = 1.1$

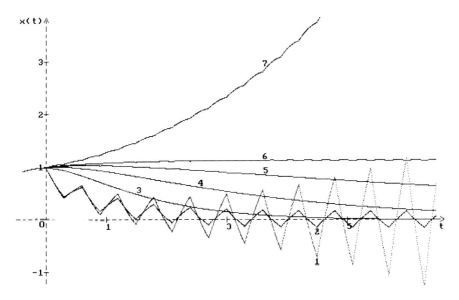

Fig. 1.30 Solutions of (1.20) for $a = 1$, $h = 0.3$, $x_0(s) = \cos(s)$, $s \in [-h, 0]$, and different values of b: (1) $b = -1.1$, (2) $b = -1$, (3) $b = 0.7$, (4) $b = 0.9$, (5) $b = 0.97$, (6) $b = 1$, (7) $b = 1.1$

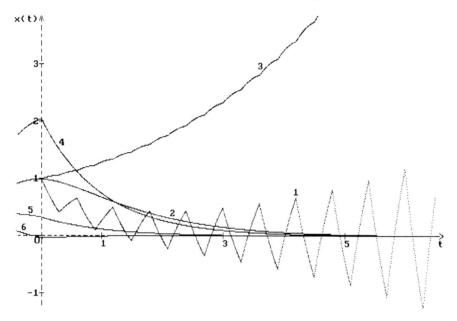

Fig. 1.31 Solutions of (1.20) for $a = 1$, $h = 0.3$, $x_0(s) = \cos(s)$, $s \in [-h, 0]$, and different values of b: (1) $b = -1.1$, (2) $b = 0.7$, (3) $b = 1.1$, and appropriate process $y(t)$: (4) $b = -1.1$, (5) $b = 0.7$, (6) $b = 1.1$

delay $h = 0.3$, the initial function $x_0(s) = \cos s$, $s \in [-h, 0]$, $a = 1$, and for different values of the parameter b: (1) $b = -1.1$, (2) $b = -1$, (3) $b = 0.7$, (4) $b = 0.9$, (5) $b = 0.97$, (6) $b = 1$, (7) $b = 1.1$.

Note that the process $y(t) = x(t) - bx(t - h)$ goes to zero for all b. In Fig. 1.31 the solutions of (1.20) are shown for the initial function $x_0(s) = \cos s$, $s \in [-h, 0]$, $a = 1$, and for different values of the parameter b: (1) $b = -1.1$, (2) $b = 0.7$, (3) $b = 1.1$ and the appropriate process $y(t)$: (4) $b = -1.1$, (5) $b = 0.7$, (6) $b = 1.1$.

One can see a similar situation for the second-order differential equation of neutral type

$$\ddot{x}(t) + a\dot{x}(t) + bx(t) = c\big[\ddot{x}(t - h) + a\dot{x}(t - h) + bx(t - h)\big],$$

$$a > 0, \ b > 0, \ c > 1. \tag{1.21}$$

If $h = 0$, then the trivial solution of (1.21) is asymptotically stable. But for any $h > 0$, the trivial solution of (1.21) is unstable since for $h > 0$, the appropriate characteristic equation $(z^2 + az + b)(1 - ce^{-hz}) = 0$ has all roots with positive real parts:

$$z_k = \frac{1}{h}(\ln c + i2k\pi), \quad i^2 = -1, \ k = 0, \pm 1, \pm 2, \ldots.$$

The solutions of (1.20) were obtained via its difference analogue in the form

$$x_{i+1} = (1 - a\Delta)x_i + b\big[x_{i+1-m} - (1 - a\Delta)x_{i-m}\big],$$

where Δ is the step of discretization, and $x_i = x(i\Delta)$, $m = h/\Delta$, $\Delta = 0.01$.

1.5 Routh–Hurwitz Conditions

Now we consider some important statements [80] for the stability of deterministic system of linear autonomous differential equations that will be essentially used below.

Linear autonomous system of ordinary differential equations has the general form

$$\dot{x}(t) = Ax(t), \tag{1.22}$$

where $x \in \mathbf{R}^n$, A is an $n \times n$ matrix with real elements a_{ij}, $i, j = 1, \dots, n$.

Theorem 1.1 (Lyapunov theorem) *The zero solution of (1.22) is asymptotically stable if and only if all roots λ of the characteristic equation*

$$\det(\lambda I - A) = 0 \tag{1.23}$$

(I is the $n \times n$ identity matrix) have negative real parts, i.e., Re $\lambda < 0$.

Definition 1.1 Let us define the trace of the kth order of a matrix A as follows:

$$S_k = \sum_{1 \le i_1 < \cdots i_k \le n} \begin{vmatrix} a_{i_1 i_1} & \cdots & a_{i_1 i_k} \\ \cdots & \cdots & \cdots \\ a_{i_k i_1} & \cdots & a_{i_k i_k} \end{vmatrix}, \quad k = 1, \dots, n.$$

Here, in particular, $S_1 = \mathrm{Tr}(A)$, $S_n = \det(A)$, $S_{n-1} = \sum_{i=1}^{n} A_{ii}$, where A_{ii} is the algebraic complement of the diagonal element a_{ii} of the matrix A.

Using the traces of the kth order, we can represent the characteristic equation (1.23) in the form

$$\lambda^n - S_1 \lambda^{n-1} + S_2 \lambda^{n-2} - \cdots + (-1)^n S_n = 0. \tag{1.24}$$

Besides, via S_k we can define the Hurwitz matrix as follows:

$$\begin{pmatrix} -S_1 & -S_3 & -S_5 & \cdots & 0 \\ 1 & S_2 & S_4 & \cdots & 0 \\ 0 & -S_1 & -S_3 & \cdots & 0 \\ 0 & 1 & S_2 & \cdots & 0 \\ \cdots & \cdots & \cdots & \cdots & \cdots \\ \cdots & \cdots & \cdots & \cdots & (-1)^n S_n \end{pmatrix}. \tag{1.25}$$

Theorem 1.2 (Routh–Hurwitz criterion) *All roots λ of the characteristic equation* (1.24) *have negative real parts if and only if*

$$\Delta_1 = -S_1 > 0, \quad \Delta_2 = \begin{vmatrix} -S_1 & -S_3 \\ 1 & S_2 \end{vmatrix} > 0,$$

$$\Delta_3 = \begin{vmatrix} -S_1 & -S_3 & -S_5 \\ 1 & S_2 & S_4 \\ 0 & -S_1 & -S_3 \end{vmatrix} > 0, \quad \dots,$$

$$\Delta_n = \begin{vmatrix} -S_1 & -S_3 & -S_5 & \dots & 0 \\ 1 & S_2 & S_4 & \dots & 0 \\ 0 & -S_1 & -S_3 & \dots & 0 \\ 0 & 1 & S_2 & \dots & 0 \\ \dots & \dots & \dots & \dots & \dots \\ \dots & \dots & \dots & \dots & (-1)^n S_n \end{vmatrix} > 0.$$

(1.26)

Definition 1.2 A symmetric $n \times n$ matrix Q is called a positive (negative) definite matrix $(Q > 0)$ $(Q < 0)$ if there exists $c > 0$ such that $x'Qx \geq c|x|^2$ $(x'Qx \leq -c|x|^2)$ for all $x \in \mathbf{R}^n$.

Theorem 1.3 (Lyapunov theorem) *All roots λ of the characteristic equation* (1.24) *have negative real parts if and only if for an arbitrary positive definite matrix Q, the matrix equation*

$$A'P + PA = -Q \tag{1.27}$$

has a positive definite solution P.

Corollary 1.1 *Let A be a 2×2 matrix. Then the zero solution of* (1.22) *is asymptotically stable if and only if*

$$\mathrm{Tr}(A) < 0 \quad and \quad \det(A) > 0. \tag{1.28}$$

Proof It is enough to note that from (1.25) and (1.26) it follows that $\Delta_1 = -S_1 = -\mathrm{Tr}(A) > 0$ and $\Delta_2 = -S_1 S_2 = -\mathrm{Tr}(A)\det(A) > 0$. □

Corollary 1.2 *Let A be a 3×3 matrix. Then the zero solution of* (1.22) *is asymptotically stable if and only if*

$$S_1 < 0, \quad S_1 S_2 < S_3 < 0.$$

Proof It is enough to note that from (1.25) and (1.26) it follows that $\Delta_1 = -S_1 > 0$, $\Delta_2 = S_3 - S_1 S_2 > 0$, $\Delta_3 = S_3(S_1 S_2 - S_3) > 0$. □

Remark 1.1 If

$$A = \begin{pmatrix} a_{11} & a_{12} \\ a_{21} & a_{22} \end{pmatrix}, \quad Q = \begin{pmatrix} q_{11} & q_{12} \\ q_{12} & q_{22} \end{pmatrix}, \quad P = \begin{pmatrix} p_{11} & p_{12} \\ p_{12} & p_{22} \end{pmatrix},$$

then

$$PA = \begin{pmatrix} p_{11}a_{11} + p_{12}a_{21} & p_{11}a_{12} + p_{12}a_{22} \\ p_{12}a_{11} + p_{22}a_{21} & p_{12}a_{12} + p_{22}a_{22} \end{pmatrix},$$

and therefore the matrix equation (1.27) can be represented in the form of the system of the equations

$$2(p_{11}a_{11} + p_{12}a_{21}) = -q_{11},$$

$$2(p_{12}a_{12} + p_{22}a_{22}) = -q_{22},$$

$$p_{11}a_{12} + p_{12}\mathrm{Tr}(A) + p_{22}a_{21} = -q_{12},$$

which, for arbitrary positive definite matrix Q, has a positive definite solution P with the elements

$$p_{11} = \frac{(a_{22}^2 + \det(A))q_{11} + a_{21}^2 q_{22} - 2a_{22}a_{21}q_{12}}{2|\mathrm{Tr}(A)|\det(A)},$$

$$p_{22} = \frac{(a_{11}^2 + \det(A))q_{22} + a_{12}^2 q_{11} - 2a_{11}a_{12}q_{12}}{2|\mathrm{Tr}(A)|\det(A)},$$

$$p_{12} = \frac{a_{12}a_{22}q_{11} + a_{21}a_{11}q_{22} - 2a_{11}a_{22}q_{12}}{2\mathrm{Tr}(A)\det(A)}.$$

If, in particular, $q_{11} = q > 0$, $q_{22} = 1$, and $q_{12} = 0$, then

$$p_{11} = \frac{(a_{22}^2 + \det(A))q + a_{21}^2}{2|\mathrm{Tr}(A)|\det(A)},$$

$$p_{22} = \frac{a_{11}^2 + \det(A) + a_{12}^2 q}{2|\mathrm{Tr}(A)|\det(A)}, \tag{1.29}$$

$$p_{12} = \frac{a_{12}a_{22}q + a_{21}a_{11}}{2\mathrm{Tr}(A)\det(A)}.$$

Remark 1.2 In the general case the elements p_{ij} of the solution P of the matrix equation (1.27) are defined as follows [21]:

$$p_{ij} = \frac{1}{2\Delta_n} \sum_{r=0}^{n-1} \gamma_{ij}^{(r)} \Delta_{1,r+1},$$

where Δ_n is the determinant (1.26) of the Hurwitz matrix (1.25), $\Delta_{1,r+1}$ is the algebraic adjunct of the element of the first line and $(r + 1)$th column of the determinant Δ_n, $\gamma_{ij}^{(r)}$ are defined by the identity

$$(-1)^{n-1} \sum_{k,m=1}^{n} q_{km} D_{ik}(\lambda) D_{jm}(-\lambda) \equiv \sum_{r=0}^{n-1} \gamma_{ij}^{(r)} \lambda^{2(n-r-1)},$$

q_{km} are the elements of the matrix Q, and $D_{ik}(\lambda)$ are the algebraic adjuncts of the determinant

$$D(\lambda) = \begin{vmatrix} a_{11} - \lambda & \cdots & a_{1n} \\ \cdots & \cdots & \cdots \\ a_{n1} & \cdots & a_{nn} - \lambda \end{vmatrix}.$$

Example 1.10 Consider the second-order scalar differential equation

$$\ddot{x}(t) + a\dot{x}(t) + bx(t) = 0. \tag{1.30}$$

Using the new variables $x_1 = x$, $x_2 = \dot{x}$, represent (1.30) in the form of (1.22) with the matrix $A = \begin{pmatrix} 0 & 1 \\ -b & -a \end{pmatrix}$. Since $\mathrm{Tr}(A) = -a$, $\det(A) = b$, then via Corollary 1.1 the inequalities $a > 0$, $b > 0$ are necessary and sufficient conditions for the asymptotic stability of the zero solution of (1.30).

Chapter 2
Stochastic Functional Differential Equations and Procedure of Constructing Lyapunov Functionals

2.1 Short Introduction to Stochastic Functional Differential Equations

Here the basic notation of the theory of stochastic differential equations [16, 17, 81, 84–87, 210] is considered.

Let $\{\Omega, F, \mathbf{P}\}$ be a probability space, $\{F_t, t \geq 0\}$ be a nondecreasing family of sub-σ-algebras of F, i.e., $F_{t_1} \subset F_{t_2}$ for $t_1 < t_2$, $\mathbf{P}\{\cdot\}$ be the probability of an event enclosed in the braces, and $\mathbf{E}$ be the mathematical expectation.

2.1.1 Wiener Process and Its Numerical Simulation

The Wiener process sometimes is also called the Brownian motion process. Originally, the Brownian motion process was posed by the English botanist Robert Brown as a model for the motion of a small particle immersed in a liquid and thus subject to molecular collisions. The Brownian motion assumes a central role in the theory of stochastic processes and statistics. It is basic to descriptions of financial markets, the construction of a large class of Markov processes called diffusions, approximations to many queuing models, and the calculation of asymptotic distributions in large sample statistical estimation problems.

Definition 2.1 A stochastic process $w(t)$ is called the standard Wiener process (relatively to the family $\{F_t, t \geq 0\}$) if it is F_t-measurable and

- $w(0) = 0$ ($\mathbf{P}$-a.s.);
- $w(t)$ is a process with stationary and mutually independent increments;
- the increments $w(t) - w(s)$ have the normal distribution with

$$\mathbf{E}\big(w(t) - w(s)\big) = 0, \quad \mathbf{E}\big(w(t) - w(s)\big)^2 = |t - s|;$$

- for almost all $\omega \in \Omega$, the functions $w(t) = w(t, \omega)$ are continuous on $t \geq 0$ [84].

L. Shaikhet, *Lyapunov Functionals and Stability of Stochastic Functional Differential Equations*, DOI 10.1007/978-3-319-00101-2_2,
© Springer International Publishing Switzerland 2013

We will also consider an m-dimensional Wiener process

$$w(t) = \big(w_1(t), \ldots, w_m(t)\big)' \in \mathbf{R}^m$$

where the components $w_i(t)$, $i = 1, \ldots, m$, are mutually independent scalar Wiener processes, and

$$\mathbf{E}\big(w(t) - w(s)\big) = 0, \quad \mathbf{E}\big(w(t) - w(s)\big)\big(w(t) - w(s)\big)' = I|t - s|.$$

Here I is the $m \times m$ identity matrix, and the prime denotes the transposition.

The trajectories of the Wiener process are nondifferentiable functions, although formally the derivative of the Wiener process $\dot{w}(t)$ is called the white noise.

There are different ways to get numerical simulation of trajectories of a Wiener process. One of them is the following [244].

Let Y_i, $i = 1, \ldots, n$, be independent random variables that are uniformly distributed on $[0, 1]$. Then $X_i = \sqrt{12}(Y_i - 0.5) = \sqrt{3}(2Y_i - 1)$, $i = 1, \ldots, n$, are independent identically distributed random variables such that $\mathbf{E}X_i = 0$ and $\mathrm{Var}(X_i) = 1$. Define the random walk S_n, $n \geq 0$, by $S_0 = 0$ and $S_n = X_1 + \cdots + X_n$ for $n > 0$. By the central limit theorem, $\frac{1}{\sqrt{n}}S_n$ converges in distribution to $N(0, 1)$, i.e., $\frac{1}{\sqrt{n}}S_n \to N(0, 1)$.

Define the continuous-time process $W_n(t) = \frac{1}{\sqrt{n}}S_{[nt]}$, $t \geq 0$, where $[t]$ is the integer part of t, i.e., the greatest integer less than or equal to t. Therefore, for any $t > 0$, we have

$$W_n(t) = \sqrt{\frac{[nt]}{n}} \, \frac{S_{[nt]}}{\sqrt{[nt]}} \to N(0, t).$$

Also, for $t > s$, we obtain

$$W_n(t) - W_n(s) = \frac{S_{[nt]} - S_{[ns]}}{\sqrt{n}} = \frac{\sum_{j=[ns]+1}^{[nt]} X_j}{\sqrt{n}} = \frac{S_{[nt]-[ns]}}{\sqrt{n}}$$

$$= \sqrt{\frac{[nt] - [ns]}{n}} \, \frac{S_{[nt]-[ns]}}{\sqrt{[nt] - [ns]}} \to N(0, t - s).$$

Since the process $\{W_n(t), t \geq 0\}$ is not continuous, let us modify it in the following way:

$$W_n^{(c)}(t) = \frac{S_{[nt]}}{\sqrt{n}} + \big(nt - [nt]\big)\frac{X_{[nt]+1}}{\sqrt{n}}, \quad t \geq 0.$$

It is easy to see that $\mathbf{E}W_n^{(c)}(t) = 0$ and

$$\lim_{n\to\infty} \mathrm{Var}\big(W_n^{(c)}(t)\big) = \lim_{n\to\infty}\left(t - \frac{nt - [nt]}{n} + \frac{(nt - [nt])^2}{n}\right) = t.$$

So, as $n \to \infty$, $W_n^{(c)}(t)$ converges in distribution to the Wiener process $w(t)$. This means that for large enough n, the process $W_n^{(c)}(t)$ approximates the Wiener process

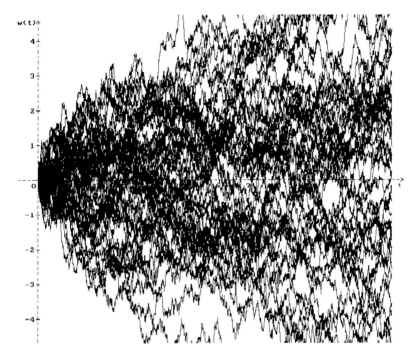

Fig. 2.1   50 trajectories of the Wiener process

well enough. 50 trajectories of the Wiener process obtained via this algorithm are
shown in Fig. 2.1.

2.1.2   Itô Integral, Itô Stochastic Differential Equation, and Itô Formula

Let $H_2[0, T]$ be the space of random functions $f(t)$ that are defined and F_t-
measurable for each $t \in [0, T]$ and for which

$$\int_0^t \mathbf{E} f^2(s)\, ds < \infty.$$

Then for all functions from $H_2[0, T]$, the Itô integral with respect to the Wiener
process $w(t)$

$$\int_0^t f(s)\, dw(s)$$

is defined and has the following properties:

$$\mathbf{E}\int_0^t f(s)\,dw(s) = 0, \quad \mathbf{E}\left(\int_0^t f(s)\,dw(s)\right)^2 = \int_0^t \mathbf{E}f^2(s)\,ds.$$

Let H_p, $p > 0$, be the space of F_0-adapted stochastic processes $\varphi(\theta)$, $\theta \le 0$, with continuous trajectories that are independent on the σ-algebra $B_{[0,\infty)}(dw)$, where $B_{[t_1,t_2]}(dw)$ is the minimal σ-algebra generated by the random variables $w(s) - w(t)$ for arbitrary $s, t : t_1 \le t \le s \le t_2$. In the space H_p two norms are defined: $\|\varphi\|_0 = \sup_{s \le 0} |\varphi(s)|$ and $\|\varphi\|_1^p = \sup_{s \le 0} \mathbf{E}|\varphi(s)|^p$.

We will consider the Itô stochastic functional differential equation [84]

$$dx(t) = a_1(t, x_t)\,dt + a_2(t, x_t)\,dw(t),$$
$$t \ge 0, \ x \in \mathbf{R}^n, \tag{2.1}$$

with the initial condition

$$x_0 = \phi \in H_p. \tag{2.2}$$

Here $x \in \mathbf{R}^n$, $x_t = x(t + s)$, $s \le 0$, $w : [0, \infty) \to \mathbf{R}^m$ is the standard Wiener process, the continuous functionals $a_1(t, \varphi)$, $a_2(t, \varphi)$ are defined on $[0, \infty) \times H_p$, $a_1 \in \mathbf{R}^n$, a_2 is an $n \times m$-dimensional matrix. It is assumed also that the functionals a_i, $i = 1, 2$, satisfy the following conditions: $a_i(t, 0) \equiv 0$, and for arbitrary functions $\varphi_1(\theta)$, $\varphi_2(\theta)$ from H_p,

$$\left|a_i(t, \varphi_1) - a_i(t, \varphi_2)\right|^2 \le \int_0^\infty \left|\varphi_1(-\theta) - \varphi_2(-\theta)\right|^2 dR_i(\theta),$$
$$\int_0^\infty dR_i(\theta) < \infty, \quad i = 1, 2, \tag{2.3}$$

where $R_i(\theta)$ are nondecreasing bounded functions.

A solution of problem (2.1)–(2.2) is a process $x(t)$ such that $x(\theta) = \phi(\theta)$ for $\theta \le 0$ and with probability 1

$$x(t) = x(0) + \int_0^t a_1(s, x_s)\,ds + \int_0^t a_2(s, x_s)\,dw(s), \quad t \ge 0,$$
$$x(0) = \phi(0).$$

The last integral is understood in the Itô sense.

Sometimes, instead of $x(t)$, we will write $x(t, \phi)$ for the solution of (2.1) with the initial function (2.2). Existence and uniqueness theorems for problem (2.1)–(2.2) are considered in [84–87, 132, 133].

To calculate the stochastic differential of the process $\eta(t) = u(t, x(t))$, where $x(t)$ is a solution of problem (2.1)–(2.2), and the function $u : [0, \infty) \times \mathbf{R}^n \to \mathbf{R}$ has continuous partial derivatives

$$u_t = \frac{\partial u(t,x)}{\partial t}, \qquad \nabla u = \left(\frac{\partial u(t,x)}{\partial x_1}, \dots, \frac{\partial u(t,x)}{\partial x_n} \right),$$

$$\nabla^2 u = \left(\frac{\partial^2 u(t,x)}{\partial x_i \partial x_j} \right), \quad i,j = 1, \dots, n,$$

the following Itô formula [84] is used:

$$d\eta(t) = Lu\big(t, x(t)\big)\, dt + \nabla u'\big(t, x(t)\big) a_2(t, x_t)\, dw(t). \tag{2.4}$$

The operator L is called the generator of (2.1) and is defined in the following way:

$$Lu\big(t, x(t)\big) = u_t\big(t, x(t)\big) + \nabla u'\big(t, x(t)\big) a_1(t, x_t)$$

$$+ \frac{1}{2} \mathrm{Tr}\big[a_2'(t, x_t) \nabla^2 u\big(t, x(t)\big) a_2(t, x_t) \big], \tag{2.5}$$

where Tr denotes the trace of a matrix.

The generator L can be applied also for some functionals $V(t, \varphi) : [0, \infty) \times H_p \to \mathbf{R}_+$. Suppose that a functional $V(t, \varphi)$ can be represented in the form $V(t, \varphi) = V(t, \varphi(0), \varphi(\theta))$, $\theta < 0$, and for $\varphi = x_t$, put

$$V_\varphi(t, x) = V(t, \varphi) = V(t, x_t) = V\big(t, x, x(t + \theta)\big),$$

$$x = \varphi(0) = x(t), \ \theta < 0. \tag{2.6}$$

Denote by D the set of the functionals for which the function $V_\varphi(t, x)$ defined by (2.6) has a continuous derivative with respect to t and two continuous derivatives with respect to x. For functionals from D, the generator L of (2.1) has the form

$$LV(t, x_t) = \frac{\partial V_\varphi(t, x(t))}{\partial t} + \nabla V_\varphi'\big(t, x(t)\big) a_1(t, x_t)$$

$$+ \frac{1}{2} \mathrm{Tr}\big[a_2'(t, x_t) \nabla^2 V_\varphi\big(t, x(t)\big) a_2(t, x_t) \big]. \tag{2.7}$$

From the Itô formula it follows that for functionals from D,

$$\mathbf{E}\big[V(t, x_t) - V(s, x_s) \big] = \int_s^t \mathbf{E} LV(\tau, x_\tau)\, d\tau, \quad t \geq s. \tag{2.8}$$

Together with (2.1), we will also consider the stochastic differential equation of neutral type [134]

$$d\big(x(t) - G(t, x_t)\big) = a_1(t, x_t)\, dt + a_2(t, x_t)\, dw(t), \quad t \geq 0,$$

$$x_0 = \phi \in H_p, \tag{2.9}$$

with the additional conditions on the functional $G(t, \varphi)$: $G(t, 0) = 0$,

$$\left|G(t, \varphi)\right| \leq \int_0^\infty \left|\varphi(-s)\right| dK(s), \quad \int_0^\infty dK(s) < 1. \tag{2.10}$$

2.2 Stability of Stochastic Functional Differential Equations

2.2.1 Definitions of Stability and Basic Lyapunov-Type Theorems

Definition 2.2 The solution $x(t)$ of (2.1) with the initial function (2.2) for some $p > 0$ is called:

- Uniformly p-bounded if $\sup_{t \geq 0} \mathbf{E}|x(t)|^p < \infty$.
- Asymptotically p-trivial if $\lim_{t \to \infty} \mathbf{E}|x(t)|^p = 0$.
- p-integrable if $\int_0^\infty \mathbf{E}|x(t)|^p \, dt < \infty$.

Definition 2.3 The trivial solution of (2.1) for some $p > 0$ is called:

- p-stable if for each $\varepsilon > 0$, there exists $\delta > 0$ such that $\mathbf{E}|x(t, \phi)|^p < \varepsilon$, $t \geq 0$, provided that $\|\phi\|_1^p < \delta$.
- Asymptotically p-stable if it is p-stable and for each initial function ϕ, the solution $x(t)$ of (2.1) is asymptotically p-trivial.
- Exponentially p-stable if it is p-stable and there exists $\lambda > 0$ such that for each initial function ϕ, there exists $C > 0$ (which may depend on ϕ) such that $\mathbf{E}|x(t, \phi)|^p \leq Ce^{-\lambda t}$ for $t > 0$.
- Stable in probability if for any $\varepsilon_1 > 0$ and $\varepsilon_2 > 0$, there exists $\delta > 0$ such that the solution $x(t, \phi)$ of (2.1) satisfies the condition $\mathbf{P}\{\sup_{t \geq 0} |x(t, \phi)| > \varepsilon_1 / F_0\} < \varepsilon_2$ for any initial function ϕ such that $\mathbf{P}\{\|\phi\|_0 < \delta\} = 1$.

In particular, if $p = 2$, then the solution of (2.1) is called respectively mean-square bounded, mean-square stable, asymptotically mean-square stable, and so on.

Definition 2.4 A nonnegative functional $V(t, \varphi)$, defined on $[0, \infty) \times H_p$, such that $V(t, 0) \equiv 0$ and $\lim_{t \to 0} \mathbf{E}V(t, x_t) = 0$ if $\lim_{t \to 0} \mathbf{E}|x(t)|^p = 0$, $p > 0$, is called an F_p-functional.

Certain stability conditions for stochastic functional differential equations can be stated in terms of Lyapunov functionals. In the sequel, c_i are different positive numbers.

Theorem 2.1 *Let $V : [t_0, \infty) \times H_p \to \mathbf{R}_+$ be a continuous functional such that for any solution $x(t)$ of problem (2.1)–(2.2) and $p \geq 2$, the following inequalities hold:*

$$\mathbf{E}V(t, x_t) \geq c_1 \mathbf{E}\left|x(t)\right|^p, \quad t \geq 0, \tag{2.11}$$

$$EV(0, \phi) \le c_2 \|\phi\|_1^p, \tag{2.12}$$

$$E\big[V(t, x_t) - V(0, \phi)\big] \le -c_3 \int_0^t E|x(s)|^p \, ds, \quad t \ge 0. \tag{2.13}$$

Then the trivial solution of (2.1) *is asymptotically p-stable.*

Proof From (2.11)–(2.13) we have

$$c_1 E|x(t)|^p \le EV(t, x_t) \le EV(0, \phi) \le c_2 \|\phi\|_1^p. \tag{2.14}$$

This proves the *p*-stability.

To prove the asymptotic *p*-stability, let us show that the solution of (2.1) is asymptotically *p*-trivial for any initial function ϕ. Note that from (2.14) we obtain

$$\sup_{t \ge 0} E|x(t)|^p \le \frac{c_2}{c_1} \|\phi\|_1^p. \tag{2.15}$$

From (2.13) and (2.12) it follows that

$$\int_0^\infty E|x(s)|^p \, ds \le \frac{1}{c_3} EV(0, \phi) \le \frac{c_2}{c_3} \|\phi\|_1^p < \infty, \tag{2.16}$$

i.e., the solution of (2.1) is *p*-integrable. Applying the generator L to the function $|x(t)|^p$ via (2.5), we obtain

$$\begin{aligned}
EL|x(t)|^p = \frac{p}{2}\Big[&2E|x(t)|^{p-2} x'(t)a_1(t, x_t) \\
&+ E|x(t)|^{p-2} \mathrm{Tr}\big[a_2'(t, x_t)a_2(t, x_t)\big] \\
&+ (p-2)E|x(t)|^{p-4}|x'(t)a_2(t, x_t)|^2\Big].
\end{aligned}$$

By the Hölder inequality, (2.3), and (2.15), there is a constant c_4 such that

$$\begin{aligned}
2E|x(t)|^{p-2}&|x'(t)a_1(t, x_t)| \\
&\le E|x(t)|^{p-2}\big[|x(t)|^2 + |a_1(t, x_t)|^2\big] \\
&\le E|x(t)|^p + \int_0^\infty E|x(t)|^{p-2}|x(t-\theta)|^2 \, dR_1(\theta) \\
&\le E|x(t)|^p + \int_0^\infty \big(E|x(t)|^p\big)^{\frac{p-2}{p}} \big(E|x(t-\theta)|^p\big)^{\frac{2}{p}} \, dR_1(\theta) \\
&\le c_4.
\end{aligned}$$

Analogously,

$$E|x(t)|^{p-2}\big|\mathrm{Tr}\big[a_2'(t, x_t)a_2(t, x_t)\big]\big| \le c_4,$$

$$\mathbf{E}\big|x(t)\big|^{p-4}\big|x'(t)a_2(t,x_t)\big|^2 \le c_4.$$

Hence, there exists a constant c_5 such that $|\mathbf{E}L|x(t)|^p| \le c_5$, and using (2.8) for $t_2 \ge t_1 \ge 0$, we obtain

$$\big|\mathbf{E}\big|x(t_2)\big|^p - \mathbf{E}\big|x(t_1)\big|^p\big| \le c_5(t_2 - t_1),$$

i.e., the function $\mathbf{E}|x(t)|^p$ satisfies the Lipschitz condition. From this, (2.15), and (2.16) it follows that $\lim_{t\to\infty} \mathbf{E}|x(t)|^p = 0$. The proof is completed. □

Remark 2.1 From (2.8) it follows that for the functional $V \in D$, condition (2.13) in Theorem 2.1 follows from the inequality

$$\mathbf{E}LV(t,x_t) \le -c_3\mathbf{E}\big|x(t)\big|^p, \quad t \ge 0. \tag{2.17}$$

Theorem 2.2 *Let there exist a functional $V(t,\varphi) \in D$ such that for any solution $x(t)$ of problem (2.1)–(2.2) and $p \ge 2$, the following inequalities hold:*

$$V(t,x_t) \ge c_1\big|x(t)\big|^p, \tag{2.18}$$

$$V(0,\phi) \le c_2\|\phi\|_0^p, \tag{2.19}$$

$$LV(t,x_t) \le 0, \quad t \ge 0, \tag{2.20}$$

$c_i > 0$, for any initial function ϕ such that $\mathbf{P}\{\|\phi\|_0 \le \delta\} = 1$, where $\delta > 0$ is small enough. Then the zero solution of (2.1) is stable in probability.

Proof Let us suppose that $\mathbf{P}\{\|\phi\|_0 \le \delta\} = 1$. From (2.20) it follows that the process $V(t,x_t)$ is a supermartingale. By (2.18), (2.19), and the inequality for supermartingales [84–87] we have

$$\mathbf{P}\left\{\sup_{t\ge t_0}\big|x(t,\phi)\big| > \varepsilon_1/F_0\right\} \le \mathbf{P}\left\{\sup_{t\ge t_0} V(t,x_t) > c_1\varepsilon_1^p/F_0\right\} \le \frac{V(t_0,\phi)}{c_1\varepsilon_1^p} \le \frac{c_2\delta^p}{c_1\varepsilon_1^p} < \varepsilon_2$$

for $\delta < \varepsilon_1(c_1\varepsilon_2/c_2)^{1/p}$. The theorem is proven. □

Theorem 2.3 *Let there exist a functional $W : [0,\infty) \times H_2 \to \mathbf{R}_+$ satisfying the condition $\mathbf{E}W(t,\varphi) \le c_1\|\varphi\|_1^2$ and such that for the functional*

$$V(t,\varphi) = W(t,\varphi) + \big|\varphi(0) - G(t,\varphi)\big|^2, \tag{2.21}$$

where $G(t,\varphi)$ satisfies condition (2.10), the following estimates are valid:

$$\mathbf{E}V(0,\phi) \le c_2\|\phi\|_1^2,$$

$$\mathbf{E}V(t,x_t) - \mathbf{E}V(0,\phi) \le -c_3\int_0^t \mathbf{E}\big|x(s)\big|^2\,ds, \quad t \ge 0, \tag{2.22}$$

where c_i, $i = 1, 2, 3$, are some positive constants. Then the zero solution of (2.9) is asymptotically mean-square stable.

The proof of Theorem 2.3 is similar to Theorem 2.1 and can be found in [132–135].

Theorem 2.4 *Let there exist a functional $V(t, \varphi) \in D$ such that for some $p > 0$ and $\lambda > 0$, the following conditions hold:*

$$\mathbf{E}V(t, x_t) \geq c_1 e^{\lambda t} \mathbf{E}|x(t)|^p, \quad t \geq 0, \tag{2.23}$$

$$\mathbf{E}V(0, \phi) \leq c_2 \|\phi\|_1^p, \tag{2.24}$$

$$\mathbf{E}LV(t, x_t) \leq 0, \quad t \geq 0. \tag{2.25}$$

Then the trivial solution of (2.1) is exponentially p-stable.

Proof Integrating (2.25) via (2.8), we obtain $\mathbf{E}V(t, x_t) \leq \mathbf{E}V(0, \phi)$. From this and from (2.23)–(2.24) it follows that

$$c_1 \mathbf{E}|x(t)|^p \leq e^{-\lambda t} \mathbf{E}V(0, \phi) \leq c_2 \|\phi\|_1^p.$$

The inequality $c_1 \mathbf{E}|x(t)|^2 \leq c_2 \|\phi\|_1^p$ means that the trivial solution of (2.1) is p-stable. Besides, from the inequality $c_1 \mathbf{E}|x(t)|^2 \leq e^{-\lambda t} \mathbf{E}V(0, \phi)$ it follows that the trivial solution of (2.1) is exponentially p-stable. The proof is completed. □

Corollary 2.1 *Let there exist a functional $V_0(t, \varphi) \in D$ such that for some $p > 0$, the following conditions hold:*

$$c_1 \mathbf{E}|x(t)|^p \leq \mathbf{E}V_0(t, x_t)$$

$$\leq c_2 \mathbf{E}|x(t)|^p + \sum_{i=0}^{m} \int_0^{\infty} \int_{t-\theta}^{t} (s - t + \theta)^i \mathbf{E}|x(s)|^p \, ds \, dK_i(\theta), \tag{2.26}$$

$$\mathbf{E}LV_0(t, x_t) \leq -c_3 \mathbf{E}|x(t)|^p, \tag{2.27}$$

where $m \geq 0$, $K_i(\theta)$, $i = 0, 1, \ldots, m$, are nondecreasing functions such that, for some small enough $\lambda > 0$,

$$\sum_{i=0}^{m} \eta_i(\lambda) < \infty, \quad \eta_i(\lambda) = \frac{1}{i+1} \int_0^{\infty} e^{\lambda \theta} \theta^{i+1} \, dK_i(\theta). \tag{2.28}$$

Then the trivial solution of (2.1) is exponentially p-stable.

Proof It is enough to show that by the conditions (2.26)–(2.28) there exists a functional $V(t, \varphi)$ that satisfies the conditions of Theorem 2.4. Indeed, put $V_1(t, \varphi) = e^{\lambda t} V_0(t, \varphi)$. By (2.26) and (2.27) we have

$$\mathbf{E}LV_1(t, x_t) = \lambda e^{\lambda t}\,\mathbf{E}V_0(t, x_t) + e^{\lambda t}\,\mathbf{E}LV_0(t, x_t)$$

$$\leq e^{\lambda t}\left(\lambda\left(c_2\mathbf{E}|x(t)|^p + \sum_{i=0}^{m}\int_0^{\infty}\int_{t-\theta}^{t}(s-t+\theta)^i\,\mathbf{E}|x(s)|^p\,ds\,dK_i(\theta)\right)\right.$$

$$\left. - c_3\mathbf{E}|x(t)|^p\right)$$

$$= e^{\lambda t}\left((\lambda c_2 - c_3)\mathbf{E}|x(t)|^p\right.$$

$$\left. + \lambda\sum_{i=0}^{m}\int_0^{\infty}\int_{t-\theta}^{t}(s-t+\theta)^i\,\mathbf{E}|x(s)|^p\,ds\,dK_i(\theta)\right).$$

Now put

$$V_2(t, x_t) = \lambda\sum_{i=0}^{m}\frac{1}{i+1}\int_0^{\infty}\int_{t-\theta}^{t}e^{\lambda(s+\theta)}(s-t+\theta)^{i+1}|x(s)|^p\,ds\,dK_i(\theta).$$

Then

$$LV_2(t, x_t) = \lambda\sum_{i=0}^{m}\left(e^{\lambda t}\eta_i(\lambda)|x(t)|^p\right.$$

$$\left. - \int_0^{\infty}\int_{t-\theta}^{t}e^{\lambda(s+\theta)}(s-t+\theta)^i|x(s)|^p\,ds\,dK_i(\theta)\right)$$

$$= \lambda e^{\lambda t}\sum_{i=0}^{m}\left(\eta_i(\lambda)|x(t)|^p\right.$$

$$\left. - \int_0^{\infty}\int_{t-\theta}^{t}e^{\lambda(s-t+\theta)}(s-t+\theta)^i|x(s)|^p\,ds\,dK_i(\theta)\right),$$

and via $s \geq t - \theta$, for the functional $V = V_1 + V_2$, it follows that

$$\mathbf{E}LV(t, x_t) \leq e^{\lambda t}\left(\left(\lambda\left(c_2 + \sum_{i=0}^{m}\eta_i(\lambda)\right) - c_3\right)\mathbf{E}|x(t)|^p\right.$$

$$\left. - \lambda\sum_{i=0}^{m}\int_0^{\infty}\int_{t-\theta}^{t}\left(e^{\lambda(s-t+\theta)} - 1\right)\mathbf{E}|x(s)|^p\,ds\,dK_i\theta\right)$$

$$\leq e^{\lambda t}\left(\lambda\left(c_1 + \sum_{i=0}^{m}\eta_i(\lambda)\right) - c_3\right)\mathbf{E}|x(t)|^p.$$

From this and from (2.28), for small enough $\lambda > 0$, we obtain (2.25). It is easy to check that conditions (2.23)–(2.24) hold too. The proof is completed.   □

Example 2.1 Consider the linear stochastic differential equation

$$\dot{x}(t) = ax(t) + bx(t - h) + \sigma x(t - \tau)\dot{w}(t). \qquad (2.29)$$

For the functional

$$V(t, x_t) = x^2(t) + |b| \int_{t-h}^{t} x^2(s)\, ds + \sigma^2 \int_{t-\tau}^{t} x^2(s)\, ds, \qquad (2.30)$$

we have

$$LV(t, x_t) = 2x(t)\big(ax(t) + bx(t - h)\big) + \sigma^2 x^2(t - \tau)$$
$$+ |b|\big(x^2(t) - x^2(t - h)\big) + \sigma^2\big(x^2(t) - x^2(t - \tau)\big)$$
$$\leq \big(2(a + |b|) + \sigma^2\big)x^2(t). \qquad (2.31)$$

So, by the condition

$$a + |b| + \frac{1}{2}\sigma^2 < 0 \qquad (2.32)$$

the functional (2.30) satisfies the conditions of Theorem 2.1 with $p = 2$, and therefore the trivial solution of (2.29) is asymptotically mean-square stable.

On the other hand, by (2.31)–(2.32) the functional (2.30) satisfies the conditions of Corollary 2.1 with $p = 2$, $m = 0$, and $dK_0(s) = (|b|\delta(s - h) + \sigma^2\delta(s - \tau))\, ds$. Therefore, the trivial solution of (2.29) is exponentially mean-square stable. By Theorems 2.1–2.4 the construction of stability conditions is reduced to the construction of some Lyapunov functionals satisfying the assumptions of these theorems. Below we will use the general method of construction of Lyapunov functionals [136–139, 146–148, 150, 265, 269–272, 278] allowing one to obtain stability conditions immediately in terms of parameters of systems under consideration. This method was successfully used for functional differential equations, for difference equations with discrete and continuous time [278], and for partial differential equations [48].

2.2.2 Formal Procedure of Constructing Lyapunov Functionals

The formal procedure of constructing Lyapunov functionals consists of four steps.

Step 1   Let us represent (2.9) in the form

$$dz(t, x_t) = \big(b_1(t, x(t)) + c_1(t, x_t)\big)\, dt + \big(b_2(t, x(t)) + c_2(t, x_t)\big)\, dw(t), \qquad (2.33)$$

where $z(t, x_t)$ is some functional of x_t, $z(t, 0) = 0$, the functionals $b_i(t, x(t))$, $i = 1, 2$, depend on t and $x(t)$ only and do not depend on the previous values $x(t + s)$, $s < 0$, of the solution, and $b_i(t, 0) = 0$.

Step 2 Consider the auxiliary differential equation without memory

$$dy(t) = b_1(t, y(t)) dt + b_2(t, y(t)) dw(t). \tag{2.34}$$

Let us assume that the zero solution of (2.34) is asymptotically mean-square stable and therefore there exists a Lyapunov function $v(t, y)$ such that $c_1|y|^2 \le v(t, y) \le c_2|y|^2$ and $L_0 v(t, y) \le -c_3|y|^2$. Here L_0 is the generator of (2.34), $c_i > 0$, $i = 1, 2, 3$.

Step 3 Replacing the second argument y of the function $v(t, y)$ by the functional $z(t, x_t)$ from left-hand part of (2.33), we obtain the main component $V_1(t, x_t) = v(t, z(t, x_t))$ of the functional $V(t, x_t)$. Then it is necessary to calculate $L V_1$, where L is the generator of (2.33), and in a reasonable way to estimate it.

Step 4 Usually, the functional V_1 almost satisfies the requirements of stability theorems. In order to satisfy these conditions completely, an auxiliary component V_2 can be easily chosen by some standard way. As a result, the desired functional $V(t, x_t)$ takes the form $V = V_1 + V_2$.

Let us make remarks on some peculiarities of this procedure.

(1) It is clear that the representation (2.33) in the first step of the procedure is not unique. Hence, for different representations, one can construct different Lyapunov functionals, and, as a result, one can get different stability conditions.
(2) In the second step, for the auxiliary equation (2.34), one can choose different Lyapunov functions $v(t, y)$. So, for the initial equation (2.9), different Lyapunov functionals can be constructed, and, as a result, different stability conditions can be obtained.
(3) It is necessary to emphasize also that by choosing different ways of estimation of $L V_1$ one can construct different Lyapunov functionals and, as a result, one can get different stability conditions.
(4) At last, some standard way of the construction of the additional functional V_2 sometimes allows one to simplify the fourth step and do not use the functional V_2 at all. Below the corresponding auxiliary Lyapunov-type theorems are considered.

2.2.3 Auxiliary Lyapunov-Type Theorem

The following theorem in some cases allows one to use the procedure of constructing Lyapunov functionals without an additional functional V_2.

Theorem 2.5 *Let there exist a functional $V_1(t, x_t) \in D$ of type (2.21) such that*

$$\mathbf{E}LV_1(t, x_t) \leq \mathbf{E}x'(t)D(t)x(t) + \sum_{i=1}^{l} \int_0^{\infty} \mathbf{E}x'(t-s)S_i(t-s)x(t-s)\,dv_i(s)$$

$$+ \sum_{i=1}^{n} \int_0^{\infty} \mathbf{E}x_i^2(t-s)\,dK_i(s)$$

$$+ \sum_{i=1}^{k} \mathbf{E}x'\big(t - \tau_i(t)\big)Q_i\big(t - \tau_i(t)\big)x\big(t - \tau_i(t)\big)$$

$$+ \sum_{j=0}^{m} \int_0^{\infty} d\mu_j(s) \int_{t-s}^{t} (\theta - t + s)^j \mathbf{E}x'(\theta)R_j(\theta)x(\theta)\,d\theta$$

$$+ \int_0^{\infty} d\mu(s) \int_{t-s}^{t} \mathbf{E}x'(\tau)R(\tau + s, t)x(\tau)\,d\tau, \qquad (2.35)$$

where L is the generator of (2.9), $D(t)$ is a negative definite matrix, $S_i(t)$, $i = 1, \ldots, l$, $Q_i(t)$, $i = 1, \ldots, k$, $R_j(t)$, $j = 0, \ldots, m$, $R(s, t)$, $s \geq t \geq 0$, are non-negative definite matrices, $\tau_i(t)$, $i = 1, \ldots, k$, $t \geq 0$, are differentiable nonnegative functions with $\dot{\tau}_i(t) \leq \hat{\tau}_i < 1$, $K_i(s)$, $i = 1, \ldots, n$, $v_i(s)$, $i = 0, \ldots, l$, $\mu_j(s)$, $j = 0, \ldots, m$, and $\mu(s)$, $s \geq 0$, are nondecreasing functions of bounded variation such that

$$k_i = \int_0^{\infty} dK_i(s) < \infty, \qquad q_i = \int_0^{\infty} dv_i(s) < \infty,$$

$$r_j = \int_0^{\infty} \frac{s^{j+1}}{j+1}\,d\mu_j(s) < \infty, \tag{2.36}$$

and the matrix

$$G(t) = D(t) + K + \sum_{i=1}^{l} q_i S_i(t) + \sum_{i=1}^{k} \frac{1}{1 - \hat{\tau}_i} Q_i(t)$$

$$+ \sum_{j=0}^{m} r_j R_j(t) + \int_0^{\infty} d\mu(s) \int_t^{t+s} R(t+s, \theta)\,d\theta, \qquad (2.37)$$

where K is the diagonal matrix with elements k_i, $i = 1, \ldots, n$, is uniformly negative definite matrix with respect to $t \geq 0$, i.e.,

$$x'G(t)x \leq -c|x|^2, \qquad c > 0, \; x \in \mathbf{R}^n. \tag{2.38}$$

Then the zero solution of (2.9) is asymptotically mean-square stable.

Proof Put

$$V_2(t, x_t) = \sum_{i=1}^{l} \int_0^\infty dv_i(s) \int_{t-s}^{t} x'(\theta) S_i(\theta) x(\theta)\, d\theta$$

$$+ \sum_{i=1}^{n} \int_0^\infty dK_i(s) \int_{t-s}^{t} x_i^2(\theta)\, d\theta + \sum_{i=1}^{k} \frac{1}{1-\hat{\tau}_i} \int_{t-\tau_i(t)}^{t} x'(s) Q_i(s) x(s)\, ds$$

$$+ \sum_{j=0}^{m} \int_0^\infty d\mu_j(s) \int_{t-s}^{t} \frac{(\theta - t + s)^{j+1}}{j+1} x'(\theta) R_j(\theta) x(\theta)\, d\theta$$

$$+ \int_0^\infty d\mu(s) \int_{t-s}^{t} \int_t^{\tau+s} x'(\tau) R(\tau + s, \theta) x(\tau)\, d\theta\, d\tau.$$

Then

$$\mathbf{E}LV_2(t, x_t) = \sum_{i=1}^{l} q_i \mathbf{E} x'(t) S_i(t) x(t) - \int_0^\infty \mathbf{E} x'(t-s) S(t-s) x(t-s)\, dv(s)$$

$$+ \sum_{i=1}^{n} k_i \mathbf{E} x_i^2(t) - \sum_{l=1}^{n} \int_0^\infty \mathbf{E} x_l^2(t-s)\, dK_l(s)$$

$$+ \sum_{i=1}^{k} \frac{1}{1-\hat{\tau}_i} \mathbf{E} x'(t) Q_i(t) x(t)$$

$$- \sum_{i=1}^{k} \frac{1 - \dot{\tau}_i(t)}{1 - \hat{\tau}_i} \mathbf{E} x'\big(t - \tau_i(t)\big) Q_i\big(t - \tau_i(t)\big) x\big(t - \tau_i(t)\big)$$

$$+ \sum_{j=0}^{m} r_j \mathbf{E} x'(t) R_j(t) x(t)$$

$$- \sum_{j=0}^{m} \int_0^\infty d\mu_j(s) \int_{t-s}^{t} (\theta - t + s)^j \mathbf{E} x'(\theta) R_j(\theta) x(\theta)\, d\theta$$

$$+ \int_0^\infty d\mu(s) \int_t^{t+s} \mathbf{E} x'(t) R(t + s, \theta) x(t)\, d\theta$$

$$- \int_0^\infty d\mu(s) \int_{t-s}^{t} \mathbf{E} x'(\tau) R(\tau + s, t) x(\tau)\, d\tau. \tag{2.39}$$

From (2.35), (2.37), and (2.39) for the functional $V(t, x_t) = V_1(t, x_t) + V_2(t, x_t)$ it follows that

$$\mathbf{E}LV(t, x_t) \le \mathbf{E} x'(t) G(t) x(t). \tag{2.40}$$

By (2.38) and Remark 2.1 this means that the functional $V(t, x_t)$ satisfies conditions (2.22) of Theorem 2.3, and therefore the zero solution of (2.9) is asymptotically mean-square stable. The proof is completed.   □

Corollary 2.2 *Let in Theorem 2.5 inequality (2.35) be the exact equality, $G(t) = G = $ const, and the functional $V = V_1 + V_2$ be F_2-functional. Then in the scalar case the condition $G < 0$ is a necessary and sufficient condition for asymptotic mean-square stability of the zero solution of (2.9).*

Proof In the considered case, for the functional $V = V_1 + V_2$, we have $\mathbf{E}LV(t, x_t) = G\mathbf{E}x^2(t)$. If $G \geq 0$, then from this and from (2.8) it follows that

$$\mathbf{E}V(t, x_t) = \mathbf{E}V(0, \phi) + G \int_0^t \mathbf{E}x^2(\tau)\,d\tau \geq \mathbf{E}V(0, \phi) > 0.$$

This means that $\lim_{t \to \infty} \mathbf{E}V(t, x_t) \neq 0$ and therefore $\lim_{t \to \infty} \mathbf{E}|x(t)|^2 \neq 0$. The proof is completed.   □

Remark 2.2 Since the functional $V(t, x_t)$ constructed in Theorem 2.5 satisfies the conditions (2.22) and $V(t, x_t) \geq 0$, we have that $c_3 \int_0^t \mathbf{E}|x(s)|^2\,ds \leq \mathbf{E}V(0, \phi) < \infty$. This means that by conditions (2.35) and (2.38) the solution of (2.9) is also mean-square integrable.

Remark 2.3 In the scalar case, from Remark 2.2 it follows that if by condition (2.35) the solution of (2.9) is mean-square nonintegrable, i.e., $\int_0^\infty \mathbf{E}x^2(t)\,dt = \infty$, then $\sup_{t \geq 0} G(t) \geq 0$.

Remark 2.4 Theorem 2.5 is a useful development and improvement of the general method of construction of Lyapunov functionals. It allows one not to use Step 4 of the procedure and get good stability conditions using much more simple Lyapunov functional than via Theorem 2.3. It can be used in different applications.

2.3 Some Useful Statements

2.3.1 Linear Stochastic Differential Equation

Consider now conditions for asymptotic mean-square stability of the trivial solution of the linear Itô stochastic differential equation

$$\dot{x}(t) = Ax(t) + Bx(t - \tau) + \sigma x(t - h)\dot{w}(t), \tag{2.41}$$

where A, B, σ, $\tau \geq 0$, $h \geq 0$ are known constants.

Lemma 2.1 *A necessary and sufficient condition for asymptotic mean-square stability of the trivial solution of (2.41) is*

$$A + B < 0, \qquad G^{-1} > \frac{1}{2}\sigma^2, \qquad (2.42)$$

where

$$G = \begin{cases} \frac{Bq^{-1}\sin(q\tau)-1}{A+B\cos(q\tau)}, & B+|A| < 0, \ q = \sqrt{B^2 - A^2}, \\ \frac{1+|A|\tau}{2|A|}, & B = A < 0, \\ \frac{Bq^{-1}\sinh(q\tau)-1}{A+B\cosh(q\tau)}, & A+|B| < 0, \ q = \sqrt{A^2 - B^2}. \end{cases} \qquad (2.43)$$

Remark 2.5 If $A = -a$ and $B = 0$, then the necessary and sufficient stability condition (2.42)–(2.43) takes the form $a > \frac{1}{2}\sigma^2$.

Note that the proof of Lemma 2.1 is based on two old enough papers [243, 290] as it was shown briefly in the author recent book [278]. Following to advices and requests of some readers of the book [278], the author took the decision to write here the proof of this lemma in more detail.

Proof of the Lemma 2.1 A necessary and sufficient stability condition (2.42) with

$$G = 2\int_0^\infty x^2(s)\,ds, \qquad (2.44)$$

where $x(t)$ is a solution of (2.41) in the deterministic case, i.e., with $\sigma = 0$, was obtained in [243]. By the Plancherel theorem the integral (2.44) coincides [243, 290] with

$$G = \frac{2}{\pi}\int_0^\infty \frac{dt}{(A + B\cos\tau t)^2 + (t + B\sin\tau t)^2}. \qquad (2.45)$$

Let us obtain for this integral the representation (2.43) in elementary functions. Following [243], consider the functional

$$V(x_t) = \frac{1}{2}Gx^2(t) + \int_{t-\tau}^t \beta(s-t)x(s)x(t)\,ds$$
$$+ \int_{t-\tau}^t \int_s^t \delta(s-t, \theta-t)x(\theta)x(s)\,d\theta\,ds, \qquad (2.46)$$

where G is a constant, and $\beta(s)$ and $\delta(s,\theta)$ are continuously differentiable functions. By (2.46) and (2.41) with $\sigma = 0$ we obtain

$$\frac{dV(x_t)}{dt} = \left(Gx(t) + \int_{t-\tau}^t \beta(s-t)x(s)\,ds\right)\left(Ax(t) + Bx(t-\tau)\right)$$
$$+ \beta(0)x^2(t) - \beta(-\tau)x(t-\tau)x(t) - \int_{t-\tau}^t \frac{d\beta(s-t)}{ds}x(s)x(t)\,ds$$

$$+ \int_{t-\tau}^{t} \delta(s-t,0)x(t)x(s)\,ds - \int_{t-\tau}^{t} \delta(-\tau,\theta-t)x(\theta)x(t-\tau)\,d\theta$$

$$- \int_{t-\tau}^{t}\int_{s}^{t}\left(\frac{\partial\delta(s-t,\theta-t)}{\partial s}+\frac{\partial\delta(s-t,\theta-t)}{\partial\theta}\right)x(\theta)x(s)\,d\theta\,ds$$

$$= (GA+\beta(0))x^2(t) + (GB-\beta(-\tau))x(t)x(t-\tau)$$

$$+ \int_{t-\tau}^{t}\left(A\beta(s-t) - \frac{d\beta(s-t)}{ds} + \delta(0,s-t)\right)x(s)x(t)\,ds$$

$$+ \int_{t-\tau}^{t}\left(B\beta(s-t) - \delta(-\tau,s-t)\right)x(s)x(t-\tau)\,ds$$

$$- \int_{t-\tau}^{t}\int_{s}^{t}\left(\frac{\partial\delta(s-t,\theta-t)}{\partial s}+\frac{\partial\delta(s-t,\theta-t)}{\partial\theta}\right)x(\theta)x(s)\,d\theta\,ds.$$

$$(2.47)$$

Let us suppose that the functions $\beta(s)$ and $\delta(s,\theta)$ satisfy the conditions

$$GA+\beta(0)=-1, \qquad GB-\beta(-\tau)=0,$$

$$A\beta(s) - \frac{d\beta(s)}{ds} + \delta(s,0)=0, \qquad B\beta(s)-\delta(-\tau,s)=0, \qquad (2.48)$$

$$\frac{\partial\delta(s,\theta)}{\partial s}+\frac{\partial\delta(s,\theta)}{\partial\theta}=0.$$

Then from (2.47) and (2.48) it follows that $\frac{dV(x_t)}{dt}=-x^2(t)$, and therefore (if the condition for asymptotic stability of the trivial solution of the considered equation holds, i.e., $\lim_{t\to\infty}V(x_t)=0$),

$$\int_0^\infty x^2(t)\,dt = -\int_0^\infty \frac{dV(x_t)}{dt}\,dt = V(x_0). \qquad (2.49)$$

Using the initial function

$$x_\varepsilon(s) = \begin{cases} 0 & \text{if } -\tau \le s \le -\varepsilon, \\ 1+\frac{s}{\varepsilon} & \text{if } -\varepsilon \le s \le 0, \end{cases}$$

and the limit $\varepsilon \to 0$, from (2.46) we obtain $V(x_0)=\frac{1}{2}G$. From this and from (2.49) it follows that G in the functional (2.46) indeed coincides with (2.44) and has the representation (2.45).

To get for G the representation (2.43), let us solve system (2.48). From the last equation of (2.48) it follows that $\delta(s,\theta)=\varphi(s-\theta)$ and, by the forth equation of (2.48), $\varphi(s)=B\beta(-\tau-s)$. Substituting this $\varphi(s)$ into the third equation of (2.48), we obtain that the function $\beta(t)$ is defined by the differential equation

$$\dot{\beta}(t)=A\beta(t)+B\beta(-t-\tau) \qquad (2.50)$$

with the conditions

$$GA + \beta(0) + 1 = 0, \qquad GB - \beta(-\tau) = 0. \tag{2.51}$$

Suppose that $q^2 = B^2 - A^2 > 0$. Then, by (2.50),

$$\ddot{\beta}(t) = A\dot{\beta}(t) - B\dot{\beta}(-t - \tau)$$
$$= A\big(A\beta(t) + B\beta(-t-\tau)\big) - B\big(A\beta(-t-\tau) + B\beta(t+\tau-\tau)\big)$$
$$= -q^2\beta(t)$$

or

$$\ddot{\beta}(t) + q^2\beta(t) = 0. \tag{2.52}$$

Substituting the general solution $\beta(t) = C_1 \cos qt + C_2 \sin qt$ of (2.52) into (2.50) and (2.51), we obtain two equations for G, C_1, and C_2

$$GA + C_1 = -1, \qquad GB = C_1 \cos q\tau - C_2 \sin q\tau, \tag{2.53}$$

and two homogeneous linear dependent equations for C_1 and C_2

$$C_1(A + B\cos q\tau) - C_2(q + B\sin q\tau) = 0,$$
$$C_1(q - B\sin q\tau) + C_2(A - B\cos q\tau) = 0. \tag{2.54}$$

By (2.54) we have

$$C_2 = C_1 \frac{A + B\cos q\tau}{q + B\sin q\tau} = -C_1 \frac{q - B\sin q\tau}{A - B\cos q\tau}. \tag{2.55}$$

Substituting the first equality (2.55) into (2.53) and excluding C_1, we obtain

$$G = \frac{A\sin q\tau - q\cos q\tau}{q(A\cos q\tau + q\sin q\tau + B)}. \tag{2.56}$$

Multiplying the numerator and the denominator of the obtained fraction by $B\sin q\tau - q$, one can convert (2.56) to the form of the first line in (2.43). Note that the same result can be obtained using the second equality (2.55).

Suppose now that $q^2 = A^2 - B^2 > 0$. Then, similarly to (2.52), we obtain the equation $\ddot{\beta}(t) - q^2\beta(t) = 0$ with the general solution $\beta(t) = C_1 e^{qt} + C_2 e^{-qt}$. Substituting this solution into (2.50) and (2.51), similarly to (2.53) and (2.54), we have

$$GA + C_1 + C_2 = -1, \qquad GB = C_1 e^{-q\tau} + C_2 e^{q\tau},$$
$$C_1(q - A) - C_2 B e^{q\tau} = 0, \qquad C_1 B + C_2(q + A)e^{q\tau} = 0. \tag{2.57}$$

By the two last equations of (2.57),

$$C_2 = C_1 \frac{q - A}{B} e^{-q\tau} = -C_1 \frac{B}{q + A} e^{-q\tau}. \tag{2.58}$$

From the first equality of (2.58) and the two first equations of (2.57) we obtain

$$G = \frac{q - A + Be^{-q\tau}}{q(q - A - Be^{-q\tau})}.$$ (2.59)

Put now $\sinh x = \frac{1}{2}(e^x - e^{-x})$ and $\cosh x = \frac{1}{2}(e^x + e^{-x})$ (respectively, hyperbolic sine and hyperbolic cosine). Multiplying the numerator and the denominator of (2.59) by $B \sinh q\tau - q$ in the denominator, we have

$$\left(q - A - Be^{-q\tau}\right)(B \sinh q\tau - q)$$

$$= \left(q - A - Be^{-q\tau}\right)\left(B \cosh q\tau - Be^{-q\tau} - q\right)$$

$$= qB \cosh q\tau - AB \cosh q\tau - B^2 e^{-q\tau} \cosh q\tau$$

$$\quad - Bqe^{-q\tau} + ABe^{-q\tau} + B^2 e^{-2q\tau} - A^2 + B^2 + Aq + Bqe^{-q\tau}$$

$$= (q - A)(A + B \cosh q\tau) + Be^{-q\tau}\left(A + B\left(e^{q\tau} + e^{-q\tau} - \cosh q\tau\right)\right)$$

$$= \left(q - A + Be^{-q\tau}\right)(A + B \cosh q\tau).$$

As a result, we obtain (2.59) in the form of the third line in (2.43). Note also that the same result can be obtained using the second equality of (2.58).

The second line of (2.43) can be obtained from the first (or the third) line in the limit as $q \to 0$. The proof is completed.   □

2.3.2 System of Two Linear Stochastic Differential Equations

Consider the system of two stochastic differential equations without delays

$$\dot{x}_1(t) = a_{11}x_1(t) + a_{12}x_2(t) + \sigma_1 x_1(t)\dot{w}_1(t),$$
$$\dot{x}_2(t) = a_{21}x_1(t) + a_{22}x_2(t) + \sigma_2 x_2(t)\dot{w}_2(t),$$ (2.60)

where $a_{ij}, \sigma_i, i, j = 1, 2$, are constants, and $w_1(t)$ and $w_2(t)$ are mutually independent standard Wiener processes.

Put $A = \|a_{ij}\|, i, j = 1, 2$, and

$$\delta_i = \frac{1}{2}\sigma_i^2, \quad i = 1, 2.$$ (2.61)

Remark 2.6 If $\sigma_1 = \sigma_2 = 0$, then (by Corollary 1.1) the trivial solution of (2.60) is asymptotically stable if and only if

$$\text{Tr}(A) = a_{11} + a_{22} < 0, \qquad \det(A) = a_{11}a_{22} - a_{12}a_{21} > 0.$$ (2.62)

If $a_{12} = a_{21} = 0$, then (by Remark 2.5) the necessary and sufficient conditions for asymptotic mean-square stability of the trivial solution of (2.60) are

$$a_{11} + \delta_1 < 0, \qquad a_{22} + \delta_2 < 0.$$ (2.63)

Lemma 2.2 *Let for some positive definite matrix $P = \|p_{ij}\|$, $i, j = 1, 2$, the parameters of system (2.60) satisfy the conditions*

$$p_{12}a_{21} + p_{11}(a_{11} + \delta_1) < 0,$$

$$p_{12}a_{12} + p_{22}(a_{22} + \delta_2) < 0,$$

$$4\big(p_{12}a_{21} + p_{11}(a_{11} + \delta_1)\big)\big(p_{12}a_{12} + p_{22}(a_{22} + \delta_2)\big) \tag{2.64}$$

$$> \big(p_{11}a_{12} + p_{22}a_{21} + p_{12}\operatorname{Tr}(A)\big)^2.$$

Then the trivial solution of system (2.60) is asymptotically mean-square stable.

Proof Let L_0 be the generator of system (2.60). Using the Lyapunov function

$$v(t) = p_{11}x_1^2(t) + 2p_{12}x_1(t)x_2(t) + p_{22}x_2^2(t) \tag{2.65}$$

and (2.61)–(2.62) for system (2.60), we have

$$L_0 v(t) = 2\big(p_{11}x_1(t) + p_{12}x_2(t)\big)\big(a_{11}x_1(t) + a_{12}x_2(t)\big) + p_{11}\sigma_1^2 x_1^2(t)$$

$$+ 2\big(p_{12}x_1(t) + p_{22}x_2(t)\big)\big(a_{21}x_1(t) + a_{22}x_2(t)\big) + p_{22}\sigma_2^2 x_2^2(t)$$

$$= 2\big(p_{12}a_{21} + p_{11}(a_{11} + \delta_1)\big)x_1^2(t) + 2\big(p_{12}a_{12} + p_{22}(a_{22} + \delta_2)\big)x_2^2(t)$$

$$+ 2\big(p_{11}a_{12} + p_{22}a_{21} + p_{12}\operatorname{Tr}(A)\big)x_1(t)x_2(t).$$

By (2.64) $L_0 v(t)$ is a negative definite square form, i.e., the function $v(t)$ satisfies (2.17) with $p = 2$. So, the trivial solution of system (2.60) is asymptotically mean-square stable. The proof is completed. □

Corollary 2.3 *Suppose that conditions (2.62) hold, $a_{12} \neq 0$, and*

$$\delta_1 < \frac{|\operatorname{Tr}(A)|\det(A)}{A_2}, \qquad \delta_2 < \frac{|\operatorname{Tr}(A)|\det(A) - A_2\delta_1}{A_1 - |\operatorname{Tr}(A)|\delta_1}, \tag{2.66}$$

where

$$A_1 = \det(A) + a_{11}^2, \qquad A_2 = \det(A) + a_{22}^2. \tag{2.67}$$

Then the trivial solution of system (2.60) is asymptotically mean-square stable.

Proof By Remark 1.1, from (1.29) and (2.64) it follows that if, for some $q > 0$,

$$-q + 2p_{11}\delta_1 < 0, \qquad -1 + 2p_{22}\delta_2 < 0, \tag{2.68}$$

then the trivial solution of (2.60) is asymptotically mean-square stable.
   Using (1.29), we can represent (2.68) in the form

$$\frac{(A_2 q + a_{21}^2)\delta_1}{|\operatorname{Tr}(A)|\det(A)} < q, \qquad \frac{(A_1 + a_{12}^2 q)\delta_2}{|\operatorname{Tr}(A)|\det(A)} < 1.$$

From this we have

$$\frac{a_{21}^2\delta_1}{|\operatorname{Tr}(A)|\det(A) - A_2\delta_1} < q < \frac{|\operatorname{Tr}(A)|\det(A) - A_1\delta_2}{a_{12}^2\delta_2}. \qquad (2.69)$$

So, if

$$\frac{a_{21}^2\delta_1}{|\operatorname{Tr}(A)|\det(A) - A_2\delta_1} < \frac{|\operatorname{Tr}(A)|\det(A) - A_1\delta_2}{a_{12}^2\delta_2}, \qquad (2.70)$$

then there exists $q > 0$ such that (2.69), and therefore (2.68) holds.

Let us show that (2.70) holds. Indeed, by the first condition (2.66) we can rewrite (2.70) in the form

$$a_{12}^2 a_{21}^2 \delta_1 \delta_2 < \left(|\operatorname{Tr}(A)|\det(A) - A_2\delta_1\right)\left(|\operatorname{Tr}(A)|\det(A) - A_1\delta_2\right)$$
$$= \left(|\operatorname{Tr}(A)|\det(A)\right)^2 - |\operatorname{Tr}(A)|\det(A)(A_1\delta_2 + A_2\delta_1) + A_1 A_2 \delta_1 \delta_2. \qquad (2.71)$$

By (2.67) we have

$$A_1 A_2 = \left(\det(A) + a_{11}^2\right)\left(\det(A) + a_{22}^2\right)$$
$$= \left(\det(A) + a_{11}^2 + a_{22}^2\right)\det(A) + a_{11}^2 a_{22}^2$$
$$= \left(|\operatorname{Tr}(A)|^2 - (a_{11}a_{22} + a_{12}a_{21})\right)\det(A) + a_{11}^2 a_{22}^2$$
$$= |\operatorname{Tr}(A)|^2 \det(A) + a_{12}^2 a_{21}^2$$
$$\geq |\operatorname{Tr}(A)|^2 \det(A). \qquad (2.72)$$

So, by (2.71) and (2.72) it is enough to show that

$$0 < |\operatorname{Tr}(A)|\det(A) - A_2\delta_1 - \left(A_1 - |\operatorname{Tr}(A)|\delta_1\right)\delta_2. \qquad (2.73)$$

Note that from (2.66) and (2.72) it follows that

$$\delta_1 < \frac{|\operatorname{Tr}(A)|\det(A)}{A_2} \leq \frac{A_1}{|\operatorname{Tr}(A)|}.$$

So, (2.73) is equivalent to (2.66). The proof is completed.   □

Remark 2.7 If $a_{12} = 0$, then conditions (2.66) coincide with (2.63). Indeed, by (2.62) from (2.66) we obtain

$$\delta_1 < \frac{|a_{11} + a_{22}|a_{11}a_{22}}{(a_{11} + a_{22})a_{22}} = -a_{11}, \qquad \delta_2 < \frac{|a_{11} + a_{22}|a_{22}(a_{11} + \delta_1)}{(a_{11} + a_{22})(a_{11} + \delta_1)} = -a_{22}.$$

Remark 2.8 From the conditions (2.63) and $a_{12}a_{21} \leq 0$ it follows that

$$|a_{11}| \leq \frac{|\operatorname{Tr}(A)|\det(A)}{A_2}, \qquad |a_{22}| \leq \frac{|\operatorname{Tr}(A)|\det(A)}{A_1}.$$

So, from the conditions (2.63) and $a_{12}a_{21} \leq 0$ it follows that

$$\delta_1 < \frac{|\operatorname{Tr}(A)|\det(A)}{A_2}, \qquad \delta_2 < \frac{|\operatorname{Tr}(A)|\det(A)}{A_1}.$$

Corollary 2.4 *Suppose that the parameters of system* (2.60) *satisfy the conditions* (2.62),

$$a_{21} > 0, \qquad A_2 > |\operatorname{Tr}(A)|\delta_2, \tag{2.74}$$

and the intervals

$$I_1 = \left(-\frac{a_{12}(a_{22} + \delta_2)}{A_2 - |\operatorname{Tr}(A)|\delta_2}, -\frac{a_{11} + \delta_1}{a_{21}} \right) \tag{2.75}$$

and

$$I_2 = \left(\frac{|\operatorname{Tr}(A)| - \sqrt{(a_{11} - a_{22})^2 + 4\det(A)}}{2a_{21}}, \frac{|\operatorname{Tr}(A)| + \sqrt{(a_{11} - a_{22})^2 + 4\det(A)}}{2a_{21}} \right) \tag{2.76}$$

have common points. Then the trivial solution of system (2.60) *is asymptotically mean-square stable.*

Proof Consider the function $v(t)$ given by (2.65) with $p_{11} = 1$, $p_{12} = \mu$, $p_{22} = \gamma$, where $\gamma = a_{21}^{-1}(\mu|\operatorname{Tr}(A)| - a_{12})$. From (2.64) it follows that $\mu \in I_1$. On the other hand, the function $v(t)$ is positive definite if and only if $\gamma > \mu^2$, which is equivalent to $\mu \in I_2$. So, the appropriate μ exists if and only if the intervals I_1 and I_2 have common points. The proof is completed. □

2.3.3 Some Useful Inequalities

Lemma 2.3 *For arbitrary vectors $a \in \mathbf{R}^n$, $b \in \mathbf{R}^n$ and an $n \times n$ matrix $R > 0$, it follows that*

$$a'b + b'a \leq a'Ra + b'R^{-1}b.$$

Proof The proof follows from the simple equality

$$0 \leq (a - R^{-1}b)'R(a - R^{-1}b) = a'Ra + b'R^{-1}b - a'b - b'a. \qquad □$$

Lemma 2.4 *For positive* P_2, *x and nonnegative* P_1, Q *such that* $P_2 > Qx$, *the following inequality holds:*

$$\frac{P_1 + Qx^{-1}}{P_2 - Qx} \geq \left(\frac{\sqrt{Q^2 + P_1 P_2} + Q}{P_2}\right)^2.$$

Proof It is enough to check that the function

$$f(x) = \frac{P_1 + Qx^{-1}}{P_2 - Qx}$$

reaches its minimum at the point

$$x_0 = \frac{P_2}{\sqrt{Q^2 + P_1 P_2} + Q}$$

and this minimum equals x_0^{-2}. The proof is completed. □

2.4 Some Unsolved Problems

In spite of the fact that the theory of stability for stochastic hereditary systems is very popular in researches, there are simply and clearly formulated problems with unknown decisions. In order to attract attention to such problems, one of them for stochastic difference equation with continuous time is represented in [277], and two unsolved stability problems for stochastic differential equations with delay are described below.

2.4.1 Problem 1

Consider the linear stochastic differential equation with delays

$$\dot{x}(t) = Ax(t) + \sum_{i=1}^{m} B_i x(t - \tau_i) + \sigma x(t - h)\dot{w}(t), \qquad (2.77)$$

where A, B_i, σ, $\tau_i > 0$, $h \geq 0$ are known constants, and $w(t)$ is the standard Wiener process.

It is known [290] that a necessary and sufficient condition for asymptotic mean-square stability of the zero solution of (2.77) can be represented in the form

$$G^{-1} > \frac{\sigma^2}{2}, \quad G = \frac{2}{\pi}\int_0^{\infty} \frac{dt}{(A + \sum_{i=1}^{m} B_i \cos \tau_i t)^2 + (t + \sum_{i=1}^{m} B_i \sin \tau_i t)^2}.$$

$$(2.78)$$

By Lemma 2.1, in the particular case $m = 1$, $B_1 = B$, $\tau_1 = \tau$ the integral (2.78) can be calculated in elementary functions, and the stability conditions take the form (2.42)–(2.43).

The problem is: *to calculate the integral* (2.78) *in elementary functions for* $m \geq 2$, *in particular, for* $m = 2$.

2.4.2 Problem 2

From (2.42) and (2.43) it follows that the zero solution of the differential equation with a constant delay $\dot{x}(t) = -bx(t - h)$ is asymptotically stable if and only if

$$0 < bh < \frac{\pi}{2}. \tag{2.79}$$

It is also known [221, 318] that the zero solution of the differential equation with a varying delay $\dot{x}(t) = -bx(t - \tau(t))$ is asymptotically stable for an arbitrary delay $\tau(t)$ such that $\tau(t) \in [0, h]$ if and only if

$$0 < bh < \frac{3}{2}. \tag{2.80}$$

Consider the stochastic differential equation with a constant delay

$$\dot{x}(t) = -bx(t - h) + \sigma x(t)\dot{w}(t). \tag{2.81}$$

From (2.42) and (2.43) it follows that the zero solution of (2.81) is asymptotically mean-square stable if and only if

$$0 < bh < \arcsin \frac{b^2 - p^2}{b^2 + p^2}, \quad p = \frac{\sigma^2}{2}. \tag{2.82}$$

In the deterministic case ($\sigma = 0$) condition (2.82) coincides with (2.79).

Consider the stochastic differential equation

$$\dot{x}(t) = -bx\big(t - \tau(t)\big) + \sigma x(t)\dot{w}(t) \tag{2.83}$$

with a varying delay $\tau(t)$ such that $\tau(t) \in [0, h]$.

The problem is: *to generalize condition* (2.80) *for* (2.83).

Chapter 3
Stability of Linear Scalar Equations

3.1 Linear Stochastic Differential Equation of Neutral Type

Using the proposed procedure of Lyapunov functionals construction, let us investigate asymptotic mean-square stability of the linear scalar stochastic differential equation of neutral type with constant coefficients

$$\dot{x}(t) + ax(t) + bx(t-h) + c\dot{x}(t-h) + \sigma x(t-\tau)\dot{w}(t) = 0, \quad |c| < 1,$$

$$x(s) = \phi(s), \quad s \in \left[-\max(h,\tau),0\right]. \tag{3.1}$$

3.1.1 The First Way of Constructing a Lyapunov Functional

Following Step 1 of the procedure, we rewrite (3.1) in the form

$$\dot{z}(x_t) = -ax(t) - bx(t-h) - \sigma x(t-\tau)\dot{w}(t),$$

$$z(x_t) = x(t) + cx(t-h).$$

Suppose that $a > 0$. Then the function $v = y^2$ is a Lyapunov function for the auxiliary differential equation $\dot{y}(t) = -ay(t)$ since $\dot{v} = -2ay^2$. Thus, the trivial solution of the auxiliary differential equation is asymptotically stable. Put $V_1 = z^2(x_t)$. Then

$$LV_1 = 2z(x_t)\left(-ax(t) - bx(t-h)\right) + \sigma^2 x^2(t-\tau)$$

$$= -2ax^2(t) - 2bcx^2(t-h) - 2(ac + b)x(t)x(t-h) + \sigma^2 x^2(t-\tau)$$

$$\leq \left(-2a + |ac+b|\right)x^2(t) + \rho x^2(t-h) + \sigma^2 x^2(t-\tau),$$

where

$$\rho = \begin{cases} 0 & \text{if } |ac+b| \leq 2bc, \\ |ac+b| - 2bc & \text{if } |ac+b| > 2bc. \end{cases} \tag{3.2}$$

L. Shaikhet, *Lyapunov Functionals and Stability of Stochastic Functional Differential Equations*, DOI 10.1007/978-3-319-00101-2_3, © Springer International Publishing Switzerland 2013

By Theorem 2.5 with $S = Q_i = R_j = R(s, t) = 0$ and

$$D = -2a + |ac + b|, \qquad dK_1(s) = \left(\rho\delta(s - h) + \sigma^2\delta(s - \tau)\right)ds$$

(here and below $\delta(s)$ is the Dirac function) we obtain that if $\sigma^2 + \rho + |ac + b| < 2a$, then the trivial solution of (3.1) is asymptotically mean-square stable.

Note that, by (3.2), $\rho \geq 0$. So, from the obtained condition it follows that $a > 0$. Using two representations for ρ in (3.2), we obtain the following two stability conditions:

$$|ac + b| \leq 2bc, \qquad \sigma^2 + |ac + b| < 2a \qquad (3.3)$$

and

$$|ac + b| > 2bc, \qquad p + |ac + b| - bc < a, \qquad p = \frac{\sigma^2}{2}. \qquad (3.4)$$

By (3.3) we have $bc = |bc|$ and $a > 0$. So, $|ac + b| = a|c| + |b|$, and inequalities (3.3) take the forms $2|bc| \geq a|c| + |b|$ and $\sigma^2 + a|c| + |b| < 2a$. The former is wrong if $2|c| < 1$. Suppose that $2|c| \geq 1$. Then

$$\frac{\sigma^2 + |b|}{2 - |c|} < a \leq \left(2 - \frac{1}{|c|}\right)|b|,$$

which is impossible since from this the contradiction $\sigma^2|c| + 2|b|(1 - |c|)^2 < 0$ follows. Thus, inequalities (3.3) are incompatible.

Consider now conditions (3.4). Suppose first that $bc \geq 0$. From this and from $a > 0$ we have $bc = |bc|$ and $|ac + b| = a|c| + |b|$, and inequalities (3.4) take the forms

$$2|bc| < a|c| + |b|, \qquad a > |b| + \frac{p}{1 - |c|}.$$

If $2|c| < 1$, then the first inequality holds for $a > 0$ and arbitrary b. If $2|c| \geq 1$, then the second inequality implies the first one. So, if $bc \geq 0$, then from (3.4) it follows that

$$bc \geq 0, \qquad a > |b| + \frac{p}{1 - |c|}. \qquad (3.5)$$

Let now $bc < 0$. Then the first inequality in (3.4) holds, and (3.4) takes the form

$$bc < 0, \qquad p + |ac + b| - bc < a. \qquad (3.6)$$

Since $bc < 0$, $|ac + b| = |a|c| - |b||$. So, if $a|c| \geq |b|$, then from (3.6) we have

$$\frac{p}{1 - |c|} - a < |b| \leq a|c|. \qquad (3.7)$$

If $a|c| < |b|$, then

$$a|c| < |b| < a - \frac{p}{1 + |c|}. \qquad (3.8)$$

Fig. 3.1 Stability regions for
(3.1) given by conditions
(3.11) for $c = -0.5$, $h = 1$
and different values of p:
(*1*) $p = 0$, (*2*) $p = 0.5$,
(*3*) $p = 1$, (*4*) $p = 1.5$

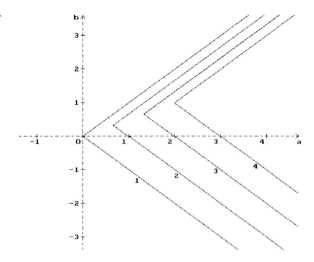

Combining (3.7) and (3.8), we obtain

$$bc < 0, \qquad \frac{p}{1 - |c|} - a < |b| < a - \frac{p}{1 + |c|}. \tag{3.9}$$

Note that since $bc < 0$, the system

$$|b| = \frac{p}{1 - |c|} - a, \qquad |b| = a - \frac{p}{1 + |c|}$$

has the solution

$$a = \frac{p}{1 - c^2}, \qquad b = -\frac{pc}{1 - c^2}. \tag{3.10}$$

Combining (3.5), (3.9), and (3.10), we obtain the stability condition in the form

$$a > \begin{cases} \frac{p}{1-c} + b, & b > -\frac{pc}{1-c^2}, \\ \frac{p}{1+c} - b, & b \le -\frac{pc}{1-c^2}. \end{cases} \tag{3.11}$$

Thus, if condition (3.11) holds, then the trivial solution of (3.1) is asymptotically
mean-square stable.

The stability regions for (3.1), given by the stability conditions (3.11), are shown
in Fig. 3.1 for $c = -0.5$, $h = 1$, and different values of p: (1) $p = 0$; (2) $p = 0.5$;
(3) $p = 1$; (4) $p = 1.5$. In Fig. 3.2 the stability regions are shown for $c = 0.5$ and
the same values of the other parameters.

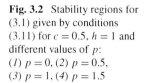

Fig. 3.2 Stability regions for
(3.1) given by conditions
(3.11) for $c = 0.5$, $h = 1$ and
different values of p:
(*1*) $p = 0$, (*2*) $p = 0.5$,
(*3*) $p = 1$, (*4*) $p = 1.5$

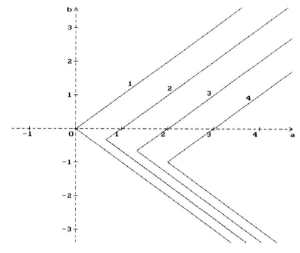

3.1.2 The Second Way of Constructing a Lyapunov Functional

To get another stability condition, rewrite (3.1) in the form

$$\dot{z}(x_t) = -(a+b)x(t) - \sigma x(t-\tau)\dot{w}(t),$$

$$z(x_t) = x(t) + cx(t-h) - b\int_{t-h}^{t} x(s)\,ds.$$

By condition (2.10) it is necessary to suppose that

$$|c| + |b|h < 1. \tag{3.12}$$

Suppose also that $a + b > 0$. Then the function $v = y^2$ is a Lyapunov function for the auxiliary differential equation $\dot{y}(t) = -(a+b)y(t)$ since $\dot{v} = -2(a+b)y^2$. Thus, the trivial solution of the auxiliary differential equation is asymptotically stable. Put $V_1 = z^2(x_t)$. Then

$$LV_1 = -2(a+b)x(t)z(x_t) + \sigma^2 x^2(t-\tau)$$

$$= -2(a+b)x^2(t) - 2(a+b)cx(t)x(t-h)$$

$$+ 2(a+b)b\int_{t-h}^{t} x(t)x(s)\,ds + \sigma^2 x^2(t-\tau)$$

$$\leq (a+b)\big(-2 + |c| + |b|h\big)x^2(t) + \sigma^2 x^2(t-\tau)$$

$$+ (a+b)\left(|c|x^2(t-h) + |b|\int_{t-h}^{t} x^2(s)\,ds\right).$$

Fig. 3.3 Stability regions for (3.1) given by conditions (3.13) for $|c| = 0.5$, $h = 0.2$, and different values of p: (1) $p = 0.2$, (2) $p = 0.6$, (3) $p = 1$, (4) $p = 1.4$

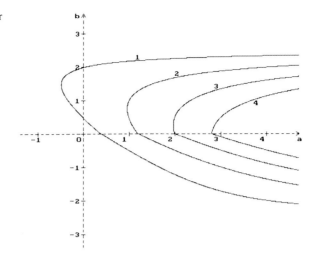

Thus, using (3.12) and Theorem 2.4 with $m = S = Q_i = R(s,t) = 0$, $n = 1$ and

$$D = (a + b)(-2 + |c| + |b|h), \qquad R_0(\theta) = (a + b)|b|, \qquad d\mu_0(s) = \delta(s - h)\,ds,$$

$$dK_1(s) = \left[(a + b)|c|\delta(s - h) + \sigma^2\delta(s - \tau)\right]ds,$$

we obtain the stability condition in the form

$$a > \frac{p}{1 - |c| - |b|h} - b, \qquad |b| < \frac{1 - |c|}{h}. \tag{3.13}$$

The stability regions for (3.1), given by the stability condition (3.13), are shown in Fig. 3.3 for $|c| = 0.5$, $h = 0.2$, and different values of p: (1) $p = 0.2$, (2) $p = 0.6$, (3) $p = 1$, (4) $p = 1.4$ and in Fig. 3.4 for $|c| = 0.5$, $p = 0.4$, and different values of h: (1) $h = 0.1$, (2) $h = 0.15$, (3) $h = 0.2$, (4) $h = 0.25$.

It is easy to see that for $b \le 0$, condition (3.11) is better than (3.13). So, condition (3.13) is better to use for $b > 0$ only in the form

$$a > \frac{p}{1 - |c| - bh} - b, \qquad 0 < b < \frac{1 - |c|}{h}. \tag{3.14}$$

The stability regions for (3.1), given by both stability conditions (3.11) and (3.14), are shown in Fig. 3.5 for $c = -0.6$, $p = 0.4$, and different values of h: (1) $h = 0.05$; (2) $h = 0.1$; (3) $h = 0.15$, (4) $h = 0.2$. In Fig. 3.6 the stability regions are shown for $c = 0.6$ and the same values of the other parameters.

Fig. 3.4 Stability regions for (3.1) given by conditions (3.13) for $|c| = 0.5$, $p = 0.4$, and different values of h: (*1*) $h = 0.1$, (*2*) $h = 0.15$, (*3*) $h = 0.2$, (*4*) $h = 0.25$

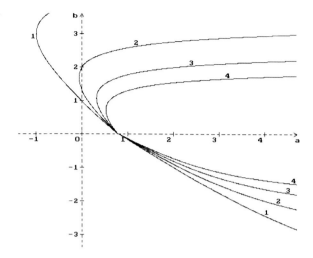

Fig. 3.5 Stability regions for (3.1) given by conditions (3.11) and (3.14) together for $c = -0.6$, $p = 0.4$, and different values of h: (*1*) $h = 0.05$, (*2*) $h = 0.1$, (*3*) $h = 0.15$, (*4*) $h = 0.2$

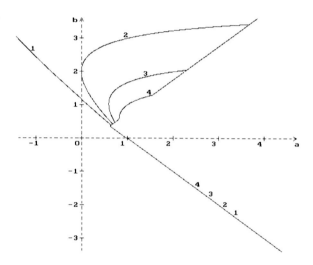

3.1.3 Some Particular Cases

(1) Note that, as $h \to 0$, condition (3.14) takes the form

$$a > \frac{p}{1 - |c|} - b, \quad b > 0. \tag{3.15}$$

On the other hand, in the case $h = 0$, for the functional

$$V = (1 + c)\left(x^2 + \int_{t-\tau}^{t} x^2(s)\, ds\right),$$

Fig. 3.6 Stability regions for (3.1) given by conditions (3.11) and (3.14) together for $c = 0.6$, $p = 0.4$, and different values of h: (1) $h = 0.05$, (2) $h = 0.1$, (3) $h = 0.15$, (4) $h = 0.2$

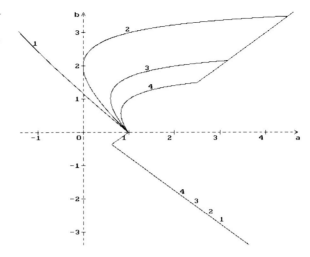

we have $LV = [-2(a + b)(1 + c) + \sigma^2]x^2(t)$. So, by Corollary 2.2 for $h = 0$ a necessary and sufficient condition for asymptotic mean-square stability of the trivial solution of (3.1) is

$$a > \frac{p}{1 + c} - b. \tag{3.16}$$

For $b > 0$ and $c > 0$, condition (3.15) is essentially worse than (3.16), but for $b > 0$ and $c \leq 0$, condition (3.15) coincides with (3.16). The second condition in (3.11) coincides with (3.16) as well.

(2) For $c = 0$, the necessary and sufficient condition for asymptotic mean-square stability of the trivial solution of (3.1) has the form (Lemma 2.1)

$$a + b > 0, \qquad G^{-1} > p, \tag{3.17}$$

where

$$p = \frac{\sigma^2}{2}, \qquad G = \begin{cases} \frac{1 + bq^{-1}\sin(qh)}{a + b\cos(qh)}, & b > |a|, q = \sqrt{b^2 - a^2}, \\[2mm] \frac{1 + ah}{2a}, & b = a > 0, \\[2mm] \frac{1 + bq^{-1}\sinh(qh)}{a + b\cosh(qh)}, & a > |b|, q = \sqrt{a^2 - b^2}. \end{cases} \tag{3.18}$$

If, in addition, $a = 0$, then the necessary and sufficient stability condition (3.17)–(3.18) takes the form

$$h < \frac{1}{b}\arcsin\frac{b^2 - p^2}{b^2 + p^2}.$$

In Fig. 3.7 the stability regions for (3.1) given by the sufficient conditions (3.11), (3.14) and the necessary and sufficient conditions (3.17)–(3.18) are shown for $c = 0$, $h = 1$, and different values of p: (1) $p = 0$; (2) $p = 0.5$; (3) $p = 1$; (2) $p = 1.5$;

Fig. 3.7 Stability regions for (3.1) given by sufficient conditions (3.11), (3.14) and necessary and sufficient conditions (3.17), (3.18) are shown for $c = 0$, $h = 1$ and different values of p: (*1*) $p = 0$, (*2*) $p = 0.5$, (*3*) $p = 1$, (*4*) $p = 1.5$, (*5*) $p = 2$

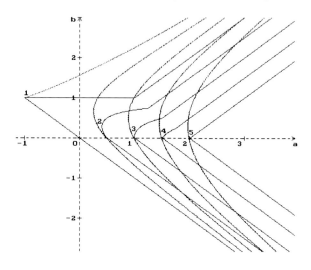

Fig. 3.8 Stability regions for (3.1) given by sufficient conditions (3.11), (3.14) and necessary and sufficient conditions (3.17), (3.18) are shown for $c = 0$, $h = 1$, and different values of p: (*1*) $p = 0$, (*2*) $p = 0.5$, (*3*) $p = 1$, (*4*) $p = 1.5$, (*5*) $p = 2$

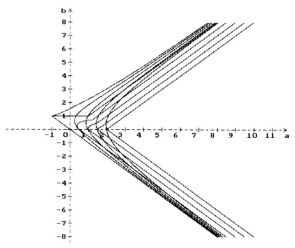

(5) $p = 2$. In Fig. 3.8 the same stability regions are shown in another scale. We can see that the sufficient conditions (3.11), (3.14) give the stability region that is sufficiently close to the exact one.

(3) For $p = 0$, from (3.11), (3.14) it follows that

$$a > \begin{cases} b, & b \geq \frac{1-|c|}{h}, \\ -b, & b < \frac{1-|c|}{h}. \end{cases} \qquad (3.19)$$

In Fig. 3.9 the exact stability regions for (3.1) are shown for $p = 0$, $h = 1$, and (1) $c = 0.5$; (2) $c = -0.5$ and the stability region given by the sufficient condition (3.19) for (3) $c = |0.5|$. In Fig. 3.10 the similar regions are shown for (1) $c = 0.85$; (2) $c = -0.85$; (3) $c = |0.85|$.

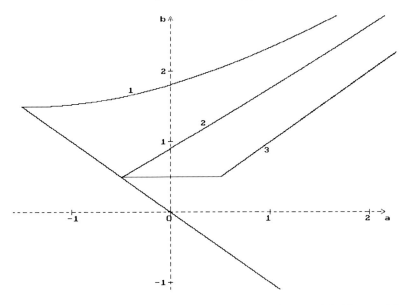

Fig. 3.9 Exact stability regions are shown for $p = 0$, $h = 1$, and (1) $c = 0.5$, (2) $c = -0.5$, and stability region, given by sufficient condition (3.19) for (3) $c = |0.5|$

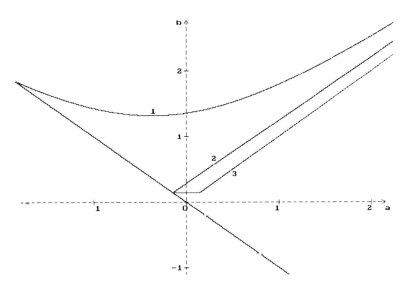

Fig. 3.10 Exact stability regions are shown for $p = 0$, $h = 1$, and (1) $c = 0.85$, (2) $c = -0.85$, and stability region, given by sufficient condition (3.19) for (3) $c = |0.85|$

3.2 Linear Differential Equation with Two Delays in Deterministic Part

Consider the scalar differential equation

$$\dot{x}(t) + ax(t - h_1) + bx(t - h_2) + \sigma x(t - \tau)\dot{w}(t) = 0, \quad 0 \le h_1 < h_2,$$

$$x(s) = \phi(s), \quad s \in \left[-\max(h_2, \tau), 0\right].$$

(3.20)

Using different representations of this equation, we will obtain different conditions for asymptotic mean-square stability of the zero solution of (3.20).

3.2.1 The First Way of Constructing a Lyapunov Functional

Rewrite (3.20) in the form

$$\dot{z}(x_t) = -ax(t) - bx(t - h_2) - \sigma x(t - \tau)\dot{w}(t),$$

$$z(x_t) = x(t) - a \int_{t-h_1}^{t} x(s)\, ds.$$

Putting $V_1 = z^2(x_t)$, we obtain

$$LV_1 = 2z(x_t)\left(-ax(t) - bx(t - h_2)\right) + \sigma^2 x^2(t - \tau)$$

$$= -2ax^2(t) - 2bx(t)x(t - h_2)$$

$$+ 2a^2 \int_{t-h_1}^{t} x(t)x(s)\, ds + 2ab \int_{t-h_1}^{t} x(t - h_2)x(s)\, ds + \sigma^2 x^2(t - \tau)$$

$$\le -2ax^2(t) + |b|\left(x^2(t) + x^2(t - h_2)\right) + a^2\left(x^2(t)h_1 + \int_{t-h_1}^{t} x^2(s)\, ds\right)$$

$$+ |ab|\left(x^2(t - h_2)h_1 + \int_{t-h_1}^{t} x^2(s)\, ds\right) + \sigma^2 x^2(t - \tau)$$

$$= \left(-2a + a^2 h_1 + |b|\right)x^2(t) + \left(a^2 + |ab|\right)\int_{t-h_1}^{t} x^2(s)\, ds$$

$$+ \left(|b| + |ab|h_1\right)x^2(t - h_2) + \sigma^2 x^2(t - \tau).$$

From this and from (2.10) and Theorem 2.5 with $m = S = Q_i = R(s, t) = 0$, $n = 1$ and

$$D = -2a + a^2 h_1 + |b|, \qquad R_0(\theta) = a^2 + |ab|, \qquad d\mu_0(s) = \delta(s - h_1)\, ds,$$

$$dK_1(s) = \left[\left(|b| + |ab|h_1\right)\delta(s - h_2) + \sigma^2 \delta(s - \tau)\right] ds$$

it follows that if

$$|b| < \frac{a(1 - ah_1) - p}{1 + ah_1}, \quad 0 < a < \frac{1}{h_1}, \tag{3.21}$$

then the trivial solution of (3.20) is asymptotically mean-square stable.

Remark 3.1 Rewriting (3.20) in the form

$$\dot{z}(x_t) = -bx(t) - ax(t - h_1) - \sigma x(t - \tau)\dot{w}(t),$$

$$z(t) = x(t) - b \int_{t-h_2}^{t} x(s)\, ds$$

and using symmetry, we obtain another sufficient condition for asymptotic mean-square stability of the trivial solution of (3.20):

$$|a| < \frac{b(1 - bh_2) - p}{1 + bh_2}, \quad 0 < b < \frac{1}{h_2}. \tag{3.22}$$

3.2.2 The Second Way of Constructing a Lyapunov Functional

Rewrite (3.20) in the form

$$\dot{z}(x_t) = -(a + b)x(t) + bx(t - h_1) - bx(t - h_2) - \sigma x(t - \tau)\dot{w}(t),$$

$$z(x_t) = x(t) - (a + b) \int_{t-h_1}^{t} x(s)\, ds.$$

Putting $V_1 = z^2(x_t)$, we obtain

$$LV_1 = 2z(x_t)\big[-(a + b)x(t) + bx(t - h_1) - bx(t - h_2)\big] + \sigma^2 x^2(t - \tau)$$

$$= -2(a + b)x^2(t) + 2bx(t)x(t - h_1) - 2bx(t)x(t - h_2)$$

$$\quad + 2(a + b)^2 \int_{t-h_1}^{t} x(t)x(s)\, ds - 2(a + b)b \int_{t-h_1}^{t} x(t - h_1)x(s)\, ds$$

$$\quad + 2(a + b)b \int_{t-h_1}^{t} x(t - h_2)x(s)\, ds + \sigma^2 x^2(t - \tau)$$

$$\leq -2(a + b)x^2(t) + |b|\big(x^2(t) + x^2(t - h_1)\big) + |b|\big(x^2(t) + x^2(t - h_2)\big)$$

$$\quad + (a + b)^2\left(x^2(t)h_1 + \int_{t-h_1}^{t} x^2(s)\, ds\right) + \sigma^2 x^2(t - \tau)$$

$$\quad + |(a + b)b|\left(x^2(t - h_1)h_1 + x^2(t - h_2)h_1 + 2\int_{t-h_1}^{t} x^2(s)\, ds\right)$$

$$= \left[-2(a+b) + (a+b)^2 h_1 + 2|b| \right] x^2(t) + \sigma^2 x^2(t-\tau)$$
$$+ |b| \left(1 + |a+b|h_1 \right) \left(x^2(t-h_1) + x^2(t-h_2) \right)$$
$$+ \left((a+b)^2 + 2|b(a+b)| \right) \int_{t-h_1}^{t} x^2(s)\,ds.$$

Thus, we obtain the representation (2.35) with $m = S = Q_i = R(s,t) = 0$, $n = 1$ and

$$D = -2(a+b) + (a+b)^2 h_1 + 2|b|,$$
$$R_0(\theta) = \left((a+b)^2 + 2|b(a+b)| \right), \qquad d\mu_0(s) = \delta(s-h_1)\,ds,$$
$$dK_1(s) = \left[|b| \left(1 + |a+b|h_1 \right) \left(\delta(s-h_1) + \delta(s-h_2) \right) + \sigma^2 \delta(s-\tau) \right] ds.$$

From this and from Theorem 2.5 it follows that if

$$2|b| < \frac{(a+b)(1-(a+b)h_1) - p}{1 + (a+b)h_1}, \qquad 0 < a+b < \frac{1}{h_1}, \tag{3.23}$$

then the trivial solution of (3.20) is asymptotically mean-square stable.

Remark 3.2 Similarly to Remark 3.1, using symmetry, we obtain another sufficient condition for asymptotic mean-square stability of the trivial solution of (3.20):

$$2|a| < \frac{(a+b)(1-(a+b)h_2) - p}{1 + (a+b)h_2}, \qquad 0 < a+b < \frac{1}{h_2}. \tag{3.24}$$

3.2.3 The Third Way of Constructing a Lyapunov Functional

Rewrite (3.20) in the form

$$\dot{z}(x_t) = -(a+b)x(t) - \sigma x(t-\tau)\dot{w}(t),$$
$$z(x_t) = x(t) - a \int_{t-h_1}^{t} x(s)\,ds - b \int_{t-h_2}^{t} x(s)\,ds. \tag{3.25}$$

Putting $V_1 = z^2(x_t)$, we obtain

$$LV_1 = 2z(x_t)\left(-(a+b) \right) x(t) + \sigma^2 x^2(t-\tau)$$
$$= -2(a+b)x^2(t) + 2a(a+b) \int_{t-h_1}^{t} x(t)x(s)\,ds$$
$$+ 2b(a+b) \int_{t-h_2}^{t} x(t)x(s)\,ds + \sigma^2 x^2(t-\tau). \tag{3.26}$$

Therefore,

$$LV_1 \leq -2(a+b)x^2(t) + |a(a+b)|\left(x^2(t)h_1 + \int_{t-h_1}^{t} x^2(s)\,ds\right)$$

$$+ |b(a+b)|\left(x^2(t)h_2 + \int_{t-h_2}^{t} x^2(s)\,ds\right) + \sigma^2 x^2(t-\tau)$$

$$= \left[-2(a+b) + |a+b|(|a|h_1 + |b|h_2)\right]x^2(t) + \sigma^2 x^2(t-\tau)$$

$$+ |a+b|\left(|a|\int_{t-h_1}^{t} x^2(s)\,ds + |b|\int_{t-h_2}^{t} x^2(s)\,ds\right).$$

Thus, we obtain the representation (2.35) with $m = S = Q_i = 0$, $n = 1$ and

$$D = -2(a+b) + |a+b|(|a|h_1 + |b|h_2), \qquad dK_1(s) = \sigma^2 \delta(s-\tau)\,ds,$$
$$R_0(\theta) = |a(a+b)|, \qquad d\mu_0(s) = \delta(s-h_1)\,ds,$$
$$R(s,t) = |b(a+b)|, \qquad d\mu(s) = \delta(s-h_2)\,ds.$$

From this and from Theorem 2.5 it follows that if

$$a+b > \frac{p}{1 - |a|h_1 - |b|h_2}, \qquad |a|h_1 + |b|h_2 < 1, \tag{3.27}$$

then the trivial solution of (3.20) is asymptotically mean-square stable.

Remark 3.3 Note that condition (3.27) follows from condition (3.22), condition (3.22) follows from condition (3.24), and condition (3.21) follows from condition (3.23).

Let us show, for instance, that condition (3.27) follows from (3.22). Indeed, by (3.22) we have $(p+|a|)b^{-1} + (|a|+b)h_2 < 1$. So, using $h_1 < h_2$, we obtain $|a|h_1 + bh_2 < (|a|+b)h_2 < 1$. To get the first condition (3.27), rewrite (3.22) and (3.27) as follows: $p < b(1-bh_2) - |a|(1+bh_2)$, $p < (a+b)(1-|a|h_1 - |b|h_2)$. So, it is enough to show that $b(1-bh_2) - |a|(1+bh_2) \leq (a+b)(1-|a|h_1 - |b|h_2)$ or $|a|(a+b)h_1 + abh_2 \leq a + |a| + |a|bh_2$. For $a = 0$, this inequality holds. If $a > 0$, then it is equivalent to $(a+b)h_1 \leq 2$. Using $h_1 < h_2$ and $ah_1 + bh_2 < 1$, we have $(a+b)h_1 < ah_1 + bh_2 < 1 < 2$. Let now $a < 0$. Then it is necessary to show that $(a+b)h_1 \leq 2bh_2$. But this follows from $(a+b)h_1 < bh_1 < bh_2 < 2bh_2$.

3.2.4 The Fourth Way of Constructing a Lyapunov Functional

Let us show that using the same representations of the initial equation but different ways of LV_1 estimation, we can get different stability conditions.

Rewrite (3.20) in the form (3.25) and put $V_1 = z^2(x_t)$ again. Using (3.26) and the condition $a + b > 0$, let us estimate LV_1 in the following way:

$$LV_1 = -2(a+b)x^2(t) + 2(a+b)^2 \int_{t-h_1}^{t} x(t)x(s)\,ds$$

$$+ 2(a+b)b \int_{t-h_2}^{t-h_1} x(t)x(s)\,ds + \sigma^2 x^2(t-\tau)$$

$$\leq -2(a+b)x^2(t) + (a+b)^2\left(h_1 x^2(t) + \int_{t-h_1}^{t} x^2(s)\,ds\right)$$

$$+ (a+b)|b|\left((h_2 - h_1)x^2(t) + \int_{t-h_2}^{t-h_1} x^2(s)\,ds\right) + \sigma^2 x^2(t-\tau)$$

$$= \left[-2(a+b) + (a+b)^2 h_1 + (a+b)|b|(h_2 - h_1)\right]x^2(t) + \sigma^2 x^2(t-\tau)$$

$$+ (a+b)^2 \int_{t-h_1}^{t} x^2(s)\,ds + (a+b)|b| \int_{t-h_2}^{t-h_1} x^2(s)\,ds.$$

From representation (2.35) with $m = S = Q_i = 0$, $n = 1$ and

$$D = -2(a+b) + (a+b)^2 h_1 + (a+b)|b|(h_2 - h_1), \qquad dK_1(s) = \sigma^2\delta(s-\tau)\,ds,$$

$$R_0(\theta) = (a+b)^2, \qquad d\mu_0(s) = \delta(s-h_1)\,ds, \qquad d\mu(s) = \delta(s-h_2)\,ds,$$

$$R(\tau + h_2, t) = \begin{cases} (a+b)|b|, & \tau \in [t-h_2, t-h_1], \\ 0, & \tau \in (t-h_1, t]. \end{cases}$$

and from Theorem 2.5 it follows that if

$$a + b > \frac{p}{1 - (a+b)h_1 - |b|(h_2 - h_1)}, \qquad (a+b)h_1 + |b|(h_2 - h_1) < 1, \quad (3.28)$$

then the trivial solution of (3.20) is asymptotically mean-square stable.

It is easy to see that condition (3.28) coincides with (3.27) for $a \geq 0$, $b \geq 0$, but in the case $ab < 0$ condition (3.28) is weaker than (3.27). Using Remark 3.3, we can conclude that the stability conditions (3.21) and (3.28) together are better than all other.

Put $p = 0$ and $h_2 = 1$. In Fig. 3.11 the stability regions for (3.20), given by conditions (3.21)–(3.24) and (3.27) (with numbers 1–5, respectively), are shown for $h_1 = 0.1$. In Fig. 3.12 the similar picture is shown for $h_1 = 0.2$ with addition of the bound of the stability region, given by the necessary and sufficient condition of asymptotic stability obtained by the characteristic equation (Example 1.3). In Fig. 3.13 we can see how the picture in Fig. 3.12 is changed for $h_1 = 0.25$.

Consider now the case $p > 0$, $h_2 = 1$. In Fig. 3.14 the stability regions, given by conditions (3.21)–(3.24), (3.27), and (3.28) (with numbers 1–6, respectively), are shown for $p = 0.1$, $h_1 = 0.1$. In Fig. 3.15 these stability regions are shown for $p = 0.1$, $h_1 = 0.5$, and in Fig. 3.16 for $p = 0.25$, $h_1 = 0.5$.

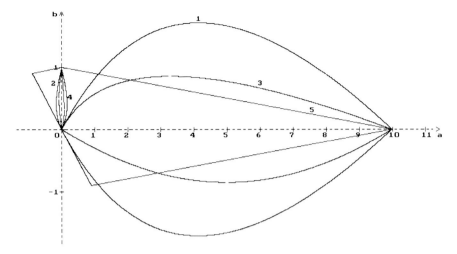

Fig. 3.11 Stability regions for (3.20), given by conditions (3.21)–(3.24), (3.27) (with numbers
1–5, respectively), are shown for $p = 0$, $h_1 = 0.1$, $h_2 = 1$

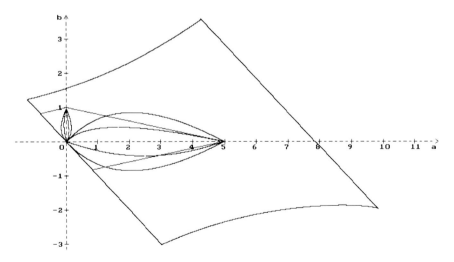

Fig. 3.12 Stability regions for (3.20), given by conditions (3.21)–(3.24), (3.27) that are similar to
Fig. 3.11, are shown for $p = 0$, $h_1 = 0.2$, $h_2 = 1$ with addition of the bound of stability region,
given by the necessary and sufficient condition

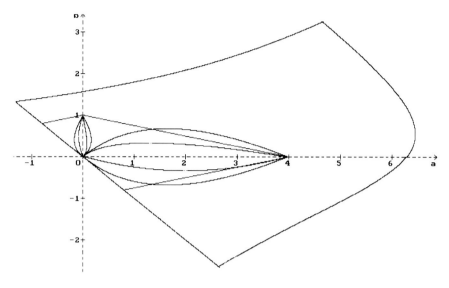

Fig. 3.13 Here it is shown how Fig. 3.12 is changed for $h_1 = 0.25$

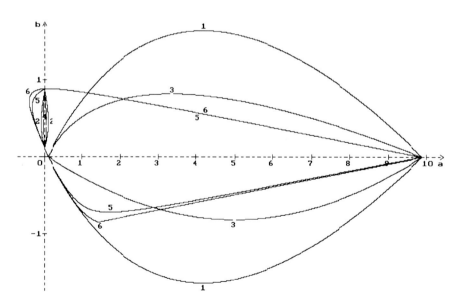

Fig. 3.14 Stability regions for (3.20), given by conditions (3.21)–(3.24), (3.27), (3.28) (with numbers *1–6*, respectively), are shown for $p = 0.1$, $h_1 = 0.1$, $h_2 = 1$

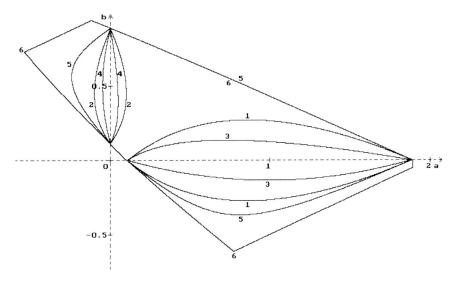

Fig. 3.15 Stability regions for (3.20), given by conditions (3.21)–(3.24), (3.27), (3.28) (with numbers *1–6*, respectively), are shown for $p = 0.1$, $h_1 = 0.5$, $h_2 = 1$

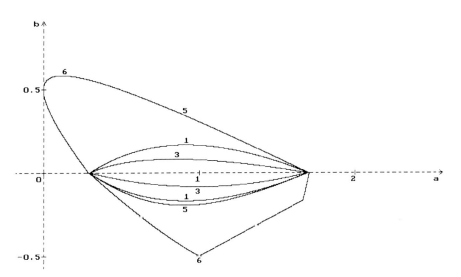

Fig. 3.16 Stability regions for (3.20), given by conditions (3.21)–(3.24), (3.27), (3.28) (with numbers *1–6*, respectively), are shown for $p = 0.25$, $h_1 = 0.5$, $h_2 = 1$

3.2.5 One Generalization for Equation with n Delays

Consider the stochastic differential equation with n delays in the deterministic part

$$\dot{x}(t) + a_0 x(t) + \sum_{i=1}^{n} a_i x(t - h_i) + \sigma x(t - \tau)\dot{w}(t) = 0,$$

$$x(s) = \phi(s), \quad s \in \left[-\max(h_n, \tau), 0\right].$$
(3.29)

Here it is supposed that

$$0 < h_1 < h_2 < \cdots < h_n.$$
(3.30)

For $V_1 = x^2(t)$, we have

$$LV_1 = -2x(t)\left(a_0 x(t) + \sum_{i=1}^{n} a_i x(t - h_i)\right) + \sigma^2 x^2(t - \tau)$$

$$\leq \left(-2a_0 + \sum_{i=1}^{n} |a_i|\right) x^2(t) + \sum_{i=1}^{n} |a_i| x^2(t - h_i) + \sigma^2 x^2(t - \tau).$$

By the representation (2.35) and Theorem 2.5 we obtain the sufficient condition for asymptotic mean-square stability of the trivial solution of (3.29) in the form

$$a_0 > \sum_{i=1}^{n} |a_i| + p, \quad p = \frac{\sigma^2}{2}.$$
(3.31)

Rewrite (3.29) as follows:

$$\dot{z}(x_t) = -S_0 x(t) - \sigma x(t - \tau)\dot{w}(t),$$

$$z(x_t) = x(t) - \sum_{i=1}^{n} a_i \int_{t-h_i}^{t} x(s)\,ds,$$
(3.32)

$$S_j = \sum_{i=j}^{n} a_i, \quad j = 0, 1, \ldots, n.$$

Suppose that $S_0 > 0$ and put $V_1 = z^2(x_t)$. Then

$$LV_1 = -2S_0 x^2(t) + 2S_0 x(t) \sum_{i=1}^{n} a_i \int_{t-h_i}^{t} x(s)\,ds + \sigma^2 x^2(t - \tau)$$

$$\leq -2S_0 x^2(t) + S_0 \sum_{i=1}^{n} |a_i| \left(x^2(t)h_i + \int_{t-h_i}^{t} x^2(s)\,ds\right) + \sigma^2 x^2(t - \tau)$$

$$= \left(-2 + \sum_{i=1}^{n} |a_i| h_i \right) S_0 x^2(t) + S_0 \sum_{i=1}^{n} |a_i| \int_{t-h_i}^{t} x^2(s) \, ds + \sigma^2 x^2(t-\tau).$$

Using Theorem 2.5 with $m = S = Q_i = R_j = 0$ and

$$D = \left(-2 + \sum_{i=1}^{n} |a_i| h_i \right) S_0, \qquad dK_1(s) = \sigma^2 \delta(s-\tau) \, ds,$$

$$d\mu(s) = \sum_{i=1}^{n} \delta(s-h_i) \, ds, \qquad R(s,t) = S_0 |a_i|,$$

we obtain the sufficient condition for asymptotic mean-square stability of the trivial solution of (3.29) in the form

$$S_0 > p \left(1 - \sum_{i=1}^{n} |a_i| h_i \right)^{-1}, \qquad \sum_{i=1}^{n} |a_i| h_i < 1, \ p = \frac{\sigma^2}{2}. \tag{3.33}$$

This condition is a generalization of condition (3.27).

Let us obtain a generalization of condition (3.28). Note that by (3.30) and $h_0 = 0$ the process $z(t)$ from (3.32) can be represented as follows:

$$z(x_t) = x(t) - \sum_{i=1}^{n} a_i \sum_{j=0}^{i-1} \int_{t-h_{j+1}}^{t-h_j} x(s) \, ds = x(t) - \sum_{j=0}^{n-1} S_{j+1} \int_{t-h_{j+1}}^{t-h_j} x(s) \, ds. \tag{3.34}$$

Using the representation (3.34), the functional $V_1 = z^2(x_t)$, and $S_0 > 0$, we get

$$LV_1 = -2S_0 x^2(t) + 2S_0 x(t) \sum_{j=0}^{n-1} S_{j+1} \int_{t-h_{j+1}}^{t-h_j} x(s) \, ds + \sigma^2 x^2(t-\tau)$$

$$\leq -2S_0 x^2(t) + S_0 \sum_{j=0}^{n-1} |S_{j+1}| \left((h_{j+1} - h_j) x^2(t) + \int_{t-h_{j+1}}^{t-h_j} x^2(s) \, ds \right)$$

$$+ \sigma^2 x^2(t-\iota)$$

$$= \left(-2 + \sum_{j=0}^{n-1} |S_{j+1}| (h_{j+1} - h_j) \right) S_0 x^2(t)$$

$$+ S_0 \sum_{j=0}^{n-1} |S_{j+1}| \int_{t-h_{j+1}}^{t-h_j} x^2(s) \, ds + \sigma^2 x^2(t-\tau).$$

By Theorem 2.5 with $m = S = Q_i = 0$ and

$$D = \left(-2 + \sum_{j=0}^{n-1} |S_{j+1}|(h_{j+1} - h_j)\right) S_0, \qquad dK_1(s) = \sigma^2 \delta(s - \tau) ds,$$

$$d\mu(s) = \sum_{j=0}^{n-1} \delta(s - h_{j+1}) ds,$$

$$R(\tau + h_{j+1}, t) = \begin{cases} S_0 |S_{j+1}|, & \tau \in [t - h_{j+1}, t - h_j], \\ 0, & \tau \in (t - h_j, t]. \end{cases}$$

from this we obtain a generalization of condition (3.28) in the form

$$S_0 > p \left(1 - \sum_{j=0}^{n-1} |S_{j+1}|(h_{j+1} - h_j)\right)^{-1}, \qquad \sum_{j=0}^{n-1} |S_{j+1}|(h_{j+1} - h_j) < 1. \quad (3.35)$$

To prove that condition (3.35) is weaker than (3.33), it is enough to show that

$$\sum_{j=0}^{n-1} |S_{j+1}|(h_{j+1} - h_j) \le \sum_{i=1}^{n} |a_i| h_i.$$

Using $h_0 = 0$, rewrite this inequality in the form

$$\sum_{j=0}^{n-1} |S_{j+1}| h_{j+1} = \sum_{j=1}^{n} |S_j| h_j \le \sum_{j=1}^{n-1} |S_{j+1}| h_j + \sum_{i=1}^{n} |a_i| h_i.$$

Now it is enough to note that $|S_j| \le |S_{j+1}| + |a_j|$, $j = 1, \ldots, n - 1$, and $S_n = a_n$.

3.3 Linear Differential Equation of nth Order

3.3.1 Case $n > 1$

Consider the scalar stochastic differential equation of the nth order

$$x^{(n)}(t) = \sum_{j=1}^{n} \int_0^\infty x^{(j-1)}(t - s) dK_j(s) + \sigma x(t - \tau)\dot{w}(t), \quad t \ge 0, \quad (3.36)$$

where $x^{(j)}(t) = \frac{d^j x(t)}{dt^j}$, $j = 1, \ldots, n$. The initial conditions for (3.36) have the form

$$x^{(j)}(\theta) = \phi^{(j)}(\theta), \quad \theta \le 0, \qquad (3.37)$$

where $\phi(\theta)$ is a given $n-1$ times continuously differentiable function. The kernels $K_j(s)$ are functions of bounded variation on $[0, \infty)$ such that

$$\alpha_{ij} = \int_0^\infty s^i |dK_j(s)| < \infty, \quad 0 \le i \le n, \ 1 \le j \le n. \tag{3.38}$$

Put $x_i(t) = x^{(i-1)}(t)$ and transform (3.36) to the system of stochastic differential equations

$$\dot{x}_i(t) = x_{i+1}(t), \quad i = 1, \ldots, n-1,$$

$$\dot{x}_n(t) = \sum_{j=1}^n \int_0^\infty x_j(t-s)\, dK_j(s) + \sigma x_1(t-\tau)\dot{w}(t). \tag{3.39}$$

Put also

$$\beta_{ij} = \int_0^\infty s^i\, dK_j(s) \tag{3.40}$$

and note that for $m = 1, \ldots, n$, we have

$$\int_0^\infty x_n(t-s)\, dK_m(s)$$

$$= x_n(t)\beta_{0m} - \frac{d}{dt}\left[\int_0^\infty dK_m(s)\int_{t-s}^t x_n(\theta)\, d\theta\right]. \tag{3.41}$$

Similarly, using (3.40), it is easy to check that for $i = 1, \ldots, n-1$,

$$\int_0^\infty x_{n-i}(t-s)\, dK_m(s)$$

$$= \sum_{j=1}^{i+1}(-1)^{j-1} x_{n-i+j-1}(t)\frac{\beta_{j-1,m}}{(j-1)!}$$

$$+ (-1)^{i+1}\frac{d}{dt}\left[\int_0^\infty dK_m(s)\int_{t-s}^t x_n(\theta)\frac{(\theta-t+s)^i}{i!}\, d\theta\right]. \tag{3.42}$$

Putting

$$z(x_t) = \sum_{l=0}^{n-1}(-1)^{l+1}\int_0^\infty dK_{n-l}(s)\int_{t-s}^t x_n(\theta)\frac{(\theta-t+s)^l}{l!}\, d\theta,$$

$$\tag{3.43}$$

$$a_l = \sum_{i=l}^{n-1}(-1)^{i-l}\frac{\beta_{i-l,n-i}}{(i-l)!}, \quad l = 0, 1, \ldots, n-1,$$

by (3.42) and (3.43) we obtain

$$\sum_{j=1}^{n} \int_0^\infty x_j(t-s)\,dK_j(s) = \sum_{i=0}^{n-1} \int_0^\infty x_{n-i}(t-s)\,dK_{n-i}(s)$$

$$= \sum_{i=0}^{n-1} \sum_{j=1}^{i+1} (-1)^{j-1} x_{n-i+j-1}(t) \frac{\beta_{j-1,n-i}}{(j-1)!} + \dot{z}(x_t)$$

$$= \sum_{i=0}^{n-1} \sum_{l=0}^{i} (-1)^{i-l} x_{n-l}(t) \frac{\beta_{i-l,n-i}}{(i-l)!} + \dot{z}(x_t)$$

$$= \sum_{l=0}^{n-1} a_l x_{n-l}(t) + \dot{z}(x_t). \tag{3.44}$$

Following the procedure of constructing Lyapunov functionals and using (3.44), rewrite (3.39) as follows:

$$\dot{x}_i(t) = x_{i+1}(t), \quad i = 1, \ldots, n-1,$$

$$\frac{d}{dt} \big[x_n(t) - z(x_t) \big] = \sum_{l=0}^{n-1} a_l x_{n-l}(t) + \sigma x_1(t-\tau)\dot{w}(t). \tag{3.45}$$

For (3.45), consider the auxiliary system of the form

$$\dot{y}_i(t) = y_{i+1}(t), \quad i = 1, \ldots, n-1, \qquad \dot{y}_n(t) = \sum_{l=0}^{n-1} a_l y_{n-l}(t). \tag{3.46}$$

Let $y = (y_1, \ldots, y_n)'$, and A be the $n \times n$ matrix

$$A = \begin{pmatrix} 0 & 1 & 0 & \cdots & 0 \\ 0 & 0 & 1 & \cdots & 0 \\ \cdots & \cdots & \cdots & \cdots & \cdots \\ 0 & 0 & 0 & \cdots & 1 \\ a_{n-1} & a_{n-2} & a_{n-3} & \cdots & a_0 \end{pmatrix}. \tag{3.47}$$

By (3.46) and (3.47) we have $\dot{y} = Ay$.

Assume that the trivial solution of (3.46) is asymptotically stable. From Theorem 1.3 it follows that for the matrix (3.47) and an arbitrary positive definite matrix Q, the matrix equation (1.27) has a unique positive definite solution P.

Consider the Lyapunov function for (3.46) of the form $v(y) = y'Py$. Because of (1.25), we have $\dot{v}(y) = -y'Qy$. According to the procedure of constructing Lyapunov functionals, we consider the functional

$$V_1(t, x_t) = \big(x_1(t), \ldots, x_{n-1}(t), x_n(t) - z(x_t) \big)' P \big(x_1(t), \ldots, x_{n-1}(t), x_n(t) - z(x_t) \big). \tag{3.48}$$

Let Q be a diagonal matrix with positive entries q_l, $l = 1, \ldots, n$. From (3.49) it follows that $LV_1(t, x_t)$ with respect to (3.45) equals

$$LV_1(t, x_t) = -\sum_{l=1}^{n} q_l x_l^2(t) - 2\sum_{l=1}^{n} z(x_t)(PA)_{nl} x_l(t) + p_{nn}\sigma^2 x_1^2(t-\tau)$$

$$\leq -\sum_{l=1}^{n} q_l x_l^2(t) + 2\sum_{l=1}^{n} \beta_l |z(x_t)x_l(t)| + p_{nn}\sigma^2 x_1^2(t-\tau), \quad (3.49)$$

where $(PA)_{nl}$ is (nl)th element of the matrix PA, $\beta_l = |(PA)_{nl}|$, and $p_{nn} = (P)_{nn}$. Put also

$$\alpha = \sum_{j=0}^{n-1} \frac{\alpha_{j+1,n-j}}{(j+1)!}, \qquad W(t, x_t) = \sum_{j=0}^{n-1} \int_0^\infty |dK_{n-j}(s)| \int_{t-s}^{t} x_n^2(\theta) \frac{(\theta-t+s)^j}{j!}\, d\theta,$$

$$(3.50)$$

and suppose that $\alpha > 0$. Then using (3.43) and some positive numbers γ_l, $l = 1, \ldots, n$, we obtain

$$2|z(x_t)x_l(t)| \leq \sum_{j=0}^{n-1} \int_0^\infty |dK_{n-j}(s)| \int_{t-s}^{t} \left(\gamma_l x_l^2(t) + \gamma_l^{-1} x_n^2(\theta)\right) \frac{(\theta-t+s)^j}{j!}\, d\theta$$

$$= \alpha\gamma_l x_l^2(t) + \gamma_l^{-1} W(t, x_t). \quad (3.51)$$

By (3.51) from (3.49) it follows that

$$LV_1(t, x_t) \leq \sum_{l=1}^{n} (\alpha\beta_l\gamma_l - q_l) x_l^2(t) + p_{nn}\sigma^2 x_1^2(t-\tau) + \sum_{l=1}^{n} \beta_l\gamma_l^{-1} W(x_t). \quad (3.52)$$

From this and from (3.50) the representation of the type of (2.35) follows, where $S(t) = Q_i(t) = R(s, t) = 0$, D is a diagonal matrix with the elements $d_{ll} = \alpha\beta_l\gamma_l - q_l$, $dK_1(s) = p_{nn}\sigma^2\delta(s-\tau)\,ds$, $dK_l(s) = 0$, $l = 2, \ldots, n$, $m = n - 1$, $d\mu_j(s) = |dK_{n-j}(s)|$, and $R_j(s)$ is the matrix with all zero elements except for $(R_j)_{nn} = \frac{1}{j!}\sum_{l=1}^{n} \beta_l\gamma_l^{-1}$.

So, by (2.37) the matrix G is the diagonal matrix with

$$G_{11} = \alpha\beta_1\gamma_1 + p_{nn}\sigma^2 - q_1, \qquad G_{ll} = \alpha\beta_l\gamma_l - q_l, \quad l = 2, \ldots, n-1,$$

$$G_{nn} - \alpha\left[\beta_n\left(\gamma_n + \frac{1}{\gamma_n}\right) + \sum_{l=1}^{n-1} \frac{\beta_l}{\gamma_l}\right] - q_n.$$

It is easy to see that G_{nn} reaches its minimum with respect to γ_n at $\gamma_n - 1$. Besides, from the matrix equation (1.27) with the matrix A given by (3.47) it follows that $\beta_1 = |a_{n-1}|p_{nn}$, $2\beta_n = q_n$. So, we can conclude that if there exist positive numbers $\gamma_1, \gamma_2, \ldots, \gamma_{n-1}$ such that

$$\gamma_1 < \frac{1}{\alpha}\left(\frac{q_1}{\beta_1} - \frac{\sigma^2}{|a_{n-1}|}\right), \qquad \gamma_l < \frac{q_l}{\alpha\beta_l}, \qquad l = 2, \ldots, n-1,$$

$$\sum_{l=1}^{n-1} \frac{\beta_l}{\gamma_l} < \left(\frac{1}{\alpha} - 1\right) q_n, \qquad \alpha < 1,$$

(3.53)

then the matrix G is a negative definite one, and therefore the zero solution of (3.36) is asymptotically mean-square stable.

Let us rewrite inequalities (3.53) in the form

$$0 < \alpha\left(\frac{q_1}{\beta_1} - \frac{\sigma^2}{|a_{n-1}|}\right)^{-1} < \frac{1}{\gamma_1}, \qquad \frac{\alpha\beta_l}{q_l} < \frac{1}{\gamma_l}, \qquad l = 2, \ldots, n-1,$$

$$\frac{\beta_1}{\gamma_1} + \sum_{l=2}^{n-1} \frac{\beta_l}{\gamma_l} < \left(\frac{1}{\alpha} - 1\right) q_n, \qquad \alpha < 1.$$

(3.54)

From the system of inequalities (3.54) it follows that

$$\alpha\beta_1\left(\frac{q_1}{\beta_1} - \frac{\sigma^2}{|a_{n-1}|}\right)^{-1} + \alpha\sum_{l=2}^{n-1} \frac{\beta_l^2}{q_l} < \frac{\beta_1}{\gamma_1} + \sum_{l=2}^{n-1} \frac{\beta_l}{\gamma_l} < \left(\frac{1}{\alpha} - 1\right) q_n. \qquad (3.55)$$

So, if the condition

$$\alpha\beta_1\left(\frac{q_1}{\beta_1} - \frac{\sigma^2}{|a_{n-1}|}\right)^{-1} + \alpha\sum_{l=2}^{n-1} \frac{\beta_l^2}{q_l} < \left(\frac{1}{\alpha} - 1\right) q_n$$

or

$$\sigma^2 < |a_{n-1}|\left(\frac{q_1}{\beta_1} - \frac{\beta_1}{\Theta q_n - \sum_{l=2}^{n-1} \beta_l^2 q_l^{-1}}\right), \qquad \sum_{l=2}^{n-1} \frac{\beta_l^2}{q_l} < \Theta q_n, \qquad \Theta = \frac{1}{\alpha^2} - \frac{1}{\alpha},$$

(3.56)

holds, then there exist positive numbers $\gamma_1, \gamma_2, \ldots, \gamma_{n-1}$ such that (3.55) also holds and the zero solution of (3.36) is asymptotically mean-square stable.

So, we have proven

Theorem 3.1 *Let there exist a diagonal matrix Q with positive entries $q_1, \ldots, q_n$ such that the matrix equation (1.27) has a positive definite solution P that satisfies inequalities (3.56). Then the zero solution of (3.36) is asymptotically mean-square stable.*

Remark 3.4 Without the loss of generality, we may assume that $q_n = 1$. Otherwise, the matrix equation (1.27) can be divided by q_n. Thus, all elements of the matrices Q and P will be divided by q_n.

Remark 3.5 Note that condition (3.56) is also correct without the assumption $\alpha > 0$. Indeed, if $\alpha = 0$ (this means also that $z(x_t) \equiv 0$), then from (3.56) we have $\Theta = \infty$ and $\sigma^2 < |a_{n-1}| q_1/\beta_1$, which also follows immediately from (3.49) and $\beta_1 = |a_{n-1}| p_{nn}$.

Remark 3.6 The stability condition obtained in Theorem 3.1 uses the representation (3.42) where the integrals in the right-hand side depend only on x_n for all i. Following the same procedure, one can try to obtain other stability conditions using the representations where the right-hand side depends on x_m for $m \leq n$. For example, for $n = 2$, we have

$$\int_0^\infty x_1(t-s)\,dK_1(s) = \beta_{01}x_1(t) - \beta_{11}x_2(t) + \frac{d}{dt}\int_0^\infty dK_1(s)$$

$$\times \int_{t-s}^t (\tau - t + s)x_2(\tau)\,d\tau,$$

$$\int_0^\infty x_i(t-s)\,dK_i(s) = \beta_{0i}x_i(t) - \frac{d}{dt}\int_0^\infty dK_i(s)\int_{t-s}^t x_i(\tau)\,d\tau, \quad i = 1, 2,$$

$$\int_0^\infty x_2(t-s)\,dK_2(s) = \frac{d}{dt}\int_0^\infty x_1(t-s)\,dK_2(s).$$

(3.57)

3.3.2 Some Particular Cases

It is easy to see that the stability condition (3.56) is the best one for those $q_1, \ldots, q_n$ for which the right-hand part of inequality (3.56) reaches its maximum. Let us consider some particular cases of condition (3.56) in which it can be formulated immediately in the terms of the parameters of the considered equation (3.36).

Let be $n = 1$. In this case, (3.36) has the form

$$\dot{x}(t) = \int_0^\infty x(t-s)\,dK(s) + \sigma x(t-\tau)\dot{w}(t), \quad t \geq 0.$$ (3.58)

Condition (3.56) cannot be used immediately since it was obtained for $n > 1$. So, note that for the functional $V_1(t, x_t) = x^2(t)$, similarly to (3.52) (for $\gamma_1 = q_1 = 1$), we have

$$LV_1(t, x_t) \leq (-1 + \alpha\beta_1)x^2(t) + p_{11}\sigma^2 x^2(t - \tau) + \beta_1 \int_0^\infty |dK(s)| \int_{t-s}^t x^2(\theta)\,d\theta,$$

where

$$\alpha = \alpha_{11} = \int_0^\infty s|dK(s)|, \qquad a_0 = \beta_{01} = \int_0^\infty dK(s) < 0,$$

$$p_{11} = -\frac{1}{2a_0} = \frac{1}{2|\beta_{01}|} > 0, \qquad \beta_1 = |p_{11}a_0| = \frac{1}{2}.$$

The stability condition for (3.58) takes the form $\sigma^2 < 2|a_0|(1 - \alpha)$.

If, in particular, $dK(s) = (-a\delta(s) - b\delta(s-h))\,ds$, then $\alpha = |b|h$, $a_0 = -a - b <$ 0, and the stability condition takes the form $\sigma^2 < 2(a + b)(1 - |b|h)$. Note that the last condition also follows immediately from (3.13) for $c = 0$.

Let be $n = 2$. In this case, (3.36) has the form

$$\ddot{x}(t) = \int_0^\infty x(t-s)\,dK_1(s) + \int_0^\infty \dot{x}(t-s)\,dK_2(s) + \sigma x(t-\tau)\dot{w}(t), \quad t \ge 0.$$
(3.59)

Following Remark 3.4, we will consider the matrix equation (1.25) with

$$A = \begin{pmatrix} 0 & 1 \\ a_1 & a_0 \end{pmatrix}, \quad P = \begin{pmatrix} p_{11} & p_{12} \\ p_{12} & p_{22} \end{pmatrix}, \quad Q = \begin{pmatrix} q & 0 \\ 0 & 1 \end{pmatrix}, \quad q > 0.$$

By (3.43), $a_0 = \beta_{02} - \beta_{11}$, $a_1 = \beta_{01}$, where β_{ij} are defined by (3.40). By Corollary 1.1 the matrix equation (1.27) has a positive definite solution if and only if $a_0 < 0$, $a_1 < 0$. From (1.29) it follows that the last element of the matrix P is $p_{22} = (q + |a_1|)(2a_0a_1)^{-1}$. Since $\beta_1 = |a_1|p_{22} = (q + |a_1|)(2|a_0|)^{-1}$, the stability condition (3.56) takes the form

$$\sigma^2 < 2|a_1|\left(\frac{q|a_0|}{q + |a_1|} - \frac{(q + |a_1|)\alpha^2}{4|a_0|(1 - \alpha)} \right),$$
(3.60)

where $\alpha = \alpha_{12} + \frac{1}{2}\alpha_{21} < 1$, α_{ij} are defined by (3.38).

The right-hand part of (3.60) reaches its maximum at $q = 2|a_0|\alpha^{-1}\sqrt{(1 - \alpha)|a_1|} - |a_1|$. So, as a result, we obtain a sufficient condition for asymptotic mean-square stability of the zero solution of (3.58) of the form

$$\sigma^2 < 2|a_1|\left(|a_0| - \alpha\sqrt{\frac{|a_1|}{1 - \alpha}} \right), \quad \alpha < 1.$$
(3.61)

Using the representations (3.57), we can get for (3.59) other stability conditions. Put, for instance,

$$z_1(t) = x_1(t) = x(t), \qquad x_2(t) = \dot{x}(t),$$

$$z_2(t) = x_2(t) + \int_0^\infty dK_1(s)\int_{t-s}^t x_1(\theta)\,d\theta + \int_0^\infty dK_2(s)\int_{t-s}^t x_2(\theta)\,d\theta.$$
(3.62)

By (3.62), (3.57), and (3.40), equation (3.59) can be represented in the form of the system

$$\dot{z}_1(t) = x_2(t),$$
$$\dot{z}_2(t) = \beta_{01}x_1(t) + \beta_{02}x_2(t) + \sigma x_1(t-\tau)\dot{w}(t),$$
(3.63)

By Corollary 1.1 the trivial solution of the auxiliary differential equation

$$\dot{y} = Ay, \quad A = \begin{pmatrix} 0 & 1 \\ \beta_{01} & \beta_{02} \end{pmatrix}, \quad y = \begin{pmatrix} y_1 \\ y_2 \end{pmatrix},$$

is asymptotically stable if and only if

$$\beta_{01} < 0, \qquad \beta_{02} < 0.$$
(3.64)

Besides, by (1.29) the matrix equation (1.27) by conditions (3.64) has the positive
definite solution P with the elements

$$p_{11} = \frac{(\beta_{01}^2 + |\beta_{02}|)q + \beta_{02}^2}{2\beta_{01}\beta_{02}}, \qquad p_{22} = \frac{|\beta_{01}| + q}{2\beta_{01}\beta_{02}}, \qquad p_{12} = -\frac{q}{2\beta_{01}}. \quad (3.65)$$

Suppose that (3.64) holds and consider the Lyapunov function $V_1 = z'Pz$ where
$z = (z_1, z_2)'$. Calculating LV_1 for (3.63), by (3.65) and (3.62) we obtain

$$LV_1 = 2\big(p_{11}z_1(t) + p_{12}z_2(t)\big)x_2(t)$$

$$+ 2\big(p_{12}z_1(t) + p_{22}z_2(t)\big)\big(\beta_{01}x_1(t) + \beta_{02}x_2(t)\big) + p_{22}\sigma^2 x_1^2(t - \tau)$$

$$= -qx_1^2(t) - x_2^2(t) + p_{22}\sigma^2 x_1^2(t - \tau)$$

$$+ 2p_{22}\beta_{01} \int_0^\infty dK_1(s) \int_{t-s}^t x_1(t)x_1(\theta)\,d\theta$$

$$- \int_0^\infty dK_1(s) \int_{t-s}^t x_2(t)x_1(\theta)\,d\theta$$

$$+ 2p_{22}\beta_{01} \int_0^\infty dK_2(s) \int_{t-s}^t x_1(t)x_2(\theta)\,d\theta$$

$$- \int_0^\infty dK_2(s) \int_{t-s}^t x_2(t)x_2(\theta)\,d\theta.$$

Using some $\gamma_1 > 0$, $\gamma_2 > 0$ and (3.38), we have

$$LV_1 \leq -qx_1^2(t) - x_2^2(t) + p_{22}\sigma^2 x_1^2(t - \tau)$$

$$+ p_{22}|\beta_{01}| \int_0^\infty |dK_1(s)| \int_{t-s}^t \big(x_1^2(t) + x_1^2(\theta)\big)\,d\theta$$

$$+ \frac{1}{2} \int_0^\infty |dK_1(s)| \int_{t-s}^t \Big(\gamma_1 x_2^2(t) + \frac{1}{\gamma_1}x_1^2(\theta)\Big)\,d\theta$$

$$+ p_{22}|\beta_{01}| \int_0^\infty |dK_2(s)| \int_{t-s}^t \Big(\gamma_2 x_1^2(t) + \frac{1}{\gamma_2}x_2^2(\theta)\Big)\,d\theta$$

$$+ \frac{1}{2} \int_0^\infty |dK_2(s)| \int_{t-s}^t \big(x_2^2(t) + x_2^2(\theta)\big)\,d\theta.$$

As a result, we obtain

$$LV_1 \leq -\big(q - p_{22}|\beta_{01}|(\alpha_{11} + \alpha_{12}\gamma_2)\big)x_1^2(t)$$

$$- \Big(1 - \frac{1}{2}(\alpha_{11}\gamma_1 + \alpha_{12})\Big)x_2^2(t) + p_{22}\sigma^2 x_1^2(t - \tau)$$

$$+ \left(p_{22} |\beta_{01}| + \frac{1}{2\gamma_1} \right) \int_0^\infty |dK_1(s)| \int_{t-s}^t x_1^2(\theta) \, d\theta$$

$$+ \left(\frac{p_{22} |\beta_{01}| 1}{\gamma_2} + \frac{1}{2} \right) \int_0^\infty |dK_2(s)| \int_{t-s}^t x_2^2(\theta) \, d\theta.$$

From Theorem 2.5 by (2.35) and (3.38) it follows that if

$$p_{22} \left(|\beta_{01}| (2\alpha_{11} + \alpha_{12}\gamma_2) + \sigma^2 \right) + \frac{\alpha_{11}}{2\gamma_1} < q,$$

$$\alpha_{12} + \frac{\alpha_{11}\gamma_1}{2} + \frac{p_{22}\alpha_{12} |\beta_{01}|}{\gamma_2} < 1,$$

(3.66)

then the trivial solution of (3.59) is asymptotically mean-square stable.

To get stability conditions immediately in the terms of (3.59), transform (3.66) by the following way. Substituting p_{22} from (3.65) into (3.66) and solving both inequalities (3.66) with respect to q, we obtain that inequalities (3.66) are equivalent to

$$0 < \left(\frac{\mu}{|\beta_{02}|} + \frac{\alpha_{11}}{\gamma_1} \right) \left(2 - \frac{\mu}{|\beta_{01}\beta_{02}|} \right)^{-1} < q$$

$$< \frac{\gamma_2 |\beta_{02}|}{\alpha_{12}} \left(2(1 - \alpha_{12}) - \alpha_{11}\gamma_1 - \frac{\alpha_{12} |\beta_{01}|}{\gamma_2 |\beta_{02}|} \right),$$

(3.67)

where

$$\mu = |\beta_{01}| (2\alpha_{11} + \alpha_{12}\gamma_2) + \sigma^2.$$

(3.68)

So, if the inequality

$$0 < \left(\frac{\mu}{|\beta_{02}|} + \frac{\alpha_{11}}{\gamma_1} \right) \left(2 - \frac{\mu}{|\beta_{01}\beta_{02}|} \right)^{-1}$$

$$< \frac{\gamma_2 |\beta_{02}|}{\alpha_{12}} \left(2(1 - \alpha_{12}) - \alpha_{11}\gamma_1 - \frac{\alpha_{12} |\beta_{01}|}{\gamma_2 |\beta_{02}|} \right)$$

holds, then there exists $q > 0$ such that (3.67) holds too.

From the last condition we have

$$\frac{\mu}{|\beta_{02}|} + \frac{\alpha_{11}}{\gamma_1} < \frac{\gamma_2 |\beta_{02}|}{\alpha_{12}} \left(2(1 - \alpha_{12}) - \alpha_{11}\gamma_1 - \frac{\alpha_{12} |\beta_{01}|}{\gamma_2 |\beta_{02}|} \right) \left(2 - \frac{\mu}{|\beta_{01}\beta_{02}|} \right)$$

$$= \frac{\gamma_2 |\beta_{02}|}{\alpha_{12}} \left(2(1 - \alpha_{12}) - \alpha_{11}\gamma_1 \right) \left(2 - \frac{\mu}{|\beta_{01}\beta_{02}|} \right) - 2|\beta_{01}| + \frac{\mu}{|\beta_{02}|}$$

or

$$\frac{\alpha_{11}}{\gamma_1} < \frac{\gamma_2 |\beta_{02}|}{\alpha_{12}} \left(2(1 - \alpha_{12}) - \alpha_{11}\gamma_1 \right) \left(2 - \frac{\mu}{|\beta_{01}\beta_{02}|} \right) - 2|\beta_{01}|.$$

Using (3.68), rewrite the obtained inequality in the form

$$\frac{2|\beta_{01}| + \alpha_{11}\gamma_1^{-1}}{2(1 - \alpha_{12}) - \alpha_{11}\gamma_1} < \gamma_2\left[2\left(\frac{|\beta_{02}|}{\alpha_{12}} - \frac{\alpha_{11}}{\alpha_{12}} - \frac{\sigma^2}{2\alpha_{12}|\beta_{01}|}\right) - \gamma_2\right]. \tag{3.69}$$

Now the numbers γ_1 and γ_2 are separated, and we have minimize the left part of this inequality with respect to $\gamma_1 > 0$ and maximize the right part of this inequality with respect to $\gamma_2 > 0$.

By Lemma 2.4 the minimum of the left part of (3.69) is reached at

$$\gamma_1 = \frac{2(1 - \alpha_{12})}{\sqrt{\alpha_{11}^2 + 4|\beta_{01}|(1 - \alpha_{12})} + \alpha_{11}}$$

and equals γ_1^{-2}. Besides, it is easy to see that the maximum of the right part of (3.69) is reached at

$$\gamma_2 = \frac{|\beta_{02}|}{\alpha_{12}} - \frac{\alpha_{11}}{\alpha_{12}} - \frac{\sigma^2}{2\alpha_{12}|\beta_{01}|}$$

and equals γ_2^2. So, condition (3.69) takes the form $\gamma_1^{-1} < \gamma_2$ or

$$\frac{\sqrt{\alpha_{11}^2 + 4|\beta_{01}|(1 - \alpha_{12})} + \alpha_{11}}{2(1 - \alpha_{12})} < \frac{|\beta_{02}|}{\alpha_{12}} - \frac{\alpha_{11}}{\alpha_{12}} - \frac{\sigma^2}{2\alpha_{12}|\beta_{01}|}.$$

As a result, from this and from (3.62) we obtain

$$\sigma^2 < 2|\beta_{01}|\left[|\beta_{02}| - \alpha_{11} - \alpha_{12}\frac{\sqrt{\alpha_{11}^2 + 4|\beta_{01}|(1 - \alpha_{12})} + \alpha_{11}}{2(1 - \alpha_{12})}\right],$$

$$\alpha_{11} + \alpha_{12} < 1. \tag{3.70}$$

So, if (3.70) holds, then the trivial solution of (3.59) is asymptotically mean-square stable.

Example 3.1 Consider the second-order differential equation

$$\ddot{x}(t) + a\dot{x}(t - h_1) + bx(t - h_2) + \sigma x(t - \tau)\dot{w}(t) = 0 \tag{3.71}$$

with $a > 0$, $b > 0$, $h_1 \geq 0$, $h_2 \geq 0$, $\tau \geq 0$. Equation (3.71) is obtained from (3.59) if $dK_1(s) = -b\delta(s - h_2)\,ds$ and $dK_2(s) = -a\delta(s - h_1)\,ds$. In this case, $\alpha_{11} = bh_2$, $\alpha_{12} = ah_1$, $\alpha_{21} = bh_2^2$, $\beta_{02} = -a < 0$, $\beta_{11} = -bh_2$, $a_0 = \beta_{02} - \beta_{11} = -a + bh_2 < 0$, $a_1 = \beta_{01} = -b < 0$. The conditions for asymptotic mean-square stability (3.61) and (3.70) give respectively

$$\sigma^2 < 2b\left(a - bh_2 - \left(ah_1 + \frac{1}{2}bh_2^2\right)\sqrt{\frac{b}{1 - ah_1 - \frac{1}{2}bh_2^2}}\right),$$

$$ah_1 + \frac{1}{2}bh_2^2 < 1, \tag{3.72}$$

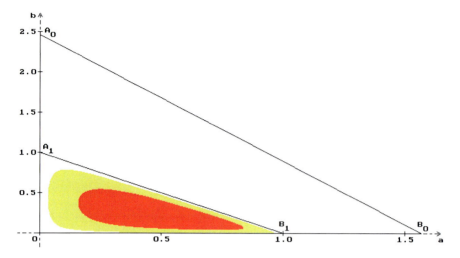

Fig. 3.17 Stability regions for (3.67), given by the conditions (3.68), (3.69), are shown for $h_1 = 1$, $h_2 = 0$, and different values of σ: $\sigma = 0.2$ (*red*), $\sigma = 0.1$ (*yellow*), and $\sigma = 0$ (*triangle $A_1 O B_1$*). The *triangle $A_0 O B_0$* gives the region of stability obtained via the characteristic equation

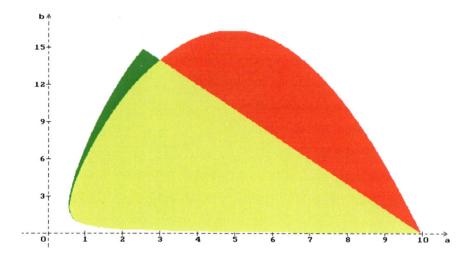

Fig. 3.18 Stability regions for (3.67), given by conditions (3.68) (*red and yellow*) and (3.69) (*green and yellow*) are shown for $h_1 = 0.2$, $h_2 = 0.5$, $\sigma = 0.25$

$$\sigma^2 < 2b\left(a - bh_2 - ah_1 \frac{\sqrt{(bh_2)^2 + 4b(1 - ah_1)} + bh_2}{2(1 - ah_1)}\right),$$

$$ah_1 + bh_2 < 1. \tag{3.73}$$

Note that for $h_2 = 0$, conditions (3.72)–(3.73) are equivalent, but for $h_2 > 0$, both conditions can give regions of stability that are different and complement each other.

In Fig. 3.17 the stability regions given by conditions (3.72)–(3.73) in the space of the parameters (a, b) are shown for $h_1 = 1$, $h_2 = 0$, and different values of σ: $\sigma = 0.2$ (red), $\sigma = 0.1$ (yellow), and $\sigma = 0$ (the triangle $A_1 O B_1$). For comparison, the triangle $A_0 O B_0$ gives a part of the region of asymptotic stability that was obtained via the characteristic equation in Example 1.6 (the case c, Fig. 1.19).

In Fig. 3.18 the stability regions given by conditions (3.72) (red and yellow) and (3.73) (green and yellow) are shown for $h_1 = 0.2$, $h_2 = 0.5$, $\sigma = 0.25$. One can see that the stability regions have a common part (yellow) but have also the different parts: red for (3.72) and green for (3.73).

Example 3.2 Consider the second-order differential equation

$$\ddot{x}(t) = ax(t) + bx(t - h_1) + bx(t - h_2) + \sigma x(t - \tau)\dot{w}(t) = 0 \qquad (3.74)$$

with $a > 0$, $h_1 \geq 0$, $h_2 \geq 0$, $\tau \geq 0$. Equation (3.74) is obtained from (3.59) if $dK_1(s) = (a\delta(s) + b_1\delta(s - h_1) + b_2\delta(s - h_2))\,ds$ and $dK_2(s) = 0$. In this case we have $a_0 = -\beta_{11} = -b_1 h_1 - b_2 h_2 < 0$, $a_1 = \beta_{01} = a + b_1 + b_2 < 0$, $\alpha = \frac{1}{2}\alpha_{21} = \frac{1}{2}(|b_1|h_1^2 + |b_2|h_2^2)$. The stability condition (3.61) takes the form

$$\sigma^2 < 2|a_1|\left(|a_0| - \alpha_{21}\sqrt{\frac{|a_1|}{2(2 - \alpha_{21})}}\right), \qquad \alpha_{21} < 2. \qquad (3.75)$$

Note that the stability condition (3.70) cannot be used here since $\beta_{02} = 0$.
Let $n = 3$. In this case, (3.36) has the form

$$\ddot{x}(t) = \int_0^\infty x(t - s)\,dK_1(s) + \int_0^\infty \dot{x}(t - s)\,dK_2(s)$$

$$+ \int_0^\infty \ddot{x}(t - s)\,dK_3(s) + \sigma x(t - \tau)\dot{w}(t). \qquad (3.76)$$

By Remark 3.4 we will consider the corresponding matrix equation (1.27) with

$$A = \begin{pmatrix} 0 & 1 & 0 \\ 0 & 0 & 1 \\ a_2 & a_1 & a_0 \end{pmatrix}, \qquad Q = \begin{pmatrix} q_1 & 0 & 0 \\ 0 & q_2 & 0 \\ 0 & 0 & 1 \end{pmatrix}, \qquad P = \begin{pmatrix} p_{11} & p_{12} & p_{13} \\ p_{12} & p_{22} & p_{23} \\ p_{13} & p_{23} & p_{33} \end{pmatrix}. \qquad (3.77)$$

By Corollary 1.2 the inequalities

$$a_0 < 0, \quad a_2 < 0, \qquad A_0 = a_0 a_1 + a_2 > 0 \qquad (3.78)$$

are necessary and sufficient conditions for the matrix equation (1.27), (3.77) to have the positive definite solution P with the elements

$$p_{11} = \frac{1}{2}\left(\frac{a_1}{a_2} + \frac{a_0^2}{A_0}\right)q_1 + \frac{a_0 a_2 q_2 + a_2^2}{2A_0}, \qquad p_{12} = \frac{a_0^2 a_1}{2a_2 A_0}q_1 + \frac{|a_2|q_2 + a_1 a_2}{2A_0},$$

$$p_{13} = \frac{q_1}{2|a_2|}, \qquad p_{22} = \frac{a_0^3 + a_2}{2a_2 A_0}q_1 + \frac{(a_0^2 + |a_1|)q_2 + a_1^2 + a_0 a_2}{2A_0},$$

$$p_{23} = \frac{a_0^2}{2|a_2|A_0}q_1 + \frac{|a_0|q_2 + |a_2|}{2A_0}, \qquad p_{33} = \frac{a_0}{2a_2 A_0}q_1 + \frac{q_2 + |a_1|}{2A_0}.$$

$$(3.79)$$

Note that from (3.78) it also follows that $a_1 < 0$. Calculating $\beta_1 = |a_2|p_{33}$, $\beta_2 = |p_{13} + a_1 p_{33}|$, we obtain the representation

$$\beta_l = p_{l1}q_1 + p_{l2}q_2 + p_{l3}, \qquad l = 1, 2, \tag{3.80}$$

where

$$\rho_{11} = \frac{|a_0|}{2A_0}, \qquad \rho_{12} = \frac{|a_2|}{2A_0}, \qquad \rho_{13} = \frac{a_1 a_2}{2A_0},$$

$$\rho_{21} = \frac{1}{2A_0}, \qquad \rho_{22} = \frac{|a_1|}{2A_0}, \qquad \rho_{23} = \frac{a_1^2}{2A_0}.$$

$$(3.81)$$

So, the stability condition (3.56) can be written in the form

$$\sigma^2 < |a_2| \sup_{q_1 > 0, q_2 > \beta_2^2 \Theta^{-1}} f(q_1, q_2), \qquad f(q_1, q_2) = \frac{q_1}{\beta_1} - \frac{\beta_1}{\Theta - \beta_2^2 q_2^{-1}}. \tag{3.82}$$

For the fixed a_i, $i = 0, 1, 2$, using (3.78)–(3.82), the supremum of the function $f(q_1, q_2)$ can be obtained numerically.

Example 3.3 Consider (3.76) with $dK_j(s) = k_j \delta(s - h_j)ds$, $\alpha_{ij} = |k_j|h_j^i$, $\beta_{ij} = k_j h_j^i$, $j = 1, 2, 3$, $i = 0, 1, 2$. Then

$$a_0 = \beta_{03} - \beta_{12} + \frac{1}{2}\beta_{21} = k_3 - k_2 h_2 + \frac{1}{2}k_1 h_1^2, \qquad a_1 = \beta_{02} - \beta_{11} = k_2 - k_1 h_1,$$

$$a_2 = \beta_{01} = k_1, \qquad A_0 = \left(k_3 - k_2 h_2 + \frac{1}{2}k_1 h_1^2\right)(k_2 - k_1 h_1) + k_1,$$

$$\alpha = \alpha_{13} + \frac{1}{2}\alpha_{22} + \frac{1}{6}\alpha_{31} = |k_3|h_3 + \frac{1}{2}|k_2|h_2^2 + \frac{1}{6}|k_1|h_1^3, \qquad \Theta = \frac{1}{\alpha^2} - \frac{1}{\alpha}.$$

Put, for example, $h_1 = h_2 = h_3 = 0.1$, $k_1 = -1$, $k_2 = -2$, $k_3 = -3$. Then $a_0 = -2.805 < 0$, $a_1 = -1.9 < 0$, $a_2 = -1 < 0$, $A_0 = 4.3295 > 0$, $\alpha \approx 0.310 < 1$, $\Theta \approx 7.171$, $\rho_{11} \approx 0.324$, $\rho_{12} \approx 0.115$, $\rho_{13} \approx 0.219$, $\rho_{21} \approx 0.115$, $\rho_{22} \approx 0.219$, $\rho_{23} \approx 0.417$. Conditions (3.74) hold. The function $f(d_1, d_2)$ reaches its supremum for $q_1 \approx 4.49$, $q_2 \approx 0.54$. The stability condition (3.82) takes the form $\sigma^2 < 2.246$.

For $h_3 = 0.2$ and the same values of all other parameters, the function $f(q_1, q_2)$ reaches its supremum for $q_1 \approx 0.75$, $q_2 \approx 0.96$, and the stability condition (3.82) takes the form $\sigma^2 < 0.1969$.

3.4 Nonautonomous Systems

3.4.1 Equations with Variable Delays

Consider the scalar stochastic differential equation with variable delays

$$\dot{x}(t) = ax(t) + \sum_{i=1}^{m} b_i x\big(t - h_i(t)\big) + \sum_{i=1}^{m} c_i \int_{t - h_i(t)}^{t} x(s)\, ds + \sigma x\big(t - \tau(t)\big) \dot{w}(t).$$

$$(3.83)$$

Suppose that in (3.83)

$$h_i(t) \le h_i^0, \qquad \dot{h}_i(t) \le \hat{h}_i < 1, \qquad \dot{\tau}(t) \le \hat{\tau} < 1 \tag{3.84}$$

and put

$$B(h) = \sum_{i=1}^{m} \frac{|b_i|}{\sqrt{1 - \hat{h}_i}}, \qquad C_0(h) = \sum_{i=1}^{m} |c_i| h_i^0. \tag{3.85}$$

Let us consider (3.83) as the representation of type (2.33) with $z(t, x_t) = x(t)$ and the auxiliary differential equation $\dot{y}(t) = ay(t)$. The zero solution of this equation is asymptotically stable if and only if $a < 0$. Using the appropriate Lyapunov function $v(y) = y^2$, we obtain the functional $V_1(t, x_t)$ in the form $V_1(t, x_t) = x^2(t)$.

Using (3.84), (3.85), and some positive numbers γ_i, $i = 1, \ldots, m$, we have

$$LV_1 = 2x(t)\left(ax(t) + \sum_{i=1}^{m} b_i x\big(t - h_i(t)\big) + \sum_{i=1}^{m} c_i \int_{t - h_i(t)}^{t} x(s)\, ds \right)$$

$$+ \sigma^2 x^2\big(t - \tau(t)\big)$$

$$\le \left(2a + C_0(h) + \sum_{i=1}^{m} \gamma_i |b_i| \right) x^2(t) + \sum_{i=1}^{m} \gamma_i^{-1} |b_i| x^2\big(t - h_i(t)\big)$$

$$+ \sum_{i=1}^{m} |c_i| \int_{t - h_i^0}^{t} x^2(s)\, ds + \sigma^2 x^2\big(t - \tau(t)\big).$$

So, we obtain the representation of type (2.35), where

$$D = 2a + C_0(h) + \sum_{i=1}^{m} \gamma_i |b_i|, \qquad k = m + 1, \qquad \tau_k(t) = \tau(t), \qquad Q_k = \sigma^2,$$

$$R_0 = 1, \qquad Q_i = \gamma_i^{-1} |b_i|, \qquad \tau_i(t) = h_i(t), \qquad i = 1, \ldots, m,$$

$$d\mu_0(s) = \sum_{i=1}^{m} |c_i| \delta\left(s - h_i^0\right) ds$$

and all other parameters are zeros. By (2.36)–(2.37) we have

$$G = 2a + 2C_0(h) + \frac{\sigma^2}{1 - \hat{\tau}} + \sum_{i=1}^{p} \left(\gamma_i + \frac{\gamma_i^{-1}}{1 - \hat{h}_i}\right) |b_i|.$$

To minimize G, put $\gamma_i = \frac{1}{\sqrt{1 - \hat{h}_i}}$. By Theorem 2.5 we obtain the following assertion: if

$$\frac{\sigma^2}{2(1 - \hat{\tau})} + B(h) + C_0(h) < |a|, \qquad a < 0, \tag{3.86}$$

then the zero solution of (3.83) is asymptotically mean-square stable.

In addition to (3.84), assume that

$$\left|\dot{h}_i(t)\right| \le \hat{h}_i^0 \tag{3.87}$$

and put

$$B_0(h) = \sum_{i=1}^{p} |b_i| h_i^0, \qquad B_1(h) = \sum_{i=1}^{p} \frac{|b_i| \hat{h}_i^0}{\sqrt{1 - \hat{h}_i}}. \tag{3.88}$$

Consider the representation (2.33) of (3.83) in the form of the differential equation of neutral type

$$\dot{z}(t, x_t) = S_0 x(t) + \sum_{i=1}^{p} \left(b_i \dot{h}_i(t) x\left(t - h_i(t)\right) + c_i \int_{t - h_i(t)}^{t} x(s)\, ds\right)$$

$$+ \sigma x\left(t - \tau(t)\right) \dot{w}(t), \tag{3.89}$$

where

$$z(t, x_t) = x(t) + \sum_{j=1}^{p} b_j \int_{t - h_j(t)}^{t} x(s)\, ds, \qquad S_0 = a + \sum_{i=1}^{p} b_i. \tag{3.90}$$

Condition (2.10) for (3.89) has the form $B_0(h) < 1$.

The auxiliary equation for (3.89) is $\dot{y}(t) = S_0 y(t)$, and the zero solution of this equation is asymptotically stable if and only if $S_0 < 0$. Using the appropriate Lyapunov function $v(y) = y^2$, we obtain the functional $V_1(t, x_t)$ in the form $V_1(t, x_t) = z^2(t, x_t)$.

By (3.84), (3.85), (3.87)–(3.90), for some positive numbers γ_{1i}, γ_{2ij}, we obtain

$$
\begin{aligned}
LV_1(t, x_t) \leq{} & 2S_0 x^2(t) + \sum_{i=1}^{p} |b_i| \hat{h}_i^0 \left(\gamma_{1i} x^2(t) + \gamma_{1i}^{-1} x^2 (t - h_i(t)) \right) \\
& + \sigma^2 x^2 (t - \tau(t)) \\
& + \sum_{i=1}^{m} \sum_{j=1}^{m} |b_j b_i| \hat{h}_i^0 \int_{t-h_j^0}^{t} \left(\gamma_{2ij} x^2(s) + \gamma_{2ij}^{-1} x^2 (t - h_i(t)) \right) ds \\
& + \sum_{i=1}^{m} \sum_{j=1}^{m} |b_j c_i| \int_{t-h_i^0}^{t} \int_{t-h_j^0}^{t} \left(x^2(\theta) + x^2(s) \right) ds \, d\theta \\
& + \sum_{j=1}^{p} |S_0 b_j + c_j| \int_{t-h_j^0}^{t} \left(x^2(t) + x^2(s) \right) ds.
\end{aligned}
$$

As a result, we have the representation of type (2.35)

$$
LV_1(t, x_t) \leq Dx^2(t) + \sigma^2 x^2 (t - \tau(t)) + \sum_{i=1}^{m} Q_i x^2 (t - h_i(t)) + \sum_{j=1}^{m} q_j \int_{t-h_j^0}^{t} x^2(s) \, ds,
$$

where

$$
D = 2S_0 + \sum_{i=1}^{m} |b_i| \hat{h}_i^0 \gamma_{1i} + \sum_{j=1}^{m} |S_0 b_j + c_j| h_j^0, \qquad d\mu_0(s) = \sum_{j=1}^{m} q_j \delta\left(s - h_j^0 \right) ds,
$$

$$
k = m + 1, \qquad \tau_k(t) = \tau(t), \qquad Q_k = \sigma^2, \qquad R_0 = 1,
$$

$$
Q_i = |b_i| \hat{h}_i^0 \gamma_{1i}^{-1} + |b_i| \hat{h}_i^0 \sum_{j=1}^{m} |b_j| h_j^0 \gamma_{2ij}^{-1}, \qquad \tau_i(t) = h_i(t), \quad i = 1, \ldots, m,
$$

$$
q_j = |S_0 b_j + c_j| + |b_j| \sum_{i=1}^{m} |b_i| \hat{h}_i^0 \gamma_{2ij} + |b_j| C_0(h) + |c_j| B_0(h).
$$

So, by (2.36)–(2.37) we obtain

$$
\begin{aligned}
G ={} & 2S_0 + 2 \sum_{j=1}^{m} |S_0 b_j + c_j| h_j^0 + 2 B_0(h) C_0(h) + \frac{\sigma^2}{1 - \hat{\tau}} \\
& + \sum_{i=1}^{m} |b_i| \hat{h}_i^0 \left(\gamma_{1i} + \frac{\gamma_{1i}^{-1}}{1 - \hat{h}_i} \right) + \sum_{j=1}^{m} |b_j| h_j^0 \sum_{i=1}^{m} |b_i| \hat{h}_i^0 \left(\gamma_{2ij} + \frac{\gamma_{2ij}^{-1}}{1 - \hat{h}_i} \right)
\end{aligned}
$$

Choosing the optimal values of $\gamma_{1i} = \gamma_{2ij} = \frac{1}{\sqrt{1-\hat{h}_i}}$, we can minimize G and by Theorem 2.5 get the following stability condition: if

$$\frac{\sigma^2}{2(1-\hat{\tau})} + \sum_{j=1}^{p} |S_0 b_j + c_j| h_j^0 + B_1(h) + B_0(h)\big(B_1(h) + C_0(h)\big) < |S_0|,$$

$$S_0 < 0. \tag{3.91}$$

then the zero solution of (3.83) is asymptotically mean-square stable.

Remark 3.7 It is easy to see that instead of condition (3.91) one can use a more rough but more simple condition of the form

$$\frac{\sigma^2}{2(1-\hat{\tau})} + \big(1 + B_0(h)\big)\big(B_1(h) + C_0(h)\big) < |S_0|\big(1 - B_0(h)\big),$$

$$S_0 < 0, \quad B_0(h) < 1. \tag{3.92}$$

Put now

$$C_1(h) = \sum_{j=1}^{m} |c_j| h_j^0 \hat{h}_j^0, \qquad C_2(h) = \sum_{i=1}^{m} |c_i| \big(h_i^0\big)^2,$$

$$A_0(h) = B_0(h) + \frac{1}{2} C_2(h), \qquad A_1(h) = B_1(h) + C_1(h), \tag{3.93}$$

and consider the representation (2.33) of (3.83) in the form of the differential equation of neutral type

$$\dot{z}(t, x_t) = S(t)x(t) + \sum_{i=1}^{m} \dot{h}_i(t)\left(b_i x\big(t - h_i(t)\big) + c_i \int_{t-h_i(t)}^{t} x(s)\,ds\right)$$

$$+ \sigma x\big(t - \tau(t)\big)\dot{w}(t), \tag{3.94}$$

where

$$z(t, x_t) = x(t) + \sum_{i=1}^{m} \int_{t-h_i(t)}^{t} \big(b_i + c_i\big(s - t + h_i(t)\big)\big)x(s)\,ds,$$

$$S(t) = a + \sum_{i=1}^{m} \big(b_i + c_i h_i(t)\big). \tag{3.95}$$

Condition (2.10) for (3.94) has the form $A_0(h) < 1$.

The auxiliary differential equation in this case is $\dot{y}(t) = S(t)y(t)$, and if $\sup_{t \geq 0} S(t) < 0$, then the zero solution of this equation is asymptotically stable. Using the appropriate Lyapunov function $v(y) = y^2$, we obtain the functional $V_1(t, x_t)$ of the form $V_1(t, x_t) = z^2(t, x_t)$.

Then by (3.84), (3.87), (3.88), (3.93)–(3.95), for some positive numbers γ_{1i}, γ_{2ij}, we obtain

$$
LV_1(t, x_t) \le 2S(t)x^2(t) + |S(t)| \sum_{i=1}^{m} \int_{t-h_i(t)}^{t} \big(|b_i| + |c_i|(s - t + h_i(t))\big)
$$

$$
\times \big(x^2(t) + x^2(s)\big)\,ds
$$

$$
+ \sum_{i=1}^{m} \hat{h}_i^0 \bigg(|b_i| \big(\gamma_{1i} x^2(t) + \gamma_{1i}^{-1} x^2(t - h_i(t))\big)
$$

$$
+ |c_i| \int_{t-h_i(t)}^{t} \big(x^2(t) + x^2(s)\big)\,ds \bigg)
$$

$$
+ \sum_{j=1}^{m} \sum_{i=1}^{m} \hat{h}_i^0 |b_j| \int_{t-h_i(t)}^{t} \big(|b_i| + |c_i|(s - t + h_i(t))\big)
$$

$$
\times \big(\gamma_{2ij} x^2(s) + \gamma_{2ij}^{-1} x^2(t - h_j(t))\big)\,ds
$$

$$
+ \sum_{j=1}^{m} \sum_{i=1}^{m} \hat{h}_i^0 |c_j| \int_{t-h_i(t)}^{t} \int_{t-h_j(t)}^{t} \big(|b_i| + |c_i|(s - t + h_i(t))\big)
$$

$$
\times \big(x^2(\theta) + x^2(s)\big)\,d\theta\,ds + \sigma^2 x^2(t - \tau(t)).
$$

Put now

$$
S_m = \inf_{t \ge 0} |S(t)|, \qquad S_M = \sup_{t \ge 0} |S(t)|,
$$

$$
I_i\big(h_i(t)\big) = \int_{t-h_i(t)}^{t} \big(|b_i| + |c_i|(s - t + h_i(t))\big)\,ds,
$$

$$
J_{0i}\big(h_i(t)\big) = \int_{t-h_i(t)}^{t} x^2(s)\,ds,
$$

$$
J_{1i}\big(h_i(t)\big) = \int_{t-h_i(t)}^{t} \big(|b_i| + |c_i|(s - t + h_i(t))\big)x^2(s)\,ds, \qquad i = 1, \ldots, m.
$$

By (3.84), (3.88), and (3.93) we have

$$
I_i\big(h_i(t)\big) \le I_i\big(h_i^0\big) = |b_i| h_i^0 + \frac{1}{2}|c_i|\big(h_i^0\big)^2, \qquad \sum_{i=1}^{m} I_i\big(h_i(t)\big) \le A_0(h),
$$

$$
J_{0i}\big(h_i(t)\big) \le J_{0i}\big(h_i^0\big), \qquad J_{1i}\big(h_i(t)\big) \le J_{1i}\big(h_i^0\big).
$$

So, we obtain the representation of type (2.35)

$$LV_1(t, x_t) \le D(t)x^2(t) + \sum_{j=1}^{m} Q_j x^2(t - h_j(t)) + \sigma^2 x^2(t - \tau(t))$$

$$+ \sum_{i=1}^{m} q_{0i} J_{0i}\left(h_i^0\right) + \sum_{i=1}^{m} q_{1i} J_{1i}\left(h_i^0\right),$$

where

$$D(t) = \left(-2 + A_0(h)\right)\left|S(t)\right| + C_1(h) + \sum_{i=1}^{m} \hat{h}_i^0 |b_i| \gamma_{1i},$$

$$k = m + 1, \quad \tau_k = \tau, \quad Q_k = \sigma^2,$$

$$Q_j = |b_j| \hat{h}_j^0 \left(\gamma_{1j}^{-1} + \sum_{i=1}^{m} \gamma_{2ij}^{-1} I_i\left(h_i^0\right)\right), \quad j = 1, \dots, m,$$

$$R_0 = R_1 = 1,$$

$$d\mu_0(s) = \sum_{i=1}^{p} \left(q_{0i} + q_{1i}|b_i|\right)\delta\left(s - h_i^0\right) ds,$$

$$d\mu_1(s) = \sum_{i=1}^{p} q_{1i}|c_i|\delta\left(s - h_i^0\right) ds,$$

$$q_{0i} = \left(1 + A_0(h)\right)|c_i| \hat{h}_i^0, \quad q_{1i} = S_M + C_1(h) + \sum_{j=1}^{p} |b_j| \hat{h}_j^0 \gamma_{2ij}.$$

As a result, we have

$$G(t) = \left(-2 + A_0(h)\right)\left|S(t)\right| + A_0(h) S_M + \frac{\sigma^2}{1 - \hat{\tau}} + 2C_1(h)\left(1 + A_0(h)\right)$$

$$+ \sum_{j=1}^{m} |b_j| \hat{h}_j^0 \left(\gamma_{1j} + \frac{\gamma_{1j}^{-1}}{1 - \hat{h}_j}\right) + \sum_{i=1}^{m} \sum_{j=1}^{m} |b_j| \hat{h}_j^0 I_i\left(h_i^0\right)\left(\gamma_{2ij} + \frac{\gamma_{2ij}^{-1}}{1 - \hat{h}_j}\right).$$

To minimize $G(t)$, put $\gamma_{1j} = \gamma_{2ij} = \frac{1}{\sqrt{1 - \hat{h}_j}}$. Then

$$G(t) = \left(-2 + A_0(h)\right)\left|S(t)\right| + A_0(h) S_M + 2A_1(h)\left(1 + A_0(h)\right) + \frac{\sigma^2}{1 - \hat{\tau}}.$$

From $\sup_{t \ge 0} S(t) < 0$ we obtain the estimation for $G(t)$:

$$\sup_{t\geq 0} G(t) \leq \left(-2 + A_0(h)\right)S_m + A_0(h)S_M + 2A_1(h)\left(1 + A_0(h)\right) + \frac{\sigma^2}{1 - \hat{\tau}}. \quad (3.96)$$

From (3.96) by Theorem 2.4 we obtain: if $\sup_{t\geq 0} S(t) < 0$ and

$$\frac{\sigma^2}{1 - \hat{\tau}} + A_0(h)S_M + 2A_1(h)\left(1 + A_0(h)\right) < \left(2 - A_0(h)\right)S_m, \quad (3.97)$$

then the zero solution of (3.83) is asymptotically mean-square stable.

3.4.2 Equations with Variable Coefficients

Consider the scalar differential equation with delays and variable coefficients

$$\dot{x}(t) = -a(t)x(t) - b(t)x(t - h) + \sigma(t)x(t - \tau)\dot{w}(t),$$
$$t \geq 0, \ h \geq 0, \ \tau \geq 0. \quad (3.98)$$

For $V(t, x_t) = x^2(t)$, we have

$$LV(t, x_t) = 2x(t)\left[-a(t)x(t) - b(t)x(t - h)\right] + \sigma^2(t)x^2(t - \tau)$$
$$\leq \left[-2a(t) + |b(t)|\right]x^2(t) + |b(t)|x^2(t - h) + \sigma^2(t)x^2(t - \tau).$$

By Theorem 2.5, if there is $c > 0$ such that

$$G(t) = -2a(t) + |b(t)| + |b(t + h)| + \sigma^2(t + \tau) \leq -c, \quad (3.99)$$

then the trivial solution of (3.98) is asymptotically mean-square stable.
To get another stability condition, rewrite (3.98) in the form

$$\dot{z}(t, x_t) = -c(t)x(t) + \sigma(t)x(t - \tau)\dot{w}(t), \quad (3.100)$$

where

$$z(t, x_t) = x(t) - \int_{t-h}^{t} b(s + h)x(s)\,ds, \qquad c(t) = a(t) + b(t + h). \quad (3.101)$$

Note that (3.100)–(3.101) is a differential equation of neutral type and suppose that

$$c(t) \geq c_0 > 0, \qquad \sup_{t \geq 0} \int_{t}^{t+h} |b(s)|\,ds < 1. \quad (3.102)$$

Consider the auxiliary differential equation without delay

$$\dot{y}(t) = -c(t)y(t). \quad (3.103)$$

Using the Lyapunov function $v(t) = y^2(t)$, by (3.99) we have $\dot{v}(t) = -2c(t)y^2(t) \leq -2c_0 y^2(t)$. So, the trivial solution of (3.103) is asymptotically stable.

Following the procedure of constructing Lyapunov functionals, we will use the Lyapunov functional $V_1(t, x_t)$ for (3.100)–(3.101) of the form $V_1(t, x_t) = z^2(t, x_t)$. Calculating $LV_1(t, x_t)$, from (3.100)–(3.101) we obtain the representation of type (2.35)

$$
LV_1(t, x_t) = c(t)\left(-2x^2(t) + 2\int_{t-h}^{t} b(s+h)x(t)x(s)\,ds\right) + \sigma^2(t)x^2(t-\tau)
$$

$$
\leq c(t)\left(-2x^2(t) + \int_{t-h}^{t} |b(s+h)|\big(x^2(t) + x^2(s)\big)\,ds\right) + \sigma^2(t)x^2(t-\tau)
$$

$$
= D(t)x^2(t) + \sigma^2(t)x^2(t-\tau) + \int_{t-h}^{t} R(s+h,t)x^2(s)\,ds,
$$

where

$$
D(t) = c(t)\left(-2 + \int_{t}^{t+h} |b(s)|\,ds\right), \qquad Q_1(t) = \sigma^2(t+\tau),
$$

$$
d\mu(\tau) = \delta(\tau - h)\,d\tau, \qquad R(s,t) = |b(s)|c(t).
$$

So, by Theorem 2.5, $G(t) = c(t)(-2 + G_0(t))$, where

$$
G_0(t) = \int_{t}^{t+h} |b(s)|\,ds + \frac{|b(t+h)|}{c(t)}\int_{t}^{t+h} c(s)\,ds + \frac{\sigma^2(t+\tau)}{c(t)},
$$

and if

$$
\sup_{t\geq 0} G_0(t) < 2, \tag{3.104}
$$

then the zero solution of (3.98) is asymptotically mean-square stable.

Note that if $h = 0$ and the functions $a(t) = a$, $b(t) = b$, $\sigma(t) = \sigma$ are constants, then condition (3.104) coincides with the necessary and sufficient condition for asymptotic mean-square stability $\sigma^2 < 2(a+b)$. But in the general case the both obtained conditions (3.99) and (3.102), (3.104) are only sufficient conditions for asymptotic mean-square stability of the trivial solution of (3.98). Indeed, consider (3.98) with $a(t) = \alpha + \sin(t)$, $b(t) = \cos(t)$, and $\sigma(t) = 0$. In Fig. 3.19 the behavior of the functions $G(t)$ (green), $c(t)$ (grey), $G_0(t)$ (blue), and the solution $x(t)$ (red) of (3.98) are shown for $x(s) = 0.1$, $s \in [-h, 0]$, $h = 0.1$, $\alpha = -0.2$. Here conditions (3.99) and (3.102), (3.104) do not hold, and the solution goes to infinity. In Fig. 3.20 the same picture is shown for $x(s) = 2.5$, $s \in [-h, 0]$, $h = 0.8$, $\alpha = 0.2$. Conditions (3.99) and (3.102), (3.104) do not hold too, but the solution $x(t)$ goes to zero. In Fig. 3.21 the same picture is shown for $x(s) = 1.5$, $s \in [-h, 0]$, $h = 1.1$, $\alpha = 1.7$. Conditions (3.99) and (3.102), (3.104) hold, and the solution $x(t)$ goes to zero.

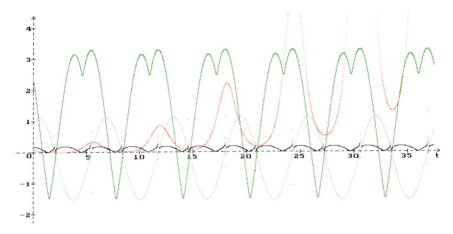

Fig. 3.19 Functions $G(t)$ (*green*), $c(t)$ (*grey*), $G_0(t)$ (*blue*), and the solution $x(t)$ (*red*) of (3.98) are shown for $x(s) = 0.1$, $s \in [-h, 0]$, $h = 0.1$, $\alpha = -0.2$

Consider the scalar stochastic differential equation with variable coefficients and unbounded delay

$$\dot{x}(t) + a(t)x(t) + \int_0^\infty b(t, s)x(t - s)\, ds + \sigma(t)x(t - \tau)\dot{w}(t) = 0. \qquad (3.105)$$

For the function $V_1(t, x_t) = x^2(t)$, we have

$$LV_1(t, x_t) = 2x(t)\left(-a(t)x(t) - \int_0^\infty b(t, s)x(t - s)\, ds\right) + \sigma^2(t)x^2(t - \tau)$$

$$\leq \left(-2a(t) + \int_0^\infty |b(t, s)|\, ds\right)x^2(t)$$

$$+ \int_0^\infty |b(t, s)|x^2(t - s)\, ds + \sigma^2(t)x^2(t - \tau).$$

To obtain a stability condition, we consider the functional

$$V_2(t, x_t) = \int_0^\infty \int_{t-s}^t |b(\tau + s, s)|x^2(\tau)\, d\tau\, ds + \int_{t-\tau}^t \sigma^2(s + \tau)x^2(s)\, ds.$$

Then

$$LV_2(t, x_t) = \int_0^\infty \left(|b(t + s, s)|x^2(t) - |b(t, s)|x^2(t - s)\right) ds$$

$$+ \sigma^2(t + \tau)x^2(t) - \sigma^2(t)x^2(t - \tau).$$

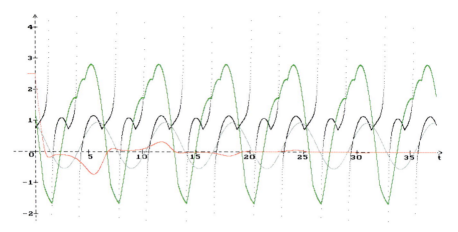

Fig. 3.20 Functions $G(t)$ (*green*), $c(t)$ (*grey*), $G_0(t)$ (*blue*) and the solution $x(t)$ (*red*) of (3.98) are shown for $x(s) = 2.5$, $s \in [-h, 0]$, $h = 0.8$, $\alpha = 0.2$

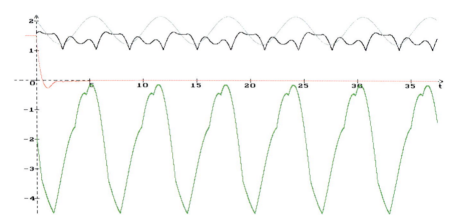

Fig. 3.21 Functions $G(t)$ (*green*), $c(t)$ (*grey*), $G_0(t)$ (*blue*), and the solution $x(t)$ (*red*) of (3.98) are shown for $x(s) = 1.5$, $s \in [-h, 0]$, $h = 1.1$, $\alpha = 1.7$

For $V = V_1 + V_2$, we obtain $LV(t, x_t) \le G(t)x^2(t)$, where

$$G(t) = -2a(t) + \int_0^\infty \left(\left| b(t, s) \right| + \left| b(t + s, s) \right| \right) ds + \sigma^2(t + \tau). \qquad (3.106)$$

So, if $\sup_{t \ge 0} G(t) < 0$, then the trivial solution of (3.105) is asymptotically mean-square stable.

Note that if $b(t, s) = b(t)\delta(s - h)$, then (3.105) coincides with (3.98), and the function (3.106) coincides with (3.99).

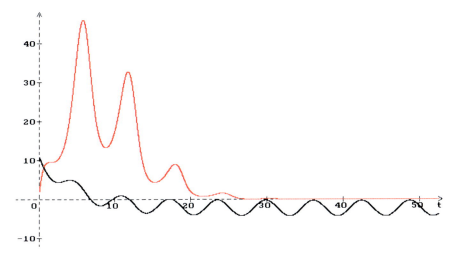

Fig. 3.22 The solution of (3.107) (*red*) and the function $G(t)$ (*blue*) are shown in the deterministic case ($\sigma(t) = 0$) for $a(t) = 1.2 + \sin t$, $x_0 = 2$, $\gamma = 0.1$, $\lambda = 0.2$

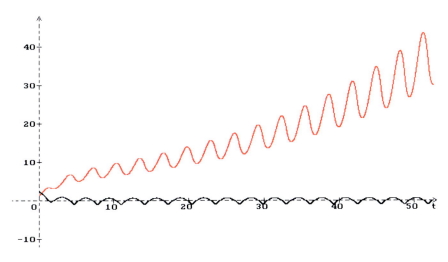

Fig. 3.23 The solution of (3.107) (*red*) and the function $G(t)$ (*blue*) are shown in the deterministic case ($\sigma(t) = 0$) for $a(t) = 0.6 - |\cos t|$, $x_0 = 1.5$, $\gamma = 1$, $\lambda = 0.6$

The negativity of the function $G(t)$ is a sufficient stability condition but not a necessary one. Put, for instance, in (3.105)

$$b(t, s) = -e^{-\lambda t - \gamma s}, \quad \gamma > 0, \ \lambda > 0, \quad x(s) = x_0, \quad s \le 0.$$

Then (3.105) and the function $G(t)$ take the forms

$$\dot{x}(t) + a(t)x(t) + \sigma(t)x(t-\tau)\dot{w}(t) = e^{-\lambda t}\left[\frac{x_0}{\gamma}e^{-\gamma t} + \int_0^t e^{-\gamma s}x(t-s)\,ds\right],$$

(3.107)

$$G(t) = -2a(t) + \sigma^2(t-\tau) + e^{-\lambda t}\left(\frac{1}{\gamma} + \frac{1}{\gamma+\lambda}\right).$$

(3.108)

If $\sup_{t\geq 0} G(t) < 0$, then the trivial solution of (3.107) is asymptotically mean-square stable.

Note that in some cases the trivial solution of (3.107) is asymptotically stable if the function $G(t)$ given by (3.108) is a negative one for large enough t only. In Fig. 3.22 the trajectory of the solution of (3.107) (red) and the function $G(t)$ (blue) are shown in the deterministic case ($\sigma(t) = 0$) for $a(t) = 1.2 + \sin t$, $x_0 = 2$, $\gamma = 0.1$, $\lambda = 0.2$. In this case the condition $G(t) \leq 0$ holds for all large enough t, and the solution of (3.107) goes to zero. In Fig. 3.23 we can see a similar picture with $a(t) = 0.6 - |\cos t|$, $x_0 = 1.5$, $\gamma = 1$, $\lambda = 0.6$. In this case the function $G(t)$ for all $t \geq 0$ has both negative and positive values, and the solution of (3.108) goes to infinity.

Chapter 4
Stability of Linear Systems of Two Equations

Below, several examples are considered where the procedure of constructing Lyapunov functionals is used for stability investigation of linear systems of two equations with constant delays, with distributed delays, and with variable coefficients.

4.1 Linear Systems of Two Equations with Constant Delays

Example 4.1 Consider the system of two stochastic differential equations with fixed delays

$$\dot{x}_1(t) = ax_2(t) - bx_1(t-h) + \sigma_1 x_1(t-\tau_1)\dot{w}_1(t),$$
$$\dot{x}_2(t) = -ax_1(t) - bx_2(t-h) + \sigma_2 x_2(t-\tau_2)\dot{w}_2(t),$$

(4.1)

where $w_1(t)$ and $w_2(t)$ are the mutually independent standard Wiener processes.
Put $x(t) = (x_1(t), x_2(t))'$, $w(t) = (w_1(t), w_2(t))'$,

$$\rho(t) = \int_{t-h}^{t} x(s)\,ds, \quad z(t) = x(t) - b\rho(t),$$

$$B(x_t) = \begin{pmatrix} \sigma_1 x(t-\tau_1) & 0 \\ 0 & \sigma_2 x(t-\tau_2) \end{pmatrix},$$

(4.2)

$$A = \begin{pmatrix} -b & a \\ -a & -b \end{pmatrix},$$

(4.3)

and rewrite (4.1) in the form

$$\dot{z}(t) = Ax(t) + B(x_t)\dot{w}(t).$$

(4.4)

By (4.3) and Corollary 1.1 the condition $b > 0$ is a necessary and sufficient condition for asymptotic stability of the zero solution of the auxiliary differential equation $\dot{y}(t) = Ay(t)$. Let P be a positive definite solution of the matrix equation (1.27),

L. Shaikhet, *Lyapunov Functionals and Stability of Stochastic Functional Differential Equations*, DOI 10.1007/978-3-319-00101-2_4,
© Springer International Publishing Switzerland 2013

where Q is (for simplicity) the identity matrix, and A is defined by (4.3). By (1.29) the elements of P are $p_{11} = p_{22} = (2b)^{-1}$, $p_{12} = 0$.

Following to the procedure of constructing Lyapunov functionals, consider the functional $V_1 = z'(t)Pz(t) = (2b)^{-1}|z(t)|^2$. Calculating LV_1 for (4.4), by (4.2) we obtain

$$LV_1 = \frac{1}{2b}\left[2z'(t)Ax(t) + \sum_{i=1}^{2}\sigma_i^2 x_i^2(t - \tau_i)\right]$$

$$= \frac{1}{2b}\left[2\big(x(t) - b\rho(t)\big)'Ax(t) + \sum_{i=1}^{2}\sigma_i^2 x_i^2(t - \tau_i)\right]$$

$$= \frac{1}{2b}\left[2x'(t)Ax(t) - 2b\int_{t-h}^{t} x'(s)Ax(t)\,ds + \sum_{i=1}^{2}\sigma_i^2 x_i^2(t - \tau_i)\right].$$

Note, that $x'(t)Ax(t) = -b|x(t)|^2$ and $2x'(s)Ax(t) \le \|A\|(|x(s)|^2 + |x(t)|^2)$, where $\|A\| = \sup_{|x|=1}|Ax|$ is the operator norm of a matrix A. Therefore,

$$LV_1 \le \frac{1}{2b}\left[-2b|x(t)|^2 + b\|A\|\int_{t-h}^{t}\big(|x(s)|^2 + |x(t)|^2\big)\,ds + \sum_{i=1}^{2}\sigma_i^2 x_i^2(t - \tau_i)\right]$$

$$= \left(-1 + \frac{1}{2}\|A\|h\right)|x(t)|^2 + \frac{1}{2}\|A\|\int_{t-h}^{t}|x(s)|^2\,ds + \frac{1}{2b}\sum_{i=1}^{2}\sigma_i^2 x_i^2(t - \tau_i).$$

So, we have got the representation of type (2.25) with $S(t) = Q_i(t) = R(s, t) = 0$, $m = 0, n = 2$ and

$$D = \left(-1 + \frac{1}{2}\|A\|h\right)I, \qquad dK_i(s) = \frac{1}{2b}\sigma_i^2\delta(s - \tau_i)\,ds, \quad i = 1, 2,$$

$$R_0(s) = \frac{1}{2}\|A\|, \qquad d\mu_0(s) = \delta(s - h)\,ds.$$

Note also that $\|A\| = \sup_{|x|=1}\sqrt{(-bx_1 + ax_2)^2 + (-ax_1 - bx_2)^2} = \sqrt{a^2 + b^2}$. From this and from Theorem 2.4 we obtain that if

$$h\sqrt{a^2 + b^2} + \frac{p}{b} < 1, \qquad p = \frac{1}{2}\max(\sigma_1^2, \sigma_2^2),$$

or

$$|a| < \sqrt{\frac{1}{h^2}\left(1 - \frac{p}{b}\right)^2 - b^2}, \tag{4.5}$$

$$\frac{1}{2h}\left(1 - \sqrt{1 - 4ph}\right) < b < \frac{1}{2h}\left(1 + \sqrt{1 - 4ph}\right), \qquad ph < \frac{1}{4}, \tag{4.6}$$

Fig. 4.1 Stability regions
for (4.1) given by
conditions (4.5), (4.6) for
$h = 1$ and different values of
p: (*1*) $p = 0$, (*2*) $p = 0.1$,
(*3*) $p = 0.2$, (*4*) $p = 0.249$

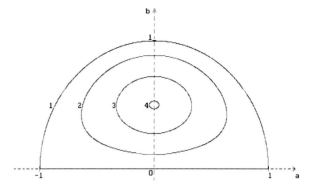

Fig. 4.2 Stability regions
for (4.1) given by
conditions (4.5), (4.6) for
$p = 0.1$ and different values
of h: (*1*) $h = 0$, (*2*) $h = 1$,
(*3*) $h = 2$, (*4*) $h = 2.45$

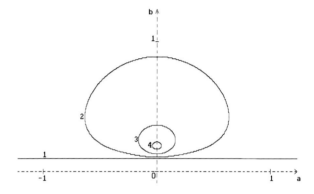

Fig. 4.3 Stability regions
for (4.1) given by
conditions (4.5), (4.6) for
$h = 0$ and different values of
p: (*1*) $p = 0$, (*2*) $p = 0.25$

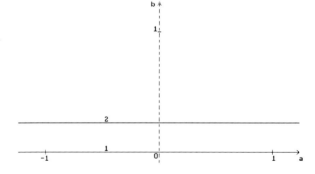

then the trivial solution of system (4.1) is asymptotically mean-square stable.

The stability regions for (4.1) given by conditions (4.5)–(4.6) are shown in Fig. 4.1 for $h = 1$ and different values of p: (1) $p = 0$; (2) $p = 0.1$; (3) $p = 0.2$; (4) $p = 0.249$; in Fig. 4.2 for $p = 0.1$ and different values of h. (1) $h = 0$; (2) $h = 1$; (3) $h = 2$; (4) $h = 2.45$; in Fig. 4.3 for $h = 0$ and different values of p: (1) $p = 0$; (2) $p = 0.25$.

Example 4.2 Consider now the system of two stochastic differential equations with fixed delays

$$\dot{x}_1(t) = ax_2(t) - bx_1(t - h) + \sigma_1 x_1(t - \tau_1)\dot{w}(t),$$
$$\dot{x}_2(t) = ax_1(t) - bx_2(t - h) + \sigma_2 x_2(t - \tau_2)\dot{w}(t), \tag{4.7}$$

which can be transformed to the form (4.4), (4.2) with the matrix

$$A = \begin{pmatrix} -b & a \\ a & -b \end{pmatrix}. \tag{4.8}$$

By (4.8) and Corollary 1.1 the condition $b > |a|$ is a necessary and sufficient condition for asymptotic stability of the zero solution of the auxiliary differential equation $\dot{y}(t) = Ay(t)$. By (1.29) the positive definite solution P of the matrix equation (1.27), where $Q = I$ (the identity matrix), and A is defined by (4.8), has the elements

$$p_{11} = p_{22} = \frac{b}{2(b^2 - a^2)}, \qquad p_{12} = \frac{a}{2(b^2 - a^2)}. \tag{4.9}$$

Put $V_1 = z'(t)Pz(t)$ and note that, by (4.8)–(4.9), $2PA = -I$. Calculating LV_1 for (4.4) with the parameters defined by (4.2), (4.8), and (4.9), we obtain

$$LV_1 = -\big(x(t) - b\rho(t)\big)'x(t) + \sum_{i=1}^{2} p_{ii}\sigma_i^2 x_i^2(t - \tau_i)$$

$$= -|x(t)|^2 + b\int_{t-h}^{t} x'(s)x(t)\,ds + \sum_{i=1}^{2} p_{ii}\sigma_i^2 x_i^2(t - \tau_i)$$

$$\leq \left(-1 + \frac{bh}{2}\right)|x(t)|^2 + \frac{b}{2}\int_{t-h}^{t} |x(s)|^2\,ds + \sum_{i=1}^{2} p_{ii}\sigma_i^2 x_i^2(t - \tau_i).$$

So, we have the representation of type (2.25) with $S(t) = Q_i(t) = R(s,t) = 0$, $m = 1$, $n = 2$ and

$$D = \left(-1 + \frac{bh}{2}\right)I, \qquad dK_i(s) = p_{ii}\sigma_i^2\delta(s - \tau_i)\,ds, \quad i = 1, 2,$$

$$d\mu_0(s) = \delta(s - h)\,ds, \qquad R_0(s) = \frac{b}{2}.$$

By Theorem 2.4 and (4.9) we obtain that if

$$b\left(h + \frac{p}{b^2 - a^2}\right) < 1, \qquad p = \frac{1}{2}\max(\sigma_1^2, \sigma_2^2),$$

Fig. 4.4 Stability regions for
(4.7) given by conditions
(4.10), (4.6) for $h = 1$ and
different values of p:
(*1*) $p = 0$, (*2*) $p = 0.1$,
(*3*) $p = 0.2$, (*4*) $p = 0.249$

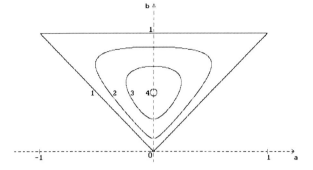

Fig. 4.5 Stability regions for
(4.7) given by conditions
(4.10), (4.6) for $p = 0.1$ and
different values of h:
(*1*) $h = 0$, (*2*) $h = 1$,
(*3*) $h = 2$, (*4*) $h = 2.45$

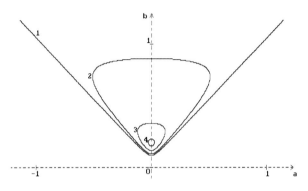

Fig. 4.6 Stability regions for
(4.7) given by conditions
(4.10), (4.6) for $h = 0$ and
different values of p:
(*1*) $p = 0$, (*2*) $p = 0.25$

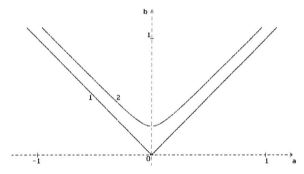

or if the conditions

$$|a| < \sqrt{b\left(b - \frac{p}{1 - bh}\right)} \tag{4.10}$$

and (4.6) hold, then the trivial solution of (4.7) is asymptotically mean-square stable.

The stability regions for (4.7) given by conditions (4.10), (4.6) are shown in
Fig. 4.4 for $h = 1$ and different values of p: (1) $p = 0$; (2) $p = 0.1$; (3) $p = 0.2$;
(4) $p = 0.249$; in Fig. 4.5 for $p = 0.1$ and different values of h: (1) $h = 0$; (2) $h = 1$;
(3) $h = 2$; (4) $h = 2.45$; in Fig. 4.6 for $h = 0$ and different values of p: (1) $p = 0$;
(2) $p = 0.25$.

4.2 Linear Systems of Two Equations with Distributed Delays

Example 4.3 Consider the system of two stochastic differential equations with distributed delays

$$\dot{x}_1(t) = ax_2(t) - b\int_{t-h}^{t} x_1(s)\,ds + \sigma_1 \int_{t-\tau_1}^{t} x_1(s)\,ds\,\dot{w}_1(t),$$

$$\dot{x}_2(t) = -ax_1(t) - b\int_{t-h}^{t} x_2(s)\,ds + \sigma_2 \int_{t-\tau_2}^{t} x_2(s)\,ds\,\dot{w}_2(t),$$
$$(4.11)$$

where $w_1(t)$ and $w_2(t)$ are the mutually independent standard Wiener processes.
Put $x(t) = (x_1(t), x_2(t))'$, $w(t) = (w_1(t), w_2(t))'$,

$$\rho(t) = \int_{t-h}^{t} (s-t+h)x(s)\,ds, \qquad z(t) = x(t) - b\rho(t),$$

$$B(x_t) = \begin{pmatrix} \sigma_1 \int_{t-\tau_1}^{t} x_1(s)\,ds & 0 \\ 0 & \sigma_2 \int_{t-\tau_2}^{t} x_2(s)\,ds \end{pmatrix},$$
$$(4.12)$$

$$A = \begin{pmatrix} -bh & a \\ -a & -bh \end{pmatrix}.$$
$$(4.13)$$

Then (4.11) can be represented in the form (4.4).

By (4.13) and Corollary 1.1 the conditions $b > 0$, $h > 0$ are necessary and sufficient conditions for asymptotic stability of the zero solution of the auxiliary differential equation $\dot{y}(t) = Ay(t)$. The positive definite solution P of the matrix differential equation (1.27) with $Q = I$ and A defined by (4.13) has the elements $p_{11} = p_{22} = (2bh)^{-1}$, $p_{12} = 0$.

Following the procedure of constructing Lyapunov functionals, put

$$V_1 = z'(t)Pz(t) = \frac{1}{2bh}|z(t)|^2.$$

Calculating LV_1 for system (4.11), by (4.12)–(4.13) we have

$$LV_1 = \frac{1}{2bh}\left[2z'(t)Ax(t) + \sum_{i=1}^{2}\left(\int_{t-\tau_i}^{t} \sigma_i x_i(s)\,ds \right)^2 \right]$$

$$= \frac{1}{2bh}\left[2(x(t) - b\rho(t))'Ax(t) + \sum_{i=1}^{2}\left(\int_{t-\tau_i}^{t} \sigma_i x_i(s)\,ds \right)^2 \right]$$

$$= \frac{1}{2bh}\left[2x'(t)Ax(t) - 2b\int_{t-h}^{t} (s-t+h)x'(s)Ax(t)\,ds \right.$$

$$+ \left. \sum_{i=1}^{2}\left(\int_{t-\tau_i}^{t} \sigma_i x_i(s)\,ds \right)^2 \right].$$

Using $x'(t)Ax(t) = -bh|x(t)|^2$ and $2x'(s)Ax(t) \leq \|A\|(|x(s)|^2 + |x(t)|^2)$, we obtain

$$LV_1 \leq \frac{1}{2bh}\left[-2bh|x(t)|^2 + b\|A\|\int_{t-h}^{t}(s-t+h)(|x(s)|^2 + |x(t)|^2)\,ds\right.$$

$$\left. + \sum_{i=1}^{2}\sigma_i^2\tau_i\int_{t-\tau_i}^{t}x_i^2(s)\,ds\right] = \left(-1 + \frac{1}{4}\|A\|h\right)|x(t)|^2$$

$$+ \frac{1}{2h}\|A\|\int_{t-h}^{t}(s-t+h)|x(s)|^2\,ds + \frac{1}{2bh}\sum_{i=1}^{2}\sigma_i^2\tau_i\int_{t-\tau_i}^{t}x_i^2(s)\,ds.$$

So, we have got the representation of type (2.25) with $S(t) = Q_i(t) = R(s,t) = 0$, $m = 1$, $n = 2$ and

$$D = \left(-1 + \frac{1}{4}\|A\|h\right)I, \qquad R_0(s) = \frac{1}{2bh}, \qquad R_1(s) = \frac{1}{2h}\|A\|,$$

$$d\mu_0(s) = \sum_{i=1}^{2}\sigma_i^2\tau_i\delta(s - \tau_i)\,ds, \qquad d\mu_1(s) = \delta(s - h)\,ds.$$

Note also that $\|A\| = \sqrt{a^2 + b^2h^2}$. From this and from Theorem 2.4 it follows that if

$$\frac{h}{2}\sqrt{a^2 + b^2h^2} + \frac{p_\tau}{bh} < 1, \qquad p_\tau = \frac{1}{2}\max(\sigma_1^2\tau_1^2, \sigma_2^2\tau_2^2),$$

or if the conditions

$$|a| < \sqrt{\frac{4}{h^2}\left(1 - \frac{p_\tau}{bh}\right)^2 - b^2h^2}, \tag{4.14}$$

$$\frac{p_\tau}{h} < \frac{1}{h^2}(1 - \sqrt{1 - 2p_\tau h}) < b < \frac{1}{h^2}(1 + \sqrt{1 - 2p_\tau h}), \qquad p_\tau h < \frac{1}{2}, \tag{4.15}$$

hold, then the trivial solution of (4.11) is asymptotically mean-square stable.

The stability regions for (4.11) given by conditions (4.14)–(4.15) are shown in Fig. 4.7 for $h = 1.1$ and different values of p_τ: (1) $p_\tau = 0$, (2) $p_\tau = 0.2$, (3) $p_\tau = 0.4$, (4) $p_\tau = 0.45$, in Figs. 4.8 and 4.9 for $p_\tau = 0.1$ and $p_\tau = 2.5$ respectively and different values of h: (1) $h = 0.085$, (2) $h = 0.1$, (3) $h = 0.12$, (4) $h = 0.16$, in Fig. 4.10 for $p_\tau = 2.5$ and the following values of h. (1) $h = 0.01$, (2) $h = 0.005$, (3) $h = 0.0025$, (4) $h = 0.00125$. In the last figure one can see that by $p_\tau > 0$, as $h \to 0$, the stability region goes to infinity.

Fig. 4.7 Stability regions for
(4.11) given by conditions
(4.14), (4.15) for $h = 1.1$ and
different values of p_τ:
(*1*) $p_\tau = 0$, (*2*) $p_\tau = 0.2$,
(*3*) $p_\tau = 0.4$, (*4*) $p_\tau = 0.45$

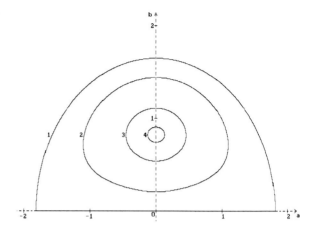

Fig. 4.8 Stability regions for
(4.11) given by conditions
(4.14), (4.15) for $p_\tau = 0.1$
and different values of h:
(*1*) $h = 0.085$, (*2*) $h = 0.1$,
(*3*) $h = 0.12$, (*4*) $h = 0.16$

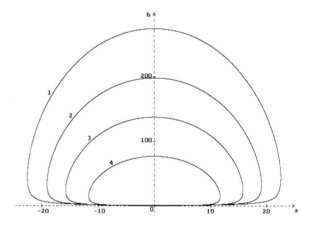

Fig. 4.9 Stability regions for
(4.11) given by conditions
(4.14), (4.15) for $p_\tau = 2.5$
and different values of h:
(*1*) $h = 0.085$, (*2*) $h = 0.1$,
(*3*) $h = 0.12$, (*4*) $h = 0.16$

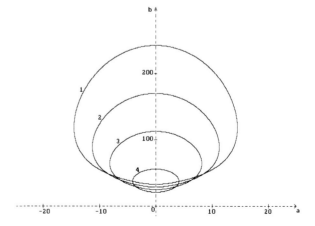

Fig. 4.10 Stability regions
for (4.11) given by conditions
(4.14), (4.15) for $p_\tau = 2.5$
and different values of h:
(*1*) $h = 0.01$, (*2*) $h = 0.005$,
(*3*) $h = 0.0025$,
(*4*) $h = 0.00125$

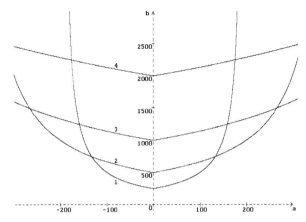

Example 4.4 Consider now the system of two stochastic differential equations with
distributed delays

$$\dot{x}_1(t) = ax_2(t) - b \int_{t-h}^{t} x_1(s)\,ds + \sigma_1 \int_{t-\tau_1}^{t} x_1(s)\,ds\,\dot{w}(t),$$
$$\dot{x}_2(t) = ax_1(t) - b \int_{t-h}^{t} x_2(s)\,ds + \sigma_2 \int_{t-\tau_2}^{t} x_2(s)\,ds\,\dot{w}(t),$$

(4.16)

which can be transformed to the form (4.4), (4.12) with the matrix

$$A = \begin{pmatrix} -bh & a \\ a & -bh \end{pmatrix}.$$

(4.17)

By (4.17) and Corollary 1.1 the condition $bh > |a|$ is a necessary and suffi-
cient condition for asymptotic stability of the zero solution of the auxiliary equation
$\dot{y}(t) = Ay(t)$. The positive definite solution P of the matrix equation (1.27) with
$Q = I$ and A defined by (4.17) has the elements

$$p_{11} = p_{22} = \frac{bh}{2(b^2h^2 - a^2)}, \qquad p_{12} = \frac{a}{2(b^2h^2 - a^2)}.$$

(4.18)

Following the procedure of constructing Lyapunov functionals, put $V_1 = z'(t)Pz(t)$ and note that $2PA = -I$. Thus, calculating LV_1 for (4.16), we obtain

$$LV_1 = -\big(x(t) - bp(t)\big)'x(t) + \sum_{i=1}^{2} p_{ii}\left(\int_{t-\tau_i}^{t} \sigma_i x_i(s)\,ds\right)^2$$

$$= -|x(t)|^2 + b \int_{t-h}^{t} (s - t + h)x'(s)x(t) + \sum_{i=1}^{2} p_{ii}\left(\int_{t-\tau_i}^{t} \sigma_i x_i(s)\,ds\right)^2$$

$$\leq -|x(t)|^2 + \frac{b}{2} \int_{t-h}^{t} (s - t + h)\left(|x(s)|^2 + |x(t)|^2\right) ds$$

$$+ \sum_{i=1}^{2} p_{ii}\sigma_i^2 \tau_i \int_{t-\tau_i}^{t} x_i^2(s)\, ds$$

$$= \left(-1 + \frac{bh^2}{4}\right)|x(t)|^2 + \frac{b}{2} \int_{t-h}^{t} (s - t + h)|x(s)|^2\, ds$$

$$+ \sum_{i=1}^{2} p_{ii}\sigma_i^2 \tau_i \int_{t-\tau_i}^{t} x_i^2(s)\, ds.$$

So, we have got the representation of type (2.25) with $S(t) = Q_i(t) = R(s, t) = 0$, $m = 1$, $n = 2$ and

$$D = \left(-1 + \frac{bh^2}{4}\right) I, \qquad dK_i(s) = p_{ii}\sigma_i^2 \delta(s - \tau_i)\, ds, \quad i = 1, 2,$$

$$d\mu_1(s) = \delta(s - h)\, ds, \qquad R_1(s) = \frac{b}{2}.$$

From this and from Theorem 2.4 it follows that if

$$bh\left(\frac{h}{2} + \frac{p_\tau}{b^2 h^2 - a^2}\right) < 1, \qquad p_\tau = \frac{1}{2}\max\left(\sigma_1^2 \tau_1^2, \sigma_2^2 \tau_2^2\right),$$

or if the conditions

$$|a| < \sqrt{bh\left(bh - \frac{2p_\tau}{2 - bh^2}\right)} \qquad\qquad (4.19)$$

and (4.15) hold, then the trivial solution of (4.16) is asymptotically mean-square stable.

The stability regions for (4.16) given by conditions (4.19), (4.15) are shown in Fig. 4.11 for $h = 1.1$ and different values of p_τ: (1) $p_\tau = 0$, (2) $p_\tau = 0.2$, (3) $p_\tau = 0.4$, (4) $p_\tau = 0.45$, in Figs. 4.12 and 4.13 for $p_\tau = 0.1$ and $p_\tau = 2.5$, respectively, and different values of h: (1) $h = 0.085$, (2) $h = 0.1$, (3) $h = 0.12$, (4) $h = 0.16$, in Fig. 4.14 for $p_\tau = 2$ and the following values of h: (1) $h = 0.01$, (2) $h = 0.005$, (3) $h = 0.0025$, (4) $h = 0.00125$. In the last figure one can see that by $p_\tau > 0$, as $h \to 0$, the stability region disappears.

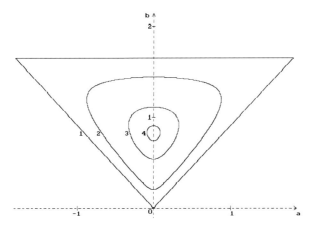

Fig. 4.11 Stability regions for (4.16) given by conditions (4.19), (4.15) for $h = 1.1$ and different values of p_τ: (*1*) $p_\tau = 0$, (*2*) $p_\tau = 0.2$, (*3*) $p_\tau = 0.4$, (*4*) $p_\tau = 0.45$

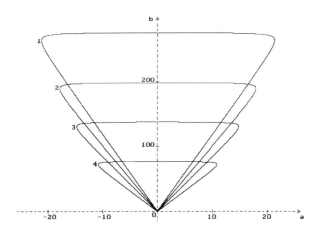

Fig. 4.12 Stability regions for (4.16) given by conditions (4.19), (4.15) for $p_\tau = 0.1$ and different values of h: (*1*) $h = 0.085$, (*2*) $h = 0.1$, (*3*) $h = 0.12$, (*4*) $h = 0.16$

4.3   Linear Systems of Two Equations with Variable Coefficients

Example 4.5 Consider the system of two stochastic differential equations with variable coefficients before the terms with delays

$$\dot{x}_1(t) = -ax_1(t) + b(t)x_1(t-h) + cx_2(t) + \sigma_1(t)x_1(t-\tau_1)\dot{w}_1(t),$$
$$\dot{x}_2(t) = -cx_1(t) + \sigma_2(t)x_2(t-\tau_2)\dot{w}_2(t), \quad t \geq 0, \tag{4.20}$$

where a, c, $h \geq 0$, $\tau_1 \geq 0$, $\tau_2 \geq 0$ are constants, and $b(t)$, $\sigma_1(t)$, $\sigma_2(t)$ are bounded functions.

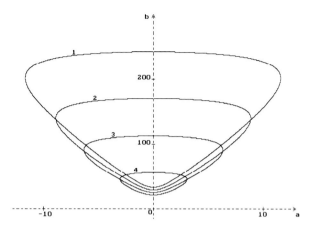

Fig. 4.13 Stability regions for (4.16) given by conditions (4.19), (4.15) for $p_\tau = 2.5$ and different values of h: (1) $h = 0.085$, (2) $h = 0.1$, (3) $h = 0.12$, (4) $h = 0.16$

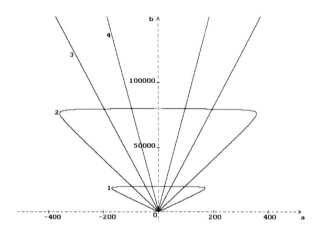

Fig. 4.14 Stability regions for (4.16) given by conditions (4.19), (4.15) for $p_\tau = 2$ and different values of h: (1) $h = 0.01$, (2) $h = 0.005$, (3) $h = 0.0025$, (4) $h = 0.00125$

Let us show that by the conditions

$$a > 0, \qquad |c| > 0, \qquad \sup_{t \geq 0} \mu(t) < a,$$

$$\delta_2(t + \tau_2) < \frac{4c^2 v(t)}{(\sqrt{\beta_1(t) + 4(c^2 + a v(t))} + \sqrt{\beta_1(t)})^2},$$ \hfill (4.21)

where

$$\mu(t) = \delta_1(t + \tau_1) + \beta_0(t + h), \qquad v(t) = a - \mu(t) > 0,$$

$$\beta_0(t) = \frac{1}{2}\big(|b(t)| + |b(t + h)|\big), \qquad \beta_1(t) = |b(t)b(t + h)|, \tag{4.22}$$

the trivial solution of (4.20) is asymptotically mean-square stable.
    Rewrite (4.20) in the form

$$\dot{x}(t) = Ax(t) + B(t)x(t - h) + \sigma(t, x_t)\dot{w}(t),$$

where $x(t) = (x_1(t), x_2(t))'$, $w(t) = (w_1(t), w_2(t))'$,

$$A = \begin{pmatrix} -a & c \\ -c & 0 \end{pmatrix}, \qquad B(t) = \begin{pmatrix} b(t) & 0 \\ 0 & 0 \end{pmatrix},$$

$$\sigma(t, x_t) = \begin{pmatrix} \sigma_1(t)x_1(t - \tau_1) & 0 \\ 0 & \sigma_2(t)x_2(t - \tau_2) \end{pmatrix}. \tag{4.23}$$

Let Q be the symmetric 2×2 matrix with the elements $q_{11} = q > 0$, $q_{22} = 1$, $q_{12} = 0$, and the matrix P be a solution of the matrix equation (1.27) where A is defined in (4.23). By (1.29) the elements of the matrix P are

$$p_{11} = \frac{q + 1}{2a}, \qquad p_{22} = \frac{a}{2c^2} + \frac{q + 1}{2a}, \qquad p_{12} = -\frac{1}{2c}. \tag{4.24}$$

For the functional $V_1 = x'(t)Px(t)$, by (1.27) we have

$$\begin{aligned}
LV_1 &= \big(Ax(t) + B(t)x(t - h)\big)' Px(t) + x'(t)P\big(Ax(t) + B(t)x(t - h)\big) \\
&\quad + p_{11}\sigma_1^2(t)x_1^2(t - \tau_1) + p_{22}\sigma_2^2(t)x_2^2(t - \tau_2) \\
&= x'(t)\big(A'P + PA\big)x(t) + 2x'(t - h)B'(t)Px(t) \\
&\quad + p_{11}\sigma_1^2(t)x_1^2(t - \tau_1) + p_{22}\sigma_2^2(t)x_2^2(t - \tau_2) \\
&= -qx_1^2(t) - x_2^2(t) + 2x'(t - h)B'(t)Px(t) \\
&\quad + p_{11}\sigma_1^2(t)x_1^2(t - \tau_1) + p_{22}\sigma_2^2(t)x_2^2(t - \tau_2). \tag{4.25}
\end{aligned}$$

By (4.23), for some $\gamma > 0$, we obtain

$$\begin{aligned}
2x'(t - h)B'(t)Px(t) &= 2b(t)x_1(t - h)\big(p_{11}x_1(t) + p_{12}x_2(t)\big) \\
&\leq |b(t)|\big[p_{11}\big(x_1^2(t) + x_1^2(t - h)\big) \\
&\quad + |p_{12}|\big(\gamma x_2^2(t) + \gamma^{-1}x_1^2(t - h)\big)\big].
\end{aligned}$$

From this and from (4.25) we have

$$LV_1 \le \left(-q + p_{11}|b(t)|\right)x_1^2(t) + (-1 + \gamma)|p_{12}b(t)|x_2^2(t)$$
$$+ \left(p_{11} + \gamma^{-1}|p_{12}|\right)|b(t)|x_1^2(t - h)$$
$$+ p_{11}\sigma_1^2(t)x_1^2(t - \tau_1) + p_{22}\sigma_2^2(t)x_2^2(t - \tau_2).$$

So, by Theorem 2.5 we obtain that if

$$p_{11}\left(|b(t)| + |b(t + h)| + \sigma_1^2(t + \tau_1)\right) + \gamma^{-1}|p_{12}b(t + h)| < q,$$
$$\gamma|p_{12}b(t)| + p_{22}\sigma_2^2(t + \tau_2) < 1, \tag{4.26}$$

then the trivial solution of (4.20) is asymptotically mean-square stable.

Note that inequalities (4.26) can be represented in the form

$$0 < \frac{|p_{12}b(t + h)|}{q - p_{11}(|b(t)| + |b(t + h)| + \sigma_1^2(t + \tau_1))} < \gamma < \frac{1 - p_{22}\sigma_2^2(t + \tau_2)}{|p_{12}b(t)|}. \tag{4.27}$$

So, if

$$0 < \frac{|p_{12}b(t + h)|}{q - p_{11}(|b(t)| + |b(t + h)| + \sigma_1^2(t + \tau_1))} < \frac{1 - p_{22}\sigma_2^2(t + \tau_2)}{|p_{12}b(t)|}$$

or (which is the same)

$$p_{12}^2|b(t)b(t + h)| < \left(1 - p_{22}\sigma_2^2(t + \tau_2)\right)$$
$$\times \left(q - p_{11}\left(|b(t)| + |b(t + h)| + \sigma_1^2(t + \tau_1)\right)\right), \tag{4.28}$$

then there exists $\gamma > 0$ that satisfies condition (4.27).

Substituting (4.24) into (4.28) and using (4.22), we obtain

$$\frac{\beta_1(t)}{4c^2} < \left(1 - \left(\frac{a}{c^2} + \frac{q + 1}{a}\right)\delta_2(t + \tau_2)\right)\left(q - \frac{q + 1}{a}\mu(t)\right).$$

From this by (4.22) it follows that

$$\frac{a^2}{4}\beta_1(t) < \left(ac^2 - \left(a^2 + (q + 1)c^2\right)\delta_2(t + \tau_2)\right)\left(qv(t) - \mu(t)\right).$$

Rewrite this inequality in the form

$$c^2v(t)\delta_2(t + \tau_2)q^2$$
$$- \left[\left(ac^2 - \left(a^2 + c^2\right)\delta_2(t + \tau_2)\right)v(t) + c^2\mu(t)\delta_2(t + \tau_2)\right]q$$
$$+ \left(ac^2 - \left(a^2 + c^2\right)\delta_2(t + \tau_2)\right)\mu(t) + \frac{a^2}{4}\beta_1(t) < 0.$$

The obtained condition holds for arbitrary $q > 0$ and each $t \geq 0$ if and only if for each $t \geq 0$, the following condition holds:

$$\left[\left(ac^2 - (a^2 + c^2)\delta_2(t + \tau_2)\right)v(t) + c^2\mu(t)\delta_2(t + \tau_2)\right]^2$$
$$> c^2 v(t)\delta_2(t + \tau_2)\left[4\left(ac^2 - (a^2 + c^2)\delta_2(t + \tau_2)\right)\mu(t) + a^2\beta_1(t)\right]$$
$$= 4\left(ac^2 - (a^2 + c^2)\delta_2(t + \tau_2)\right)\mu(t)c^2 v(t)\delta_2(t + \tau_2)$$
$$+ a^2 c^2 v(t)\beta_1(t)\delta_2(t + \tau_2),$$

or

$$\left[\left(ac^2 - (a^2 + c^2)\delta_2(t + \tau_2)\right)v(t) - c^2\mu(t)\delta_2(t + \tau_2)\right]^2 > a^2 c^2 v(t)\beta_1(t)\delta_2(t + \tau_2),$$

or

$$\left[c^2 v(t) - \left(c^2 + av(t)\right)\delta_2(t + \tau_2)\right]^2 > c^2 v(t)\beta_1(t)\delta_2(t + \tau_2). \qquad (4.29)$$

From (4.21) it follows that

$$\delta_2(t + \tau_2) < \frac{c^2 v(t)}{c^2 + av(t)}.$$

So, (4.29) can be written in the form

$$c^2 v(t) - \left(c^2 + av(t)\right)\delta_2(t + \tau_2) > c\sqrt{v(t)\beta_1(t)\delta_2(t + \tau_2)}$$

or

$$\left(c^2 + av(t)\right)\delta_2(t + \tau_2) + c\sqrt{v(t)\beta_1(t)\delta_2(t + \tau_2)} - c^2 v(t) < 0.$$

From this we obtain

$$\sqrt{\delta_2(t + \tau_2)} < \frac{c\sqrt{v(t)}\left(\sqrt{\beta_1(t) + 4(c^2 + av(t))} - \sqrt{\beta_1(t)}\right)}{2(c^2 + av(t))}$$

$$= \frac{2c\sqrt{v(t)}}{\sqrt{\beta_1(t) + 4(c^2 + av(t))} + \sqrt{\beta_1(t)}},$$

which is equivalent to (4.21).

Note that in the case $b(t) = 0$, $\sigma_i(t) = \sigma_i = \text{const}$, $i = 1, 2$, conditions (4.21) take the form

$$\delta_1 < a, \qquad \delta_2 < \frac{c^2(a - \delta_1)}{c^2 + a(a - \delta_1)},$$

which immediately follows from (2.56)–(2.57).

Chapter 5
Stability of Systems with Nonlinearities

The idea of the method of investigation that is used below is similar to the method of the first approximation. Namely, a linear part of the considered differential equation is interpreted as being undisturbed, and the other part as the disturbance. Assuming that the undisturbed equation is stable and the nonlinear disturbance in the right-hand side can be majorized in a certain sense, this method makes it possible to obtain stability conditions.

5.1 Systems with Nonlinearities in Stochastic Part

5.1.1 Scalar First-Order Differential Equation

Let us consider the scalar differential equation

$$\dot{x}(t) = -ax(t) + \int_0^\infty x(t-s)\,dK(s) + \sigma(t, x_t)\dot{w}(t), \quad t \geq 0, \qquad (5.1)$$

where

$$|\sigma(t, \varphi)| \leq \int_0^\infty |\varphi(-s)|\,dR(s), \qquad (5.2)$$

the function $K(s)$ and the nondecreasing function $R(s)$ are functions of bounded variation such that

$$k_0 = \int_0^\infty |dK(s)| < \infty, \qquad R = \int_0^\infty dR(s) < \infty. \qquad (5.3)$$

L. Shaikhet, *Lyapunov Functionals and Stability of Stochastic Functional Differential Equations*, DOI 10.1007/978-3-319-00101-2_5,
© Springer International Publishing Switzerland 2013

5.1.1.1 The First Way of Constructing a Lyapunov Functional

We will consider the auxiliary differential equation (2.34) for (5.1) in the form $\dot{y}(t) = -ay(t)$, $a > 0$. Put $V_1 = x^2$. Calculating LV_1, by (5.2) and (5.3) we have

$$
LV_1 = 2x(t)\left(-ax(t) + \int_0^\infty x(t-s)\,dK(s)\right) + \sigma^2(t, x_t)
$$

$$
\leq -2ax^2(t) + \int_0^\infty \left(x^2(t) + x^2(t-s)\right)|dK(s)| + R\int_0^\infty x^2(t-s)\,dR(s)
$$

$$
= (-2a + k_0)x^2(t) + \int_0^\infty x^2(t-s)|dK(s)| + R\int_0^\infty x^2(t-s)\,dR(s).
$$

By the representation (2.35) with $D = -2a + k_0$ and $dK_1(s) = |dK(s)| + R\,dR(s)$ (all other parameters are zeros) and by Theorem 2.5 we obtain: if

$$
a > k_0 + p, \qquad p = \frac{1}{2}R^2, \tag{5.4}
$$

then the trivial solution of (5.1) is asymptotically mean-square stable.

Example 5.1 Consider the scalar differential equation

$$
\dot{x}(t) = -ax(t) + b\int_0^h (h-s)x(t-s)\,ds + \sigma(t, x_t)\dot{w}(t), \qquad t \geq 0, \tag{5.5}
$$

which is a particular case of (5.1) with

$$
dK(s) = f(s)\,ds, \qquad f(s) = \begin{cases} b(h-s), & s \in [0, h], \\ 0, & s > h. \end{cases} \tag{5.6}
$$

In this case condition (5.4) takes the form

$$
a > |b|\frac{h^2}{2} + p. \tag{5.7}
$$

In Fig. 5.1 the regions of asymptotic mean-square stability for (5.5), given by condition (5.7), are shown for $h = 0.9$ and different values of the parameter p: (1) $p = 0$, (2) $p = 0.5$, (3) $p = 1$, (4) $p = 1.5$, (5) $p = 2$, (6) $p = 2.5$, (7) $p = 3$, (8) $p = 3.5$.

5.1.1.2 The Second Way of Constructing a Lyapunov Functional

Put

$$
k = \int_0^\infty dK(s), \qquad k_1 = \int_0^\infty s|dK(s)| \tag{5.8}
$$

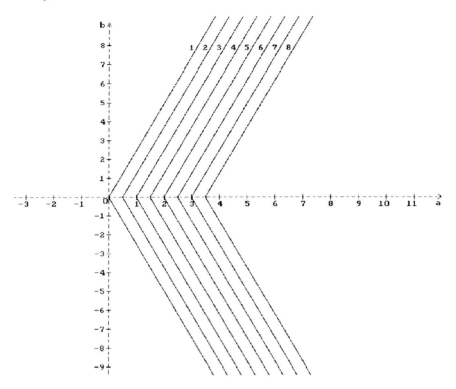

Fig. 5.1 Stability regions for (5.5) given by conditions (5.7) for $h = 0.9$ and different values of p:
(*1*) $p = 0$, (*2*) $p = 0.5$, (*3*) $p = 1$, (*4*) $p = 1.5$, (*5*) $p = 2$, (*6*) $p = 2.5$, (*7*) $p = 3$, (*8*) $p = 3.5$

and represent (5.1) in the form

$$\dot{z}(x_t) = -(a - k)x(t) + \sigma(t, x_t)\dot{w}(t),$$

$$z(x_t) = x(t) + \int_0^\infty \int_{t-s}^t x(\theta)\, d\theta\, dK(s). \tag{5.9}$$

Note that

$$\int_0^\infty \int_{t-s}^t x(\theta)\, d\theta\, dK(s) = \int_0^\infty \int_0^s x(t - \theta)\, d\theta\, dK(s)$$

$$= \int_0^\infty x(t - \theta) \int_\theta^\infty dK(s)\, d\theta$$

and

$$\int_0^\infty \int_\theta^\infty |dK(s)|\, d\theta = \int_0^\infty \int_0^s d\theta\, |dK(s)| = k_1.$$

By (5.9) and (2.10) we have to suppose that $k_1 < 1$, and by (2.34) we will consider the auxiliary differential equation $\dot{y}(t) = -(a - k)y(t)$ with the necessary and sufficient condition $a > k$ for asymptotic stability of the trivial solution.

Put $V_1 = z^2(x_t)$. Calculating LV_1 and using (5.8), (5.9), (5.2), and (5.3), we have

$$LV_1 = -2z(x_t)(a - k)x(t) + \sigma^2(t, x_t)$$

$$\leq (a - k)\left[-2x^2(t) + \int_0^\infty \int_{t-s}^t \left(x^2(t) + x^2(\theta)\right) d\theta \, |dK(s)|\right]$$

$$+ R\int_0^\infty x^2(t - s) \, dR(s)$$

$$= -(a - k)(2 - k_1)x^2(t) + (a - k)\int_0^\infty |dK(s)| \int_{t-s}^t x^2(\theta) \, d\theta$$

$$+ R\int_0^\infty x^2(t - s) \, dR(s).$$

So, we obtain the representation of type (2.35) with $D = -(a - k)(2 - k_1)$, $d\mu_0 = (a - k)|dK(s)|$, $n = 1$, and $dK_1(s) = R \, dR(s)$. From Theorem 2.5 it follows that if

$$a > k + \frac{p}{1 - k_1}, \quad k_1 < 1, \quad p = \frac{1}{2}R^2, \tag{5.10}$$

then the trivial solution of (5.1) is asymptotically mean-square stable.

Example 5.2 In the case (5.6), condition (5.10) takes the form

$$a > b\frac{h^2}{2} + p\left(1 - |b|\frac{h^3}{6}\right)^{-1}, \quad |b|\frac{h^3}{6} < 1. \tag{5.11}$$

In Fig. 5.2 the regions of asymptotic mean-square stability for (5.5), given by condition (5.11) are shown for $h = 0.9$ and different values of the parameter p: (1) $p = 0$, (2) $p = 0.1$, (3) $p = 0.3$, (4) $p = 0.7$, (5) $p = 1.2$, (6) $p = 1.7$, (7) $p = 2.5$, (8) $ph = 3$.

5.1.1.3 The Third Way of Constructing a Lyapunov Functional

Let us show that in some particular cases using a special way of LV_1 estimation we can get new stability conditions.

Consider (5.5) and suppose that $b \leq 0$. Putting $V_1 = x^2$, by (5.5), (5.2), and (5.3) we obtain

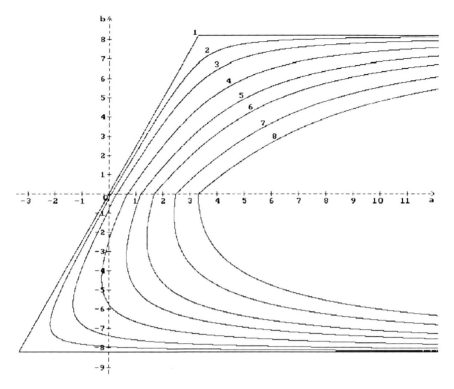

Fig. 5.2 Stability regions for (5.5) given by conditions (5.11), for $h = 0.9$ and different values of p: (*1*) $p = 0$, (*2*) $p = 0.1$, (*3*) $p = 0.3$, (*4*) $p = 0.7$, (*5*) $p = 1.2$, (*6*) $p = 1.7$, (*7*) $p = 2.5$, (*8*) $ph = 3$

$$LV_1 = 2x(t)\left(-ax(t) + b\int_0^h (h - s)x(t - s)\,ds\right) + \sigma^2(t, x_t)$$

$$\leq -2ax^2(t) - 2|b|x(t)\int_0^h (h - s)x(t - s)\,ds + R\int_0^\infty x^2(t - s)\,dR(s).$$

$$(5.12)$$

Choosing V_2 in the form

$$V_2 = |b|\int_0^h \left(\int_{t-s}^t x(\theta)\,d\theta\right)^2 ds + R\int_0^\infty \int_{t-s}^t x^2(\theta)\,d\theta\,dR(s),$$

we have

$$LV_2 = 2|b|\int_0^h \int_{t-s}^t x(\theta)\,d\theta\left(x(t) - x(t - s)\right)ds$$

$$+ R^2 x^2(t) - R\int_0^\infty x^2(t - s)\,dR(s). \qquad (5.13)$$

Note that

$$\int_0^h \int_{t-s}^t x(\theta) \, d\theta \left(x(t) - x(t-s) \right) ds$$

$$= \int_0^h \int_0^s x(t-\theta) \, d\theta \left(x(t) - x(t-s) \right) ds$$

$$= x(t) \int_0^h \int_0^s x(t-\theta) \, d\theta \, ds - \int_0^h x(t-s) \int_0^s x(t-\theta) \, d\theta \, ds \quad (5.14)$$

and

$$\int_0^h \int_0^s x(t-\theta) \, d\theta \, ds = \int_0^h \int_\theta^h ds \, x(t-\theta) \, d\theta = \int_0^h (h-s)x(t-s) \, ds. \quad (5.15)$$

Besides, changing the order of integration, we have

$$\int_0^h x(t-s) \int_0^s x(t-\theta) \, d\theta \, ds = \int_0^h x(t-\theta) \int_\theta^h x(t-s) \, ds \, d\theta$$

$$= \int_0^h x(t-s) \int_s^h x(t-\theta) \, d\theta \, ds.$$

Therefore,

$$2 \int_0^h x(t-s) \int_0^s x(t-\theta) \, d\theta \, ds = \int_0^h x(t-s) \int_0^s x(t-\theta) \, d\theta \, ds$$

$$+ \int_0^h x(t-s) \int_s^h x(t-\theta) \, d\theta \, ds$$

$$= \left(\int_0^h x(t-s) \, ds \right)^2$$

$$\geq 0,$$

and from (5.13)–(5.15) we obtain

$$LV_2 \leq R^2 x^2(t) + 2|b|x(t) \int_0^h (h-s)x(t-s) \, ds - R \int_0^\infty x^2(t-s) \, dR(s). \quad (5.16)$$

As a result, for the functional $V = V_1 + V_2$, from (5.12) and (5.16) it follows that

$$LV \leq -2(a-p)x^2(t), \quad p = \frac{1}{2}R^2.$$

So, if

$$a > p, \qquad b \leq 0, \qquad\qquad\qquad (5.17)$$

then the trivial solution of (5.5) is asymptotically mean-square stable.

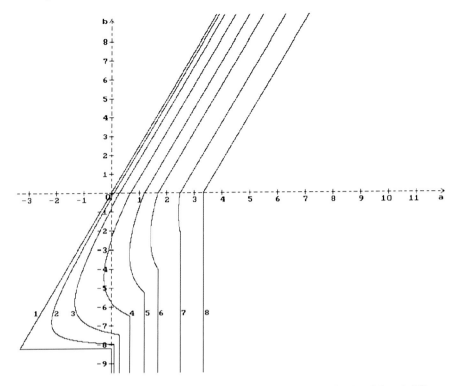

Fig. 5.3 Stability regions for (5.5) given by conditions (5.18), (5.19), for $h = 0.9$ and different values of p: (*1*) $p = 0$, (*2*) $p = 0.1$, (*3*) $p = 0.3$, (*4*) $p = 0.7$, (*5*) $p = 1.2$, (*6*) $p = 1.7$, (*7*) $p = 2.5$, (*8*) $ph = 3$

Using (5.7), (5.11), and (5.17), we can get a sufficient condition for asymptotic mean-square stability of the trivial solution of (5.5) in the following form.

If $ph < 3$, then

$$a > \begin{cases} b\frac{h^2}{2} + p, & b \geq 0, \\ b\frac{h^2}{2} + p(1 + b\frac{h^3}{6})^{-1}, & -2(3 - ph)h^{-3} \leq b < 0, \\ p, & b < -2(3 - ph)h^{-3}. \end{cases} \qquad (5.18)$$

If $ph \geq 3$, then

$$a > \begin{cases} b\frac{h^2}{2} + p, & b \geq 0, \\ p, & b < 0. \end{cases} \qquad (5.19)$$

In Fig. 5.3 the regions of asymptotic mean-square stability for (5.5), given by conditions (5.18) and (5.19), are shown for $h = 0.9$ and different values of the parameter p: (1) $p = 0$, (2) $p = 0.1$, (3) $p = 0.3$, (4) $p = 0.7$, (5) $p = 1.2$, (6) $p = 1.7$, (7) $p = 2.5$, (8) $ph = 3$.

5.1.2 Scalar Second-Order Differential Equation

Consider the scalar linear second-order differential equation with distributed delays

$$\ddot{x}(t) = -a\dot{x}(t) - \int_0^\infty \dot{x}(t-s)\,dK_0(s) - \int_0^\infty x(t-s)\,dK_1(s)$$

$$+\sigma(t, x_t, \dot{x}_t)\dot{w}(t), \quad x_0 = \phi, \ \dot{x}_0 = \dot{\phi}. \tag{5.20}$$

Here $K_i(s)$, $i = 0, 1$, are functions of bounded variation, and the functional $\sigma(t, \varphi, \psi)$ satisfies the condition

$$\sigma^2(t, \varphi, \psi) \le \int_0^\infty \varphi^2(-s)\,dr_1(s) + \int_0^\infty \psi^2(-s)\,dr_2(s), \tag{5.21}$$

where $r_i(s)$ are nondecreasing functions such that

$$r_i = \int_0^\infty dr_i(s) < \infty, \quad i = 1, 2. \tag{5.22}$$

Let us obtain sufficient conditions for asymptotic mean-square stability of the trivial solution of (5.20) using the proposed procedure of Lyapunov functionals construction.

Put $x_1 = x$, $x_2 = \dot{x}$,

$$\alpha_{ij} = \int_0^\infty s^i |dK_j(s)|, \qquad \beta_{ij} = \int_0^\infty s^i\,dK_j(s).$$

Following the first step of the procedure, represent (5.20) in the form

$$\dot{z}_1(t) = x_2(t),$$
$$\dot{z}_2(t) = -bx_1(t) - ax_2(t) + \sigma(t, x_t, \dot{x}_t)\dot{w}(t), \tag{5.23}$$

where $b = \beta_{01}$, $z(t) = (z_1(t), z_2(t))$,

$$z_1(t) = x_1(t),$$

$$z_2(t) = x_2(t) + \int_0^\infty x_1(t-\theta)\,dK_0(\theta) - \int_0^\infty \int_{t-\theta}^t x_1(s)\,ds\,dK_1(\theta). \tag{5.24}$$

In this case the auxiliary system has the form

$$\dot{y}(t) = Ay(t), \quad y = \begin{pmatrix} y_1 \\ y_2 \end{pmatrix}, \quad A = \begin{pmatrix} 0 & 1 \\ -b & -a \end{pmatrix}. \tag{5.25}$$

Note that the matrix Lyapunov equation (1.27) with the matrix A from (5.25) and the matrix Q with the elements $q_{11} = q > 0$, $q_{22} = 1$, $q_{12} = 0$ is equivalent to the system

$$2bp_{12} = q, \qquad 2(ap_{22} - p_{12}) = 1, \qquad p_{11} - ap_{12} - bp_{22} = 0 \tag{5.26}$$

and has (see Corollary 1.1) the positive definite solution P with elements p_{ij} if and only if $a > 0, b > 0$. So, the function $v(y) = y'Py$ is a Lyapunov function for (5.25). We will construct a Lyapunov functional V for (5.23) in the form $V = V_1 + V_2$, where $V_1 = z'(t)Pz(t)$. Calculating LV_1, where L is the generator of (5.23), and using (5.24) and (5.26), we have

$$LV_1 = 2\big(p_{11}x_1(t) + p_{12}z_2(t)\big)x_2(t)$$

$$- 2\big(p_{12}x_1(t) + p_{22}z_2(t)\big)\big(bx_1(t) + ax_2(t)\big) + p_{22}\sigma^2(t, x_{1t}, x_{2t})$$

$$= -qx_1^2(t) - x_2^2(t) + p_{22}\sigma^2(t, x_{1t}, x_{2t})$$

$$- 2bp_{22}\left(\int_0^\infty x_1(t)x_1(t-\theta)\,dK_0(\theta) - \int_0^\infty\int_{t-\theta}^t x_1(t)x_1(s)\,ds\,dK_1(\theta)\right)$$

$$- \int_0^\infty x_2(t)x_1(t-\theta)\,dK_0(\theta) + \int_0^\infty\int_{t-\theta}^t x_2(t)x_1(s)\,ds\,dK_1(\theta).$$

Putting $\alpha = \alpha_{00} + \alpha_{11}$, for some $\gamma > 0$, we obtain

$$LV_1 \le -qx_1^2(t) - x_2^2(t) + p_{22}\left(\int_0^\infty x_1^2(t-s)\,dr_1(s) + \int_0^\infty x_2^2(t-s)\,dr_2(s)\right)$$

$$+ bp_{22}\Bigg[\int_0^\infty \big(x_1^2(t) + x_1^2(t-\theta)\big)\big|dK_0(\theta)\big|$$

$$+ \int_0^\infty\int_{t-\theta}^t \big(x_1^2(t) + x_1^2(s)\big)\,ds\,\big|dK_1(\theta)\big|\Bigg]$$

$$+ \frac{1}{2}\Bigg[\int_0^\infty\left(\frac{x_2^2(t)}{\gamma} + \gamma x_1^2(t-\theta)\right)\big|dK_0(\theta)\big|$$

$$+ \int_0^\infty\int_{t-\theta}^t\left(\frac{x_2^2(t)}{\gamma} + \gamma x_1^2(s)\right)\,ds\,\big|dK_1(\theta)\big|\Bigg]$$

$$= (-q + \alpha bp_{22})x_1^2(t) + \left(-1 + \frac{\alpha}{2\gamma}\right)x_2^2(t)$$

$$+ \int_0^\infty x_1^2(t-\theta)\left(p_{22}\,dr_1(\theta) + \left(bp_{22} + \frac{\gamma}{2}\right)\big|dK_0(\theta)\big|\right)$$

$$+ p_{22}\int_0^\infty x_2^2(t-s)\,dr_2(s) + \left(bp_{22} + \frac{\gamma}{2}\right)\int_0^\infty\int_{t-\theta}^t x_1^2(s)\,ds\,\big|dK_1(\theta)\big|.$$

By Theorem 2.5 the stability condition takes the form

$$(2\alpha b + r_1)p_{22} + \frac{\alpha\gamma}{2} < q, \qquad \frac{\alpha}{2\gamma} + p_{22}r_2 < 1. \tag{5.27}$$

From (5.26) it follows that $p_{22} = (q + b)(2ab)^{-1}$. Substituting p_{22} into (5.27) and using the condition

$$\frac{2\alpha b + r_1}{2ab} < 1, \tag{5.28}$$

we obtain

$$\left(\frac{\alpha\gamma}{2} + \frac{\alpha b}{a} + \frac{r_1}{2a}\right)\left(1 - \frac{2\alpha b + r_1}{2ab}\right)^{-1} < q < \frac{2ab}{r_2}\left(1 - \frac{\alpha}{2\gamma} - \frac{r_2}{2a}\right). \tag{5.29}$$

So, if for some $\gamma > 0$,

$$\left(\frac{\alpha\gamma}{2} + \frac{\alpha b}{a} + \frac{r_1}{2a}\right)\left(1 - \frac{2\alpha b + r_1}{2ab}\right)^{-1} < \frac{2ab}{r_2}\left(1 - \frac{\alpha}{2\gamma} - \frac{r_2}{2a}\right), \tag{5.30}$$

then there exists $q > 0$ such that inequalities (5.29) hold. From (5.30) it follows that

$$\frac{\alpha}{2}\left(\gamma + \frac{2b(a - \alpha) - r_1}{\gamma r_2}\right) < \frac{2b(a - \alpha) - r_1}{r_2} - b.$$

To minimize the left-hand part of this inequality, put $\gamma = \sqrt{(2b(a - \alpha) - r_1)r_2^{-1}}$. Then we have $\alpha\gamma < \gamma^2 - b$. From this and using the fact that system (5.23)–(5.24) is a system of neutral type, we obtain

$$2b(a - \alpha) > r_1 + r_2\left[b + \frac{\alpha^2 + \alpha\sqrt{\alpha^2 + 4b}}{2}\right], \quad \alpha < 1. \tag{5.31}$$

So, if conditions (5.31) hold, then the trivial solution of (5.20) is asymptotically mean-square stable.

Remark 5.1 Note that instead of the representation (5.23)–(5.24) for $z_2(t)$, we can use some other representations:

$$\dot{z}_2(t) = -\beta_{01}x_1(t) - (a + \beta_{00})x_2(t) + \sigma^2(t, x_t, \dot{x}_t)\dot{w}(t),$$

$$z_2(t) = x_2(t) - \int_0^\infty \int_{t-\theta}^t x_1(s)\,ds\,dK_1(\theta) - \int_0^\infty \int_{t-\theta}^t x_2(s)\,ds\,dK_0(\theta),$$

or

$$\dot{z}_2(t) = -\beta_{01}x_1(t) - (a - \beta_{11})x_2(t) + \sigma^2(t, x_t, \dot{x}_t)\dot{w}(t),$$

$$z_2(t) = x_2(t) + \int_0^\infty x_1(t - \theta)\,dK_0(\theta) + \int_0^\infty \int_{t-\theta}^t (s - t + \theta)x_2(s)\,ds\,dK_1(\theta),$$

or

$$\dot{z}_2(t) = -\beta_{01}x_1(t) - (a + \beta_{00} - \beta_{11})x_2(t) + \sigma^2(t, x_t, \dot{x}_t)\dot{w}(t),$$

$$z_2(t) = x_2(t) - \int_0^\infty \int_{t-\theta}^t x_2(s)\,ds\,dK_0(\theta) + \int_0^\infty \int_{t-\theta}^t (s - t + \theta)x_2(s)\,ds\,dK_1(\theta).$$

Using these different representations, we can obtain different stability conditions for asymptotic mean-square stability of the trivial solution of (5.20).

Remark 5.2 Consider some particular cases of condition (5.31).

(1) Let $r_2 = 0$. Then (5.31) coincides with (5.28).
(2) Let $dK_0(\theta) = 0$ and $dK_1(\theta) = b\delta(\theta)\,d\theta$. Then $\alpha_{00} = \alpha_{11} = 0$ and (5.20), (5.31) are respectively

$$\ddot{x}(t) = -a\dot{x}(t) - bx(t) + \sigma(t, x_t, \dot{x}_t)\dot{w}(t),$$

$$2a > r_2 + \frac{r_1}{b}, \quad b > 0.$$

(3) Let $dK_0(\theta) = 0$, $dK_1(\theta) = b\delta(\theta - h)\,d\theta$, and $\sigma(t, x_t, \dot{x}_t) = \sigma x(t - \tau)$. Then $\alpha_{00} = 0$, $\alpha_{11} = bh$, $r_1 = \sigma^2$, $r_2 = 0$, and (5.20) and the stability conditions (5.31) respectively take the forms

$$\ddot{x}(t) = -a\dot{x}(t) - bx(t - h) + \sigma x(t - \tau)\dot{w}(t), \tag{5.32}$$

$$a > bh + \frac{\sigma^2}{2b}, \quad 0 < bh < 1. \tag{5.33}$$

Note that condition (3.70) for (5.32) coincides with (5.33), and condition (3.61) for (5.32) takes the form

$$a > bh + bh^2\sqrt{\frac{b}{2(2 - bh^2)}} + \frac{\sigma^2}{2b}, \quad 0 < bh^2 < 2. \tag{5.34}$$

Note also that conditions (5.33)–(5.34) follow from (3.73), (3.72), respectively, if $h_1 = 0$, $h_2 = h$.

It is clear that if $0 < b < h^{-1}$, then condition (5.33) is weaker than (5.34). But, on the other hand, if $h^{-1} \le b < 2h^{-2}$, $h < 2$, then (5.33) cannot be used, and (5.34) gives some additional stability region. In Fig. 5.4 the regions of asymptotic mean-square stability for (5.32), given by conditions (5.33)–(5.34), are shown for $\sigma = 0.7$ and different values of the delay h: (1) $h = 1$, (2) $h = 1.2$, (3) $h = 1.4$.

Fig. 5.4 Stability regions for (5.32), given by conditions (5.33), (5.34) for $\sigma = 0.7$ and different values of the delay h: (1) $h = 1$, (2) $h = 1.2$, (3) $h = 1.4$

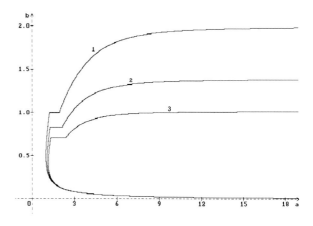

5.2 Systems with Nonlinearities in Both Deterministic and Stochastic Parts

Consider the nonlinear scalar differential equation of neutral type

$$dz(t) = (b - ax(t))z(t)(dt + \sigma\, dw(t)),$$

$$z(t) = x(t) - cx(t - h),$$

$$(5.35)$$

where $a > 0$, $b > 0$, $h \geq 0$, $|c| < 1$, σ are some known constants. Let us investigate asymptotic mean-square stability of the equilibrium point $x(t) \equiv x^* = a^{-1}b$ with respect to perturbations of the initial function ϕ satisfying the following condition:

$$z(0) = \phi(0) - c\phi(-h) > 0. \qquad (5.36)$$

Put $h = 0$. Then (5.35) takes the form

$$dx(t) = -a(x(t) - x^*)x(t)(dt + \sigma\, dw(t)). \qquad (5.37)$$

Let us show that if

$$0 < pb < 1, \quad p = \frac{\sigma^2}{2}, \qquad (5.38)$$

then the function

$$v(x(t)) = x(t) - x^* - x^* \ln \frac{x(t)}{x^*} \qquad (5.39)$$

is a Lyapunov function for the auxiliary equation (5.37).

First at all, note that $v(x^*) = 0$, and since $x - 1 - \ln x \geq 0$ for $x > 0$, we have

$$v\big(x(t)\big) = x^*\left(\frac{x(t)}{x^*} - 1 - \ln\frac{x(t)}{x^*}\right) \geq 0, \quad x(t) \geq 0.$$

Using that $ax^* = b$, we obtain

$$Lv\big(x(t)\big) = -a\big(x(t) - x^*\big)x(t)\left(1 - \frac{x^*}{x(t)}\right) + pa^2\big(x(t) - x^*\big)^2 x^2(t)\frac{x^*}{x^2(t)}$$

$$= -a(1 - bp)\big(x(t) - x^*\big)^2.$$

So, by condition (5.38) the function (5.39) is a Lyapunov function for the solution x^* of (5.37).

Using the procedure of constructing Lyapunov functionals, we will construct a Lyapunov functional V for (5.35) in the form $V = V_1 + V_2$, where

$$V_1 = v\big(z(t)\big) = z(t) - z^* - z^* \ln\frac{z(t)}{z^*},$$

$$z(t) = x(t) - cx(t - h), \quad z^* = (1 - c)x^*.$$

Calculating LV_1, by (5.35) and $b = ax^*$ we have

$$LV_1 = \big(b - ax(t)\big)z(t)\left(1 - \frac{z^*}{z(t)}\right) + p\big(b - ax(t)\big)^2 z^2(t)\frac{z^*}{z^2(t)}$$

$$= -a\big(x(t) - x^*\big)\big(z(t) - z^*\big) + abp(1 - c)\big(x(t) - x^*\big)^2$$

$$= -a\big(x(t) - x^*\big)\big(x(t) - cx(t - h) - x^* + cx^*\big) + abp(1 - c)\big(x(t) - x^*\big)^2$$

$$= -a\big[1 - bp(1 - c)\big]\big(x(t) - x^*\big)^2 + ac\big(x(t) - x^*\big)\big(x(t - h) - x^*\big)$$

$$\leq -a\big[1 - bp(1 - c)\big]\big(x(t) - x^*\big)^2 + \frac{a|c|}{2}\left[\big(x(t) - x^*\big)^2 + \big(x(t - h) - x^*\big)^2\right]$$

$$= -a\left(1 - bp(1 - c) - \frac{|c|}{2}\right)\big(x(t) - x^*\big)^2 + \frac{a|c|}{2}\big(x(t - h) - x^*\big)^2.$$

Putting

$$V_2 = \frac{a|c|}{2}\int_{t-h}^{t}\big(x(s) - x^*\big)^2 ds,$$

for the functional $V = V_1 + V_2$, we obtain

$$LV \leq -a\big(1 - bp(1 - c) - |c|\big)\big(x(t) - x^*\big)^2.$$

Fig. 5.5 Stability regions for
(5.35) given by conditions
(5.40) for different values
of p: (1) $p = 0$, (2) $p = 0.2$,
(3) $p = 0.4$, (4) $p = 0.6$,
(5) $p = 0.8$

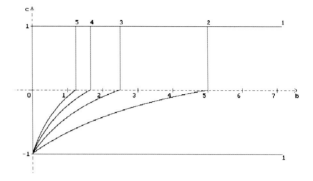

From this it follows that if conditions (5.36) and

$$0 < bp < \begin{cases} \frac{1+c}{1-c}, & -1 < c < 0, \\ 1, & 0 \le c < 1, \end{cases} \tag{5.40}$$

hold then the equilibrium point $x^*(t) \equiv a^{-1}b$ of (5.35) is asymptotically mean-square stable. Note that condition (5.40) does not depend on a and h.

In Fig. 5.5 the stability regions, given by condition (5.40), are shown for different values of p: (1) $p = 0$, (2) $p = 0.2$, (3) $p = 0.4$, (4) $p = 0.6$, (5) $p = 0.8$.

5.3 Stability in Probability of Nonlinear Systems

Consider the nonlinear stochastic differential equation

$$dx(t) = \left(\int_0^\infty dK_0(s)\, x(t-s) + g_0(t, x_t) \right) dt$$
$$+ \sum_{i=1}^{m} \left(\int_0^\infty dK_i(s) x(t-s) + g_i(t, x_t) \right) dw_i(t),$$
$$x_0 = \phi_0. \tag{5.41}$$

Here $x(t) \in \mathbf{R}^n$, $w_1(t), \dots, w_m(t)$ are the mutually independent scalar Wiener processes, $K_i(s)$, $i = 0, \dots, m$, are $n \times n$-matrices such that

$$\alpha_i = \int_0^\infty \| dK_i(s) \| < \infty, \tag{5.42}$$

and $\|A\|$ is the operator norm of a matrix A. It is assumed also that the functionals $g_i(t, \varphi)$, $i = 0, \dots, m$, satisfy the condition

$$|g_i(t, \varphi)| \le \int_0^\infty |\varphi(-s)|^{\nu_i}\, dr_i(s), \quad \|\varphi\|_0 \le \delta, \ \nu_i > 1, \tag{5.43}$$

where δ is small enough, $\|\varphi\|_0 = \sup_{s \le 0} |\varphi(s)|$, and $r_i(s)$ are nondecreasing functions satisfying the condition

$$r_i = \int_0^\infty dr_i(s) < \infty, \quad i = 0, \dots, m. \tag{5.44}$$

The generator L of (5.41) has the form

$$LV(t, x_t) = L_0 V(t, x_t) + g_0'(t, x_t) \frac{\partial V_\varphi(t, x)}{\partial x} + \frac{1}{2} \sum_{i=1}^m g_i'(t, x_t) \frac{\partial^2 V_\varphi(t, x)}{\partial x^2} g_i'(t, x_t)$$

$$+ \sum_{i=1}^m g_i'(t, x_t) \frac{\partial^2 V_\varphi(t, x)}{\partial x^2} \int_0^\infty dK_i(s) x(t-s), \tag{5.45}$$

where

$$L_0 V(t, x_t) = \frac{\partial V_\varphi(t, x)}{\partial t} + \left(\int_0^\infty dK_0(s) x(t-s) \right)' \frac{\partial V_\varphi(t, x)}{\partial x}$$

$$+ \frac{1}{2} \sum_{i=1}^m \left(\int_0^\infty dK_i(s) x(t-s) \right)' \frac{\partial^2 V_\varphi(t, x)}{\partial x^2} \int_0^\infty dK_i(s) x(t-s) \tag{5.46}$$

is the generator of the "linear part" of (5.41), i.e., of the linear differential equation

$$dx(t) = \int_0^\infty dK_0(s) x(t-s) dt + \sum_{i=1}^m \int_0^\infty dK_i(s) x(t-s) dw_i(t). \tag{5.47}$$

Theorem 5.1 *Let there exist a functional* $V_0(t, \varphi) \in D$ *that satisfies conditions* (2.18)–(2.19) *for* $p = 2$ *and*

$$L_0 V_0(t, \varphi) \le -c_0 |\varphi(0)|^2,$$

$$\left| \frac{\partial V_\varphi^0(t, x)}{\partial x} \right| \le c_1 |x| + \int_0^\infty dq(\tau) \int_{-\tau}^0 |\varphi(s)| ds, \quad \left\| \frac{\partial^2 V_\varphi^0(t, x)}{\partial x^2} \right\| \le c_2, \tag{5.48}$$

$$c_i > 0, \ i = 0, 1, 2, \quad \gamma = \int_0^\infty \tau \, dq(\tau) < \infty.$$

Then the trivial solution of (5.41) *is stable in probability.*

Proof We will construct a functional V that satisfies the conditions of Theorem 2.2 (for $p = 2$). The functional V will be constructed in the form $V = V_0 + V_1$, where

V_0 satisfies conditions (5.48). Calculating LV, by (5.45) we have

$$LV = LV_0 + LV_1 = L_0V_0 + LV_1 + g_0'(t, x_t)\frac{\partial V_\varphi^0(t, x)}{\partial x}$$

$$+ \sum_{i=1}^{m} g_i'(t, x_t)\frac{\partial^2 V_\varphi^0(t, x)}{\partial x^2}\int_0^\infty dK_i(s)\,x(t - s)$$

$$+ \frac{1}{2}\sum_{i=1}^{m} g_i'(t, x_t)\frac{\partial^2 V_\varphi^0(t, x)}{\partial x^2}g_i'(t, x_t). \tag{5.49}$$

Supposing that $|x(s)| \le \delta$ for $s \le t$, by inequalities (5.43), (5.48) we estimate the terms in (5.49). So, we get

$$\left|g_0'(t, x_t)\frac{\partial V_\varphi^0(t, x)}{\partial x}\right| \le \int_0^\infty |x(t - s)|^{\nu_0}\,dr_0(s)\left|\frac{\partial V_\varphi^0(t, x)}{\partial x}\right|$$

$$\le \delta^{\nu_0-1}\int_0^\infty |x(t - s)|\,dr_0(s)$$

$$\times \left(c_1|x(t)| + \int_0^\infty dq(\tau)\int_{t-\tau}^t |x(s)|\,ds\right)$$

$$= c_1\delta^{\nu_0-1}\int_0^\infty |x(t - s)||x(t)|\,dr_0(s)$$

$$+ \delta^{\nu_0-1}\int_0^\infty dr_0(\theta)\int_0^\infty dq(\tau)\int_{t-\tau}^t |x(t - \theta)||x(s)|\,ds$$

$$\le \frac{1}{2}c_1\delta^{\nu_0-1}\int_0^\infty \left(|x(t - s)|^2 + |x(t)|^2\right)dr_0(s)$$

$$+ \frac{1}{2}\delta^{\nu_0-1}\int_0^\infty dr_0(\theta)\int_0^\infty dq(\tau)$$

$$\times \int_{t-\tau}^t \left(|x(t - \theta)|^2 + |x(s)|^2\right)ds$$

$$= \frac{1}{2}c_1\delta^{\nu_0-1}\left(r_0|x(t)|^2 + \int_0^\infty |x(t - s)|^2\,dr_0(s)\right)$$

$$+ \frac{1}{2}\delta^{\nu_0-1}\left(\gamma\int_0^\infty |x(t - \theta)|^2\,dr_0(\theta)\right.$$

$$\left. + r_0\int_0^\infty dq(\tau)\int_{t-\tau}^t |x(s)|^2\,ds\right)$$

$$= \frac{1}{2} c_1 \delta^{\nu_0 - 1} r_0 |x(t)|^2$$

$$+ \frac{1}{2} \delta^{\nu_0 - 1} \left((c_1 + \gamma) \int_0^\infty |x(t-s)|^2 \, dr_0(s) \right.$$

$$\left. + r_0 \int_0^\infty dq(\tau) \int_{t-\tau}^t |x(s)|^2 \, ds \right). \tag{5.50}$$

Similarly, we have

$$\left| \sum_{i=1}^m g_i'(t, x_t) \frac{\partial^2 V_\varphi^0(t, x)}{\partial x^2} \int_0^\infty dK_i(s) x(t-s) \right|$$

$$\leq c_2 \sum_{i=1}^m \int_0^\infty dr_i(s) |x(t-s)|^{\nu_i} \int_0^\infty \|dK_i(\tau)\| |x(t-\tau)|$$

$$\leq \frac{c_2}{2} \sum_{i=1}^m \delta^{\nu_i - 1} \int_0^\infty dr_i(s) \int_0^\infty \|dK_i(\tau)\| \left(|x(t-s)|^2 + |x(t-\tau)|^2 \right)$$

$$= \frac{c_2}{2} \sum_{i=1}^m \delta^{\nu_i - 1} \left(\alpha_i \int_0^\infty dr_i(s) |x(t-s)|^2 + \beta_i \int_0^\infty \|dK_i(\tau)\| |x(t-\tau)|^2 \right) \tag{5.51}$$

and

$$\left| \sum_{i=1}^m g_i'(t, x_t) \frac{\partial^2 V_\varphi^0(t, x)}{\partial x^2} g_i(t, x_t) \right|$$

$$\leq c_2 \sum_{i=1}^m |g_i(t, x_t)|^2$$

$$\leq c_2 \sum_{i=1}^m \left(\int_0^\infty dr_i(s) |x(t-s)|^{\nu_i} \right)^2$$

$$\leq c_2 \sum_{i=1}^m \delta^{2(\nu_i - 1)} r_i \int_0^\infty dr_i(s) |x(t-s)|^2. \tag{5.52}$$

Let us define the functional V_1 as follows:

$$V_1 = \frac{1}{2} \delta^{\nu_0 - 1} (c_1 + \gamma) \int_0^\infty dr_0(\tau) \int_{t-\tau}^t |x(s)|^2 \, ds$$

$$+ \frac{1}{2} \delta^{\nu_0 - 1} r_0 \int_0^\infty dq(\tau) \int_{t-\tau}^t (s - t + \tau) |x(s)|^2 \, ds$$

$$+ \frac{c_2}{2} \sum_{i=1}^{m} \delta^{v_i-1} r_i \int_0^{\infty} \|dK_i(\tau)\| \int_{t-\tau}^{t} |x(s)|^2 \, ds$$

$$+ \frac{c_2}{2} \sum_{i=1}^{m} \delta^{v_i-1} (\alpha_i + 2\delta^{v_i-1} r_i) \int_0^{\infty} dr_i(\tau) \int_{t-\tau}^{t} |x(s)|^2 \, ds. \qquad (5.53)$$

It is easy to see that $0 \le V_1(t, \varphi) \le c\|\varphi\|^2$, $c > 0$. Therefore, the functional $V = V_0 + V_1$ satisfies conditions (2.18)–(2.19). Calculating LV_1 and using (5.48)–(5.53), we obtain

$$LV(t, x_t) \le - \left[c_0 - \delta^{v_0-1} \beta_0 \left(\frac{c_1}{2} + \gamma \right) - c_2 \sum_{i=1}^{m} \delta^{v_i-1} r_i \left(\alpha_i + \delta^{v_i-1} r_i \right) \right] |x(t)|^2.$$

For small enough δ, the term in the square brackets is positive. Therefore, $LV \le 0$, and the functional V satisfies the conditions of Theorem 2.2. So, the trivial solution of (5.41) is stable in probability. The proof is completed. $\qquad \Box$

Remark 5.3 The asymptotic mean-square stability of the trivial solution of the linear differential equation (5.47) follows from the existence of the functional V_0 that satisfies the conditions of Theorem 5.1. Therefore, in order to obtain sufficient conditions for stability in probability of the trivial solution of the nonlinear differential equation (5.41) with the order of nonlinearity higher than one, it is enough to obtain, by virtue of some Lyapunov functional, sufficient conditions for asymptotical mean-square stability of the trivial solution of "the linear part" of (5.41), i.e., of (5.47). For example, it is easy to show [261] that the functional

$$V_0(t, x_t) = |x(t)|^2 + v \left| x(t) + \int_{+0}^{\infty} dK_0(\tau) \int_{t-\tau}^{t} x(s) \, ds \right|^2$$

$$+ v \int_{+0}^{\infty} \left| \left(\int_0^{\infty} dK_0(\theta) \right)' dK_0(\tau) \right| \int_{t-\tau}^{t} (s - t + \tau) |x(s)|^2 \, ds$$

$$+ \int_{+0}^{\infty} |dK_0(\tau)| \int_{t-\tau}^{t} |x(s)|^2 \, ds$$

$$+ (v+1) \sum_{i=1}^{N} \int_0^{\infty} |dK_i(\theta)| \int_0^{\infty} |dK_i(\tau)| \int_{t-\tau}^{t} |x(s)|^2 \, ds,$$

$v \ge 0$, satisfies the conditions of Theorem 2.1 if the matrix

$$Q = \int_0^{\infty} dK_0(s) + \inf_{v \ge 0} \frac{1}{v+1} \left[\int_{+0}^{\infty} (|dK_0(s)|I - dK_0(s)) \right.$$

$$\left. + v \int_{+0}^{\infty} \tau \left| \left(\int_0^{\infty} dK_0(s) \right)' dK_0(\tau) \right| I \right] + \frac{1}{2} \sum_{i=1}^{N} \left(\int_0^{\infty} |dK_i(s)| \right)^2 I \qquad (5.54)$$

is a negative definite matrix, i.e., $x'Qx \le -c|x|^2$. Here I is the identity matrix, and $c > 0$.

So, we obtain the following theorem.

Theorem 5.2 *Let conditions* (5.42)–(5.44) *hold, and the matrix* (5.54) *be a negative definite matrix. Then the trivial solution of* (5.47) *is asymptotically mean-square stable, and the trivial solution of* (5.41) *is stable in probability.*

Example 5.3 Consider the well-known Volterra population equation

$$\dot{x}(t) = ax(t)\left(1 - \frac{1}{K}\int_0^\infty x(t-s)\,dH(s)\right), \tag{5.55}$$

where

$$a > 0, \quad K > 0, \quad dH(s) \ge 0, \quad \int_0^\infty dH(s) = 1, \quad \int_0^\infty s\,dH(s) < \infty.$$

Let us assume that the parameter a is susceptible to stochastic perturbations of the type of white noise $\dot{w}(t)$. Then (5.55) is transformed to the stochastic integro–differential equation

$$\dot{x}(t) = a\left(1 + \sigma\dot{w}(t)\right)x(t)\left(1 - \frac{1}{K}\int_0^\infty x(t-s)\,dH(s)\right). \tag{5.56}$$

Substituting $x(t) = K(1 + y(t))$ into (5.56) and keeping only the linear part of the obtained equation, we get the linearization of (5.56) in the neighborhood of the steady-state solution $x(t) = K$

$$\dot{y}(t) = -a\int_0^\infty y(t-s)\,dH(s) - a\sigma\int_0^\infty y(t-s)\,dH(s)\,\dot{w}(t). \tag{5.57}$$

By Theorem 5.2 and (5.54) we obtain that the inequality

$$\min\left[2\int_{+0}^\infty dH(s), a\int_{+0}^\infty s\,dH(s)\right] + \frac{a\sigma^2}{2} < 1$$

is a sufficient condition for asymptotic mean-square stability of the trivial solution of (5.57) and for stability in probability of the steady-state solution $x(t) = K$ of (5.55).

5.4 Systems with Fractional Nonlinearity

Here a nonlinear differential equation with a fractional nonlinearity is considered, and it is supposed that this equation has an equilibrium point and is exposed to additive stochastic perturbations of the type of white noise that are directly proportional to the deviation of the system state from the equilibrium point. Stochastic perturbations of such a form were first proposed by the author in [27, 267] and successfully used later by other researchers (see, for instance, [19, 38, 39, 49, 117, 214, 257]).

The results from the previous paragraph are used below for investigation of stability in probability of the equilibrium points of a stochastic fractional differential equation. Numerous graphical illustrations of stability regions and trajectories of solutions are plotted.

5.4.1 Equilibrium Points

Nonlinear delay differential equation of type $\dot{x}(t) = -ax(t) + f(x(t - \tau))$ is rather popular among researchers. See, for example, [101, 183, 248], the famous Nicholson blowflies equation [99, 102, 224] $\dot{x}(t) = -ax(t) + bx(t-\tau)e^{-\gamma x(t-\tau)}$, the Mackey–Glass model [190]

$$\dot{x}(t) = -ax(t) + \frac{bx(t - \tau)}{1 + x^n(t - \tau)}.$$

On the other hand, recently there is a very large interest in studying the behavior of solutions of nonlinear difference equations with fractional nonlinearity of the type

$$x_{n+1} = \frac{\mu + \sum_{j=0}^{k} a_j x_{n-j}}{\lambda + \sum_{j=0}^{k} b_j x_{n-j}}, \quad n = 0, 1, \ldots,$$

(see [95, 162–164, 232, 278] and a long list of the references therein).

Here, similarly to [232, 278], the stability of equilibrium points of the nonlinear differential equation with fractional nonlinearity

$$\dot{x}(t) = -ax(t) + \frac{\mu + \sum_{j=0}^{k} a_j x(t - \tau_j)}{\lambda + \sum_{j=0}^{k} b_j x(t - \tau_j)}, \quad t > 0, \tag{5.58}$$

and the initial condition

$$x(s) = \phi(s), \quad s \in [-\tau, 0], \ \tau = \max\{\tau_1, \ldots, \tau_k\}, \tag{5.59}$$

is investigated. Here μ, λ, a_j, b_j, $j = 0, \ldots, k$, $\tau_0 = 0$, $\tau_j > 0$, $j > 0$, are known constants.

Put

$$A = \sum_{j=0}^{k} a_j, \qquad B = \sum_{j=0}^{k} b_j, \tag{5.60}$$

and suppose that (5.58) has an equilibrium point $\hat{x}$ (not necessarily a positive one). By (5.58), (5.60), and the assumption

$$\lambda + B\hat{x} \neq 0 \tag{5.61}$$

the equilibrium point $\hat{x}$ is defined by the algebraic equation

$$a\hat{x} = \frac{\mu + A\hat{x}}{\lambda + B\hat{x}}. \tag{5.62}$$

If $aB \neq 0$, then by condition (5.61), (5.62) can be transformed to the form

$$aB\hat{x}^2 - (A - a\lambda)\hat{x} - \mu = 0. \tag{5.63}$$

Thus, if

$$(A - a\lambda)^2 + 4aB\mu > 0, \tag{5.64}$$

then (5.58) has two equilibrium points

$$\hat{x}_1 = \frac{A - a\lambda + \sqrt{(A - a\lambda)^2 + 4aB\mu}}{2aB} \tag{5.65}$$

and

$$\hat{x}_2 = \frac{A - a\lambda - \sqrt{(A - a\lambda)^2 + 4aB\mu}}{2aB}; \tag{5.66}$$

if

$$(A - a\lambda)^2 + 4aB\mu = 0, \tag{5.67}$$

then (5.58) has only one equilibrium point,

$$\hat{x} = \frac{A - a\lambda}{2aB}. \tag{5.68}$$

Finally, if

$$(A - a\lambda)^2 + 4aB\mu < 0, \tag{5.69}$$

then (5.58) has no equilibrium points.

Remark 5.4 Assume that $aB \neq 0$ and $\mu = 0$. If $A \neq 0$ and $A \neq a\lambda$, then (5.58) has two equilibrium points,

$$\hat{x}_1 = \frac{A - a\lambda}{aB} \quad \text{and} \quad \hat{x}_2 = 0; \tag{5.70}$$

if $A = 0$ or $A = a\lambda$, then (5.58) has only one equilibrium point $\hat{x} = 0$.

Remark 5.5 Assume that $aB = 0$. If $A \neq a\lambda$, then (5.58) has only one equilibrium point

$$\hat{x} = -\frac{\mu}{A - a\lambda}.$$

Remark 5.6 Consider the case $\mu = B = 0$, $\lambda \neq 0$. If $A \neq a\lambda$, then (5.58) has only one equilibrium point $\hat{x} = 0$; if $A = a\lambda$, then each solution $\hat{x} = \mathrm{const}$ is an equilibrium point of (5.58).

5.4.2 Stochastic Perturbations, Centering, and Linearization

Suppose that (5.58) is exposed to stochastic perturbations of the type of white noise $\dot{w}(t)$ that are proportional to the deviation of the state $x(t)$ of (5.58) from the equilibrium point $\hat{x}$. Then (5.58) takes the form

$$\dot{x}(t) = -ax(t) + \frac{\mu + \sum_{j=0}^{k} a_j x(t - \tau_j)}{\lambda + \sum_{i=0}^{k} b_i x(t - \tau_i)} + \sigma \left(x(t) - \hat{x} \right) \dot{w}(t). \tag{5.71}$$

Note that the equilibrium point $\hat{x}$ of (5.58) is also an equilibrium point of (5.71). Putting $x(t) = y(t) + \hat{x}$ and

$$\gamma_j = \frac{a_j - ab_j \hat{x}}{\lambda + B\hat{x}}, \quad j = 0, \dots, k, \tag{5.72}$$

we will center (5.71) in the neighborhood of the equilibrium point $\hat{x}$. From (5.71)–(5.72) it follows that $y(t)$ satisfies the equation

$$\dot{y}(t) = -ay(t) + \frac{\gamma_0 y(t) + \sum_{j=1}^{k} \gamma_j y(t - \tau_j)}{1 + \sum_{i=0}^{k} b_i (\lambda + B\hat{x})^{-1} y(t - \tau_i)} + \sigma y(t)\dot{w}(t). \tag{5.73}$$

It is clear that the stability of the trivial solution of (5.73) is equivalent to the stability of the equilibrium point of (5.71).

Together with the nonlinear differential equation (5.73), we will consider the linear part (in a neighborhood of the zero) of (5.73)

$$\dot{z}(t) = -(a - \gamma_0)z(t) + \sum_{j=1}^{k} \gamma_j z(t - \tau_j) + \sigma z(t)\dot{w}(t). \tag{5.74}$$

Below, the following method for stability investigation is used. Conditions for asymptotic mean-square stability of the trivial solution of the constructed linear differential equation (5.74) were obtained by the procedure of constructing Lyapunov functionals. Since the order of nonlinearity of (5.73) is higher than one, the obtained sufficient conditions for asymptotic mean-square stability at the same time are (Theorem 5.2) sufficient conditions for stability in probability of the trivial solution of the nonlinear differential equation (5.73) and therefore for stability in probability of the equilibrium point of (5.71).

5.4.3 Stability of Equilibrium Points

Note that the differential equation (5.74) is an equation of type (3.29). So, from (3.31) and (3.33) we obtain two following sufficient conditions for asymptotic mean-square stability of the trivial solution of (5.74):

$$a > \gamma_0 + \sum_{j=1}^{k} |\gamma_j| + p, \quad p = \frac{\sigma^2}{2}, \tag{5.75}$$

and

$$\left(a - \sum_{j=0}^{k} \gamma_j \right) \left(1 - \sum_{j=1}^{k} |\gamma_j| \tau_j \right) > p, \quad \sum_{j=1}^{k} |\gamma_j| \tau_j < 1. \tag{5.76}$$

Remark 5.7 If the delays are absent, i.e., $\tau_j = 0$, $j = 0, \ldots, k$, then condition (5.76) is not worse than (5.75) that does not depend on delays.

Suppose at first that condition (5.67) holds. In this case, (5.71) has only one equilibrium point $\hat{x}$ that is defined by (5.68), and by (5.72), (5.68) we have

$$\sum_{j=0}^{k} \gamma_j = \frac{A - a B \hat{x}}{\lambda + B \hat{x}} = \frac{A - \frac{1}{2}(A - a\lambda)}{\lambda + \frac{1}{2a}(A - a\lambda)} = a.$$

Thus, the stability condition (5.76) for the equilibrium point (5.68) does not hold. Moreover,

$$a = \sum_{j=0}^{k} \gamma_j \le \gamma_0 + \sum_{j=1}^{k} |\gamma_j|.$$

Thus, the stability condition (5.75) for equilibrium point (5.68) does not hold too.

Suppose now that condition (5.64) holds. Then (5.71) has two equilibrium points $\hat{x}_1$ and $\hat{x}_2$ that are defined in (5.65) and (5.66), respectively. Put

$$S = \sqrt{(A - a\lambda)^2 + 4a B \mu}, \tag{5.77}$$

$$\gamma_j^{(l)} = \frac{a_j - a b_j \hat{x}_l}{\lambda + B \hat{x}_l}, \quad j = 0, \ldots, k, \ l = 1, 2. \tag{5.78}$$

Corollary 5.1 *Assume that condition (5.64) holds and $\gamma_0^{(l)} \geq 0, l = 1, 2$. Then for fixed μ and λ, condition (5.75) cannot be true for both equilibrium points $\hat{x}_1$ and $\hat{x}_2$ together.*

Proof By (5.75), (5.78), and (5.65), for $\hat{x}_1$, we obtain

$$1 > \frac{1}{a} \sum_{j=0}^{k} |\gamma_j^{(1)}| \geq \frac{1}{a} \left| \sum_{j=0}^{k} \gamma_j^{(1)} \right| = \frac{1}{a} \left| \frac{A - aB\hat{x}_1}{\lambda + B\hat{x}_1} \right|$$

$$= \left| \frac{A - \frac{1}{2}(A - a\lambda + S)}{a\lambda + \frac{1}{2}(A - a\lambda + S)} \right| = \left| \frac{A + a\lambda - S}{A + a\lambda + S} \right|.$$

Similarly, for $\hat{x}_2$, we have

$$1 > \frac{1}{a} \sum_{j=0}^{k} |\gamma_j^{(2)}| \geq \frac{1}{a} \left| \sum_{j=0}^{k} \gamma_j^{(2)} \right| = \frac{1}{a} \left| \frac{A - aB\hat{x}_2}{\lambda + B\hat{x}_2} \right|$$

$$= \left| \frac{A - \frac{1}{2}(A - a\lambda - S)}{a\lambda + \frac{1}{2}(A - a\lambda - S)} \right| = \left| \frac{A + a\lambda + S}{A + a\lambda - S} \right|.$$

Thus, we obtain two conflicting conditions. The proof is completed. □

Corollary 5.2 *Assume that condition (5.64) holds and $a \neq 0$. If*

$$\frac{2aS}{S + A + a\lambda} \left(1 - \sum_{j=1}^{k} |\gamma_j^{(1)}| \tau_j \right) > p, \quad \sum_{j=1}^{k} |\gamma_j^{(1)}| \tau_j < 1, \tag{5.79}$$

then the equilibrium point $\hat{x} = \hat{x}_1$ (defined in (5.65)) of (5.71) is stable in probability. If

$$\frac{2aS}{S - A - a\lambda} \left(1 - \sum_{j=1}^{k} |\gamma_j^{(2)}| \tau_j \right) > p, \quad \sum_{j=1}^{k} |\gamma_j^{(2)}| \tau_j < 1, \tag{5.80}$$

then the equilibrium point $\hat{x} = \hat{x}_2$ (defined in (5.66)) of (5.71) is stable in probability. Assume now that $a = 0$. If

$$\frac{A^2}{Q} \left(1 - \frac{|A|}{Q} \tau \right) > p, \quad \tau = \sum_{j=1}^{k} |a_j| \tau_j < \frac{Q}{|A|}, \quad Q = B\mu - A\lambda, \tag{5.81}$$

then the equilibrium point $\hat{x} = -\mu A^{-1}$ is stable in probability.

Proof For $a \neq 0$, by (5.76) it is enough to note that, for $\hat{x}_1$,

$$a - \sum_{j=0}^{k} \gamma_j^{(1)} = a - \frac{A - a B \hat{x}_1}{\lambda + B \hat{x}_1} = a - \frac{A - \frac{1}{2}(A - a\lambda + S)}{\lambda + \frac{1}{2a}(A - a\lambda + S)} = \frac{2a S}{S + A + a\lambda},$$

and, similarly, for $\hat{x}_2$,

$$a - \sum_{j=0}^{k} \gamma_j^{(2)} = a - \frac{A - a B \hat{x}_2}{\lambda + B \hat{x}_2} = a - \frac{A - \frac{1}{2}(A - a\lambda - S)}{\lambda + \frac{1}{2a}(A - a\lambda - S)} = \frac{2a S}{S - A - a\lambda}.$$

For $a = 0$, by Remark 5.5 from (5.72), (5.76) we obtain

$$-\sum_{j=0}^{k} \gamma_j = -\frac{A}{\lambda - B\mu A^{-1}} = \frac{A^2}{Q}, \qquad \sum_{j=0}^{k} |\gamma_j| \tau_j = \frac{\tau}{|\lambda + B(-\mu) A^{-1}|} = \frac{|A|}{Q} \tau.$$

So, (5.76) implies (5.81). The proof is completed. $\Box$

Remark 5.8 If $\tau_j = 0$, $j = 1, \ldots, k$, then conditions (5.79), (5.80) take the forms

$$\frac{2a S}{S + A + a\lambda} > p, \qquad \frac{2a S}{S - A - a\lambda} > p, \tag{5.82}$$

respectively. Moreover, inequalities (5.82) are necessary conditions for implementation of conditions (5.79), (5.80) with arbitrary τ_j, $j = 1, \ldots, k$. If $a < 0$ or $p \geq 2a > 0$, then conditions (5.79), (5.80) cannot be true for the same μ and λ. Really, if $a < 0$, then from (5.82) the contradiction follows: $0 < S < A + a\lambda < -S < 0$. If $p \geq 2a > 0$, then from (5.82) another contradiction follows: $0 \leq -(2a/p - 1)S < A + a\lambda < (2a/p - 1)S \leq 0$.

Corollary 5.3 *Put*

$$q = \begin{cases} \frac{p}{2a - p} & \text{if } p < 2a, \\ \frac{p}{p - 2a} & \text{if } p > 2a, \\ +\infty & \text{if } p = 2a, \end{cases} \tag{5.83}$$

assume that $a \neq 0$, $S > 0$, $\tau_j = 0$, $j = 1, \ldots, k$, and consider the following four cases.

Case 1: $a > 0$, $B > 0$.

If

$$p < 2a, \qquad \mu > \begin{cases} \frac{q^2(A + a\lambda)^2 - (A - a\lambda)^2}{4a B} & \text{for } \lambda \geq -\frac{A}{a}, \\ \frac{A}{B} \lambda & \text{for } \lambda < -\frac{A}{a}, \end{cases} \tag{5.84}$$

or

$$p \geq 2a, \qquad \frac{A}{B}\lambda < \mu < \frac{q^2(A+a\lambda)^2 - (A-a\lambda)^2}{4aB}, \qquad \lambda < -\frac{A}{a}, \qquad (5.85)$$

then the equilibrium point $\hat{x}_1$ is stable in probability.
   If

$$p < 2a, \qquad \mu > \begin{cases} \dfrac{q^2(A+a\lambda)^2-(A-a\lambda)^2}{4aB} & \text{for } \lambda < -\frac{A}{a}, \\[2ex] \dfrac{A}{B}\lambda & \text{for } \lambda \geq -\frac{A}{a}, \end{cases} \qquad (5.86)$$

or

$$p \geq 2a, \qquad \frac{A}{B}\lambda < \mu < \frac{q^2(A+a\lambda)^2 - (A-a\lambda)^2}{4aB}, \qquad \lambda \geq -\frac{A}{a}, \qquad (5.87)$$

then the equilibrium point $\hat{x}_2$ is stable in probability.

Case 2: $a > 0$, $B < 0$.

   If

$$p < 2a, \qquad \mu < \begin{cases} \dfrac{q^2(A+a\lambda)^2-(A-a\lambda)^2}{4aB} & \text{for } \lambda \geq -\frac{A}{a}, \\[2ex] \dfrac{A}{B}\lambda & \text{for } \lambda < -\frac{A}{a}, \end{cases} \qquad (5.88)$$

or

$$p \geq 2a, \qquad \frac{A}{B}\lambda > \mu > \frac{q^2(A+a\lambda)^2 - (A-a\lambda)^2}{4aB}, \qquad \lambda < -\frac{A}{a}, \qquad (5.89)$$

then the equilibrium point $\hat{x}_1$ is stable in probability.
   If

$$p < 2a, \qquad \mu < \begin{cases} \dfrac{q^2(A+a\lambda)^2-(A-a\lambda)^2}{4aB} & \text{for } \lambda < -\frac{A}{a}, \\[2ex] \dfrac{A}{B}\lambda & \text{for } \lambda \geq -\frac{A}{a}, \end{cases} \qquad (5.90)$$

or

$$p \geq 2a, \qquad \frac{A}{B}\lambda > \mu > \frac{q^2(A+a\lambda)^2 - (A-a\lambda)^2}{4aB}, \qquad \lambda \geq -\frac{A}{a}, \qquad (5.91)$$

then the equilibrium point $\hat{x}_2$ is stable in probability.

Case 3: $a < 0$, $B > 0$.

If

$$\frac{A}{B}\lambda < \mu < \frac{q^2(A+a\lambda)^2 - (A-a\lambda)^2}{4aB}, \quad \lambda > -\frac{A}{a}, \tag{5.92}$$

then the equilibrium point $\hat{x}_1$ is stable in probability.

If

$$\frac{A}{B}\lambda < \mu < \frac{q^2(A+a\lambda)^2 - (A-a\lambda)^2}{4aB}, \quad \lambda < -\frac{A}{a}, \tag{5.93}$$

then the equilibrium point $\hat{x}_2$ is stable in probability.

Case 4: $a < 0$, $B < 0$.

If

$$\frac{A}{B}\lambda > \mu > \frac{q^2(A+a\lambda)^2 - (A-a\lambda)^2}{4aB}, \quad \lambda > -\frac{A}{a}, \tag{5.94}$$

then the equilibrium point $\hat{x}_1$ is stable in probability.

If

$$\frac{A}{B}\lambda > \mu > \frac{q^2(A+a\lambda)^2 - (A-a\lambda)^2}{4aB}, \quad \lambda < -\frac{A}{a}, \tag{5.95}$$

then the equilibrium point $\hat{x}_2$ is stable in probability.

Proof It is enough to prove Case 1; the proofs of the others cases are similar.

Consider the equilibrium point $\hat{x}_1$. Assume first that $p < 2a$. If $A + a\lambda \geq 0$, then by (5.77) from the first line of (5.84) it follows that $S > q(A + a\lambda)$. By (5.83) and $\tau_j = 0$, $j = 1, \ldots, k$, this inequality coincides with (5.79). If $A + a\lambda < 0$, then from the second line of (5.84) we have $B\mu > A\lambda$. So, by (5.77) we obtain $S > |A + a\lambda|$ and, therefore, $S > S + A + a\lambda > 0$. From this and from $2a > p$ condition (5.79) with $\tau_j = 0$, $j = 1, \ldots, k$, follows.

Let now $p > 2a$. Then, by (5.85), $A + a\lambda < 0$ and $B\mu > A\lambda$. Thus, from (5.77), (5.85) it follows that $q|A + a\lambda| > S > |A + a\lambda|$. From this by (5.83) condition (5.79) with $\tau_j = 0$, $j = 1, \ldots, k$, follows. Finally, if $p = 2a$, then (5.85) is equivalent to $B\mu > A\lambda$ and $S > |A + a\lambda|$, which implies (5.79) with $\tau_j = 0$, $j = 1, \ldots, k$.

Consider the equilibrium point $\hat{x}_2$. Assume first that $p < 2a$. If $A + a\lambda \geq 0$, then from the second line of (5.86) it follows that $B\mu > A\lambda$. In view of (5.77), this means that $S > A + a\lambda$, and therefore $S > S - A - a\lambda$. From this and from $p < 2a$ condition (5.80) with $\tau_j = 0$, $j = 1, \ldots, k$, follows. If $A + a\lambda < 0$, then from the first line of (5.86) we obtain $S > q|A + a\lambda|$. From this and from (5.83) condition (5.80) with $\tau_j = 0$, $j = 1, \ldots, k$, follows.

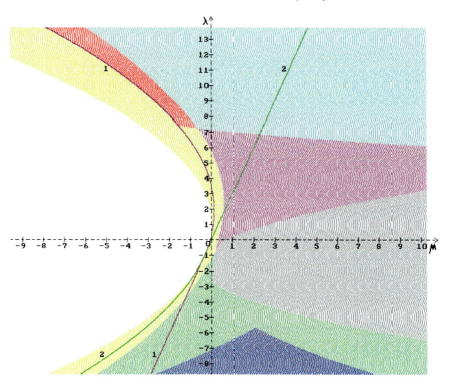

Fig. 5.6 Stability regions for the equilibrium points $\hat{x}_1$ and $\hat{x}_2$ of (5.96) by $a = 1$, $a_1 = 1.5$, $a_2 = -0.5$, $b_1 = 1.2$, $b_2 = 1.8$, $\tau_1 = 0.4$, $\tau_2 = 0.3$, $\sigma = 1.2$

Let now $p > 2a$. Then, by (5.87), $A + a\lambda \geq 0$ and $B\mu > A\lambda$. Thus, from (5.77), (5.87) it follows that $q(A + a\lambda) > S > A + a\lambda \geq 0$. From this and from (5.83) condition (5.80) with $\tau_j = 0$, $j = 1, \ldots, k$, follows. Finally, if $p = 2a$, then (5.87) is equivalent to $B\mu > A\lambda$, and, by (5.77), $S > A + a\lambda$, which implies (5.80) with $\tau_j = 0$, $j = 1, \ldots, k$. The proof is completed. $\square$

Corollary 5.4 *Put* $\tau = \sum_{j=1}^{k} |a_j| \tau_j$, $Q = B\mu - A\lambda$ *and assume that* $a = 0$, $AB \neq 0$, $\tau < Q|A|^{-1}$. *If*

$$\frac{A\lambda}{B} + \frac{A^2(1 - \sqrt{1 - 4p\tau|A^{-1}|})}{2pB} < \mu < \frac{A\lambda}{B} + \frac{A^2(1 + \sqrt{1 - 4p\tau|A^{-1}|})}{2pB}, \quad B > 0,$$

or

$$\frac{A\lambda}{B} + \frac{A^2(1 + \sqrt{1 - 4p\tau|A^{-1}|})}{2pB} < \mu < \frac{A\lambda}{B} + \frac{A^2(1 - \sqrt{1 - 4p\tau|A^{-1}|})}{2pB}, \quad B < 0,$$

then the equilibrium point $\hat{x} = -\mu A^{-1}$ *is stable in probability.*

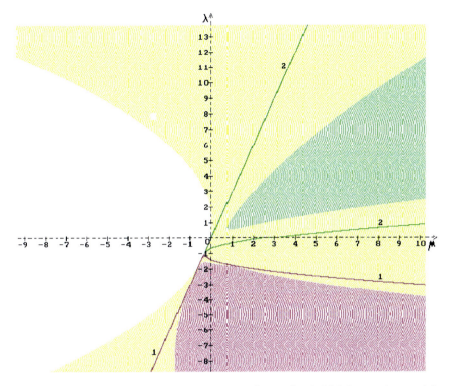

Fig. 5.7 Stability regions for the equilibrium points $\hat{x}_1$ and $\hat{x}_2$ of (5.96) by $a = 1$, $a_1 = 1.5$, $a_2 = -0.5$, $b_1 = 1.2$, $b_2 = 1.8$, $\tau_1 = 0.15$, $\tau_2 = 0.01$, $\sigma = 2.2$

Proof It is enough to note that the given conditions are the solution of the inequality

$$pQ^2 - A^2Q + A^2|A|\tau < 0, \quad \text{where } Q = B\mu - A\lambda > 0,$$

which is equivalent to (5.81).   □

5.4.4 Numerical Analysis

Below, for numerical simulation of the solutions of the equations of type (5.71), we use difference analogues of the considered equations [278] and the algorithm of numerical simulation of the Wiener process trajectories (Sect. 2.1.1).

Example 5.4 Consider the differential equation

$$\dot{x}(t) = -ax(t) + \frac{\mu + a_1x(t - \tau_1) + a_2x(t - \tau_2)}{\lambda + b_1x(t - \tau_1) + b_2x(t - \tau_2)} + \sigma\left(x(t) - \hat{x}\right)\dot{w}(t), \quad (5.96)$$

which is an equation of type (5.71) with $k = 2$, $a_0 = b_0 = 0$.

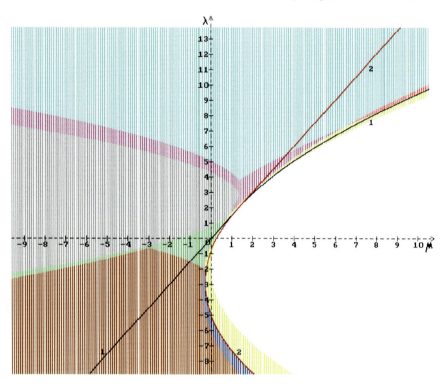

Fig. 5.8 Stability regions for the equilibrium points $\hat{x}_1$ and $\hat{x}_2$ of (5.96) by $a = 1$, $a_1 = -1.5$, $a_2 = -0.5$, $b_1 = -1.2$, $b_2 = -1.8$, $\tau_1 = 0.3$, $\tau_2 = 0.4$, $\sigma = 1.1$

Case 1. Put $a = 1$, $a_1 = 1.5$, $a_2 = -0.5$, $b_1 = 1.2$, $b_2 = 1.8$, $\tau_1 = 0.4$, $\tau_2 = 0.3$, $\sigma = 1.2$. Thus, $a > 0$, $B = 1.2 + 1.8 = 3 > 0$, $p = 0.72 < 2a = 2$, $A = 1.5 - 0.5 = 1 > 0$.

In Fig. 5.6 the regions of stability in probability for the equilibrium points $\hat{x}_1$ and $\hat{x}_2$ are shown in the space of parameters (μ, λ): in the white region there are no equilibrium points; in the yellow region there are possible unstable equilibrium points; red, cyan, magenta, and grey regions are the regions for stability in probability of the equilibrium point $\hat{x} = \hat{x}_1$ given by condition (5.75) (red and cyan) and condition (5.79) (cyan, magenta, and grey); blue, green, and grey regions are the regions for stability in probability of the equilibrium point $\hat{x} = \hat{x}_2$ given by condition (5.75) (blue) and condition (5.80) (blue, green, and grey); in the grey region both equilibrium points $\hat{x} = \hat{x}_1$ and $\hat{x} = \hat{x}_2$ are stable in probability. Curves 1 and 2 are the bounds of the equilibrium points $\hat{x}_1$ and $\hat{x}_2$ stability regions, respectively, given by conditions (5.84) and (5.86) for the case $\tau_1 = \tau_2 = 0$. One can see that the stability regions obtained for positive delays are placed inside of the regions with zero delays.

Put now $\tau_1 = 0.15$, $\tau_2 = 0.01$, $\sigma = 2.2$. Then $p = 2.42 > 2a = 2$, and condition (5.75) does not hold. Appropriate stability regions obtained with the same values

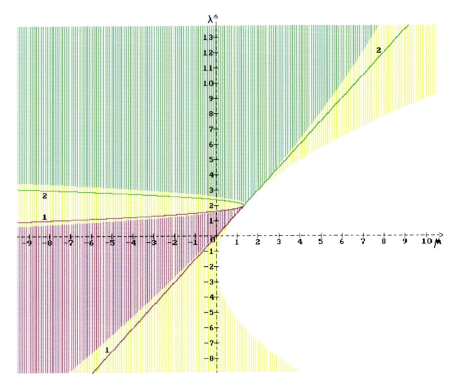

Fig. 5.9 Stability regions for the equilibrium points $\hat{x}_1$ and $\hat{x}_2$ of (5.96) by $a = 1$, $a_1 = -1.5$, $a_2 = -0.5$, $b_1 = -1.2$, $b_2 = -1.8$, $\tau_1 = 0.02$, $\tau_2 = 0.03$, $\sigma = 2.1$

of all other parameters by conditions (5.85), (5.87) for the equilibrium points $\hat{x}_1$ (magenta) and $\hat{x}_2$ (green) are shown in Fig. 5.7. As it was shown also in Fig. 5.6, the stability regions obtained for positive delays are placed inside the regions with zero delays (bounds 1 and 2).

   Case 2. Put $a = 1$, $a_1 = -1.5$, $a_2 = -0.5$, $b_1 = -1.2$, $b_2 = -1.8$, $\tau_1 = 0.3$, $\tau_2 = 0.4$, $\sigma = 1.1$. Thus, $a > 0$, $B = -1.2 - 1.8 = -3 < 0$, $p = 0.605 < 2a$, $A = -1.5 - 0.5 = -2 < 0$.

   In Fig. 5.8 the regions of stability in probability for the equilibrium points $\hat{x}_1$ and $\hat{x}_2$ are shown in the space of parameters (μ, λ): in the white region there are no equilibrium points; in the yellow region there are possible unstable equilibrium points; the red, cyan, magenta, and grey regions are the regions for stability in probability of the equilibrium point $\hat{x} = \hat{x}_1$ given by condition (5.75) (red and cyan) and condition (5.79) (cyan, magenta, and grey); blue, brown, green, and grey regions are the regions for stability in probability of the equilibrium point $\hat{x} = \hat{x}_2$ given by condition (5.75) (blue, brown) and condition (5.80) (blue, brown, green, and grey); in the grey region both equilibrium points $\hat{x} = \hat{x}_1$ and $\hat{x} = \hat{x}_2$ are stable in probability. Curves 1 and 2 are the bounds of the equilibrium points $\hat{x}_1$ and $\hat{x}_2$ stability regions, respectively, given by conditions (5.88) and (5.90) for the case $\tau_1 = \tau_2 = 0$.

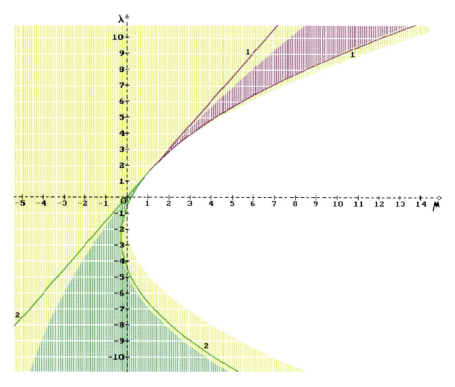

Fig. 5.10 Stability regions for the equilibrium points $\hat{x}_1$ and $\hat{x}_2$ of (5.96) by $a = -1.2$, $a_1 = 1.5$, $a_2 = 0.5$, $b_1 = 1.2$, $b_2 = 1.8$, $\tau_1 = 0.04$, $\tau_2 = 0.03$, $\sigma = 2$

One can see that the stability regions obtained for positive delays are placed inside the regions with zero delays.

Put now $\tau_1 = 0.02$, $\tau_2 = 0.03$, $\sigma = 2.1$. Then $p = 2.205 > 2a = 2$, and condition (5.75) does not hold. Appropriate stability regions obtained with the same values of all other parameters by conditions (5.89), (5.91) for the equilibrium points $\hat{x}_1$ (magenta) and $\hat{x}_2$ (green) are shown in Fig. 5.9. As it was shown also in Fig. 5.8, the stability regions obtained for positive delays are placed inside the regions with zero delays (bounds 1 and 2).

Case 3. Put $a = -1.2$, $a_1 = 1.5$, $a_2 = 0.5$, $b_1 = 1.2$, $b_2 = 1.8$, $\tau_1 = 0.04$, $\tau_2 = 0.03$, $\sigma = 2$. Thus, $a < 0$, $B = 1.2 + 1.8 = 3 > 0$, $A = 1.5 + 0.5 = 2 > 0$, and condition (5.75) does not hold. In Fig. 5.10 the regions of stability in probability for the equilibrium points $\hat{x}_1$ and $\hat{x}_2$ are shown in the space of parameters (μ, λ): in the white region there are no equilibrium points; in the yellow region there are possible unstable equilibrium points; magenta region is the region for stability in probability of the equilibrium point $\hat{x}_1$ given by condition (5.79), green region is the region for stability in probability of the equilibrium point $\hat{x}_2$ given by the condition (5.80). Curves 1 and 2 are the bounds of the equilibrium points $\hat{x}_1$ and $\hat{x}_2$ stability regions, respectively, given by conditions (5.92) and (5.93) for the case $\tau_1 = \tau_2 = 0$.

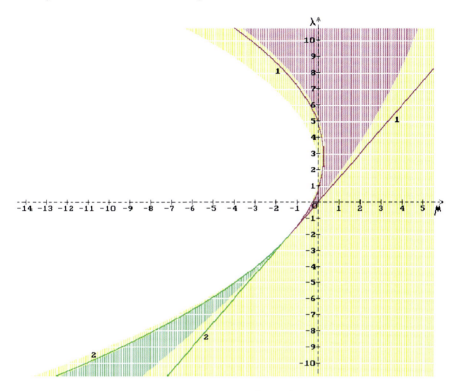

Fig. 5.11 Stability regions for the equilibrium points $\hat{x}_1$ and $\hat{x}_2$ of (5.96) by $a = -1$, $a_1 = -1.5$, $a_2 = -0.5$, $b_1 = -1.2$, $b_2 = -1.8$, $\tau_1 = 0.04$, $\tau_2 = 0.05$, $\sigma = 1.7$

One can see that the stability regions obtained for positive delays are placed inside the regions with zero delays.

Case 4. Put $a = -1$, $a_1 = -1.5$, $a_2 = -0.5$, $b_1 = -1.2$, $b_2 = -1.8$, $\tau_1 = 0.04$, $\tau_2 = 0.05$, $\sigma = 1.7$. Thus, $a < 0$, $B = -1.2 - 1.8 = -3 < 0$, $A = -1.5 - 0.5 = -2 < 0$. Appropriate regions of stability in probability for the equilibrium points $\hat{x}_1$ and $\hat{x}_2$ obtained by conditions (5.79), (5.80), (5.94), and (5.95) are shown in Fig. 5.11.

Example 5.5 Consider the differential equation

$$\dot{x}(t) = -ax(t) + \frac{\mu + a_0x(t) + a_1x(t - \tau_1)}{\lambda + b_0x(t) + b_1x(t - \tau_1)} + \sigma\left(x(t) - \hat{x}\right)\dot{w}(t), \qquad (5.97)$$

which is a particular case of (5.71) with $k = 1$. The linear part of type (5.74) for this equation has the form

$$\dot{z}(t) = \hat{\gamma}_0z(t) + \gamma_1z(t - \tau_1) + \sigma z(t)\dot{w}(t), \qquad (5.98)$$

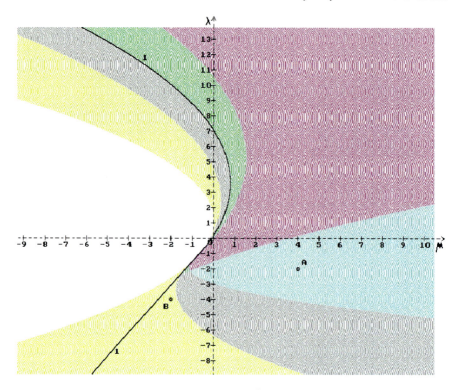

Fig. 5.12 Stability regions for the equilibrium point $\hat{x}_1 = 2.696$ of (5.97) by $a = 1$, $a_0 = -0.4$, $b_0 = 0.2$, $a_1 = 1.5$, $b_1 = 1.5$, $\tau_1 = 0.4$, $\sigma = 1.3$

where

$$\hat{\gamma}_0 = \gamma_0 - a, \qquad \gamma_j = \frac{a_j - ab_j\hat{x}}{\lambda + B\hat{x}}, \quad j = 0, 1, \ B = b_0 + b_1.$$

By Lemma 2.1 a necessary and sufficient condition for asymptotic mean-square stability of the trivial solution of (5.98) is

$$\hat{\gamma}_0 + \gamma_1 < 0, \qquad G^{-1} > p, \tag{5.99}$$

where

$$G = \begin{cases} \frac{\gamma_1 q^{-1} \sin(q\tau) - 1}{\hat{\gamma}_0 + \gamma_1 \cos(q\tau)}, & \gamma_1 + |\hat{\gamma}_0| < 0, \ q = \sqrt{\gamma_1^2 - \hat{\gamma}_0^2}, \\[2mm] \frac{1 + |\hat{\gamma}_0|\tau}{2|\hat{\gamma}_0|}, & \gamma_1 = \hat{\gamma}_0 < 0, \\[2mm] \frac{\gamma_1 q^{-1} \sinh(q\tau) - 1}{\hat{\gamma}_0 + \gamma_1 \cosh(q\tau)}, & \hat{\gamma}_0 + |\gamma_1| < 0, \ q = \sqrt{\hat{\gamma}_0^2 - \gamma_1^2}. \end{cases}$$

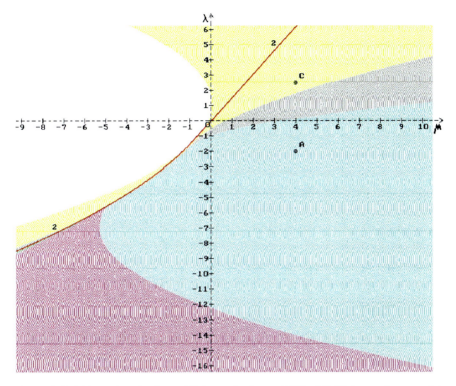

Fig. 5.13 Stability regions for the equilibrium point $\hat{x}_2 = -0.873$ of (5.97) by $a = 1$, $a_0 = -0.4$, $b_0 = 0.2$, $a_1 = 1.5$, $b_1 = 1.5$, $\tau_1 = 0.4$, $\sigma = 1.3$

Note that if, in (5.98), $\tau_1 = 0$, then the sufficient condition (5.76) for asymptotic mean-square stability of the trivial solution of (5.98) takes the form $a > \gamma_0 + \gamma_1 + p$ and coincides with the necessary and sufficient condition (5.99) for asymptotic mean-square stability of the trivial solution of (5.98). Let us show that for small enough delay, the sufficient stability conditions (5.75) and (5.76) together are sufficiently close to the necessary and sufficient stability condition (5.99).

In Fig. 5.12 stability regions for the equilibrium point $\hat{x}_1$ given by condition (5.75) (green and magenta), by condition (5.79) (magenta and cyan), and by condition (5.99) (grey, green, magenta, and cyan) are shown for the following values of the parameters:

$$a = 1, \quad a_0 = -0.4, \quad b_0 = 0.2, \quad a_1 = b_1 = 1.5, \quad \tau_1 = 0.4, \quad \sigma = 1.3.$$
$$(5.100)$$

One can see that both stability conditions (5.75) and (5.79) complement each other and both these conditions together give the region of stability (green, magenta, and cyan) that is sufficiently close to the exact stability region obtained by the necessary and sufficient stability condition (5.99).

In Fig. 5.13 the similar picture for the same values of parameters (5.100) is shown for the equilibrium point $\hat{x}_2$: both stability conditions (5.75) (magenta and green

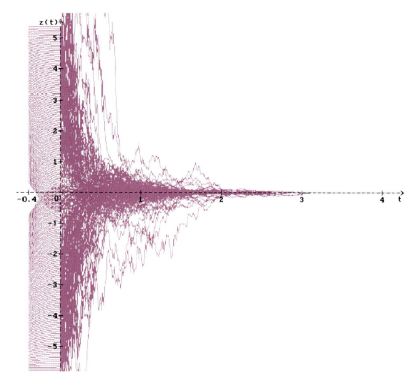

Fig. 5.14   200 trajectories of solutions of (5.98) at the point $A(4, -2)$

(small region placed between magenta and bound 2)) and (5.79) (magenta and cyan) complement each other, and both these conditions together give the region of stability (green, magenta, and cyan) that is sufficiently close to the exact stability region (green, magenta, cyan, and grey) obtained by the necessary and sufficient stability condition (5.99).

Consider the point A with $\mu = 4$ and $\lambda = -2$. This point belongs to stability regions for both equilibrium points $\hat{x}_1 = 2.696$ (Fig. 5.12) and $\hat{x}_2 = -0.873$ (Fig. 5.13). At the point $A(4, -2)$ the trivial solution of (5.98) is asymptotically mean-square stable. Thus, at the point A, the trajectories of all solutions of (5.97) with different given initial functions and the values of the parameters (5.100) converge to zero as $t \to \infty$. 200 such trajectories are shown in Fig. 5.14 by the initial functions $(s \leq 0)$

$$x(s) = \hat{x}_1 + \frac{j}{33} \cos\left(\frac{10}{7} s\right) - 8.5, \quad j = 0, 2, 4, \ldots, 198,$$

$$x(s) = \frac{25}{28} \hat{x}_1 - \frac{j}{33} \cos\left(\frac{10}{7} s\right) + 3, \quad j = 1, 3, 5, \ldots, 199.$$

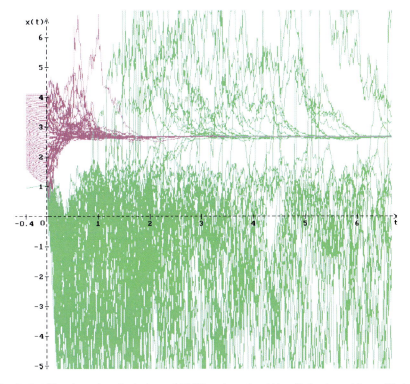

Fig. 5.15 100 trajectories of solutions of (5.97) at the point $A(4, -2)$ for the stable equilibrium point $\hat{x}_1 = 2.696$

In Fig. 5.15 trajectories of solutions of the nonlinear equation (5.97) are shown at the point A for the values of the parameters (5.100). At the point A the equilibrium point $\hat{x}_1 = 2.696$ of (5.97) is stable in probability. Thus, at the point A, 50 trajectories of solutions of (5.97) with the initial functions

$$x(s) = \hat{x}_1 - \frac{2j}{33} \cos\left(\frac{10}{7}s\right) + 1.5, \quad s \le 0, \ j = 1, 2, \ldots, 50,$$

which belong to some neighborhood of the equilibrium point $\hat{x}_1$, converge to $\hat{x}_1$ as $t \to \infty$ (magenta trajectories), but other 50 trajectories of solution with one initial function

$$x(s) = \hat{x}_1 + \frac{6}{11} \cos\left(\frac{10}{7}s\right) - 2.2, \quad s \le 0,$$

which is placed out of the neighborhood of $\hat{x}_1$, fill the whole space (green trajectories). Only some of these trajectories, which come to a neighborhood of $\hat{x}_1$, converge to $\hat{x}_1$ as $t \to \infty$.

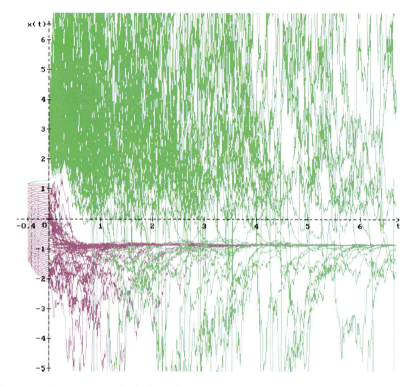

Fig. 5.16 100 trajectories of solutions of (5.97) at the point $A(4, -2)$ for the stable equilibrium point $\hat{x}_2 = -0.873$

Figure 5.16 is similar to Fig. 5.15, but it shows 100 trajectories for the equilibrium point $\hat{x}_2 = -0.873$: 50 trajectories (magenta) with the initial functions

$$x(s) = \hat{x}_2 - \frac{j}{15} \cos\left(\frac{5}{3}s\right) + 2.1, \quad s \leq 0, \ j = 1, 2, \ldots, 50,$$

which belong to a small enough neighborhood of the equilibrium point $\hat{x}_2$, converge to this equilibrium, and 50 trajectories (green) with one initial function

$$x(s) = \hat{x}_2 + \frac{4}{11} \cos(2s) + 1.8, \quad s \leq 0,$$

which is placed out of this neighborhood of $\hat{x}_2$, fill the whole space.

Consider now the point B with $\mu = -2$ and $\lambda = -4$ (Fig. 5.12). This point does not belong to stability region for the equilibrium point $\hat{x}_1 = 2.536$, and thus, at the point $B(-2, -4)$ the equilibrium point $\hat{x}_1$ is unstable. In Fig. 5.17 five hundred trajectories of the solution of (5.97) are shown with the initial function

$$x(s) = \hat{x}_1 + 0.015 \sin\left(\frac{10}{3}s\right), \quad s \leq 0,$$

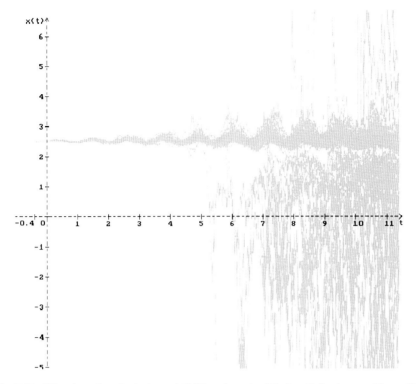

Fig. 5.17 500 trajectories of solutions of (5.97) at the point $B(-2, -4)$ for the unstable equilibrium point $\hat{x}_1 = 2.536$

which is placed close enough to the equilibrium point $\hat{x}_1$. One can see that the trajectories do not converge to $\hat{x}_1$ and fill the whole space.

In Fig. 5.18 a similar picture is shown for the unstable equilibrium point $\hat{x}_2 = -2$ at the point C with $\mu = 4$ and $\lambda = 2.5$ (Fig. 5.13) with the initial function

$$x(s) = \hat{x}_2 - 0.025 \cos\left(\frac{5}{3}s\right), \quad s \le 0.$$

Note that simulations of the solutions of (5.97) and (5.98) were obtained via its difference analogues respectively in the forms

$$x_{i+1} = (1 - a\Delta)x_i + \frac{\mu + a_0 x_i + a_1 x_{i-m}}{\lambda + b_0 x_i + b_1 x_{i-m}}\Delta + \sigma(x_i - \hat{x})(w_{i+1} - w_i)$$

and

$$z_{i+1} = (1 + \hat{\gamma}_0 \Delta)z_i + \gamma_1 \Delta z_{i-m} + \sigma z_i (w_{i+1} - w_i).$$

Here Δ is the step of discretization (that was chosen as $\Delta = 0.01$), $x_i = x(i\Delta)$, $z_i = z(i\Delta)$, $w_i = w(i\Delta)$, $m = \tau_1/\Delta$, and the trajectories of the Wiener process are simulated by the algorithm described in Sect. 2.1.1.

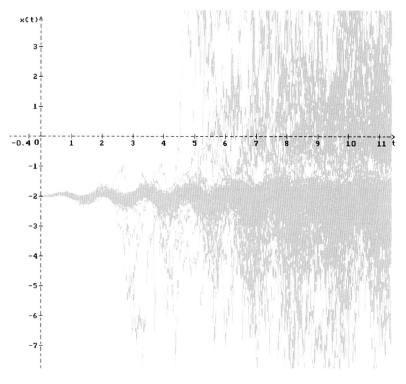

Fig. 5.18 500 trajectories of solutions of (5.97) at the point $C(4, 2.5)$ for the unstable equilibrium point $\hat{x}_2 = -2$

Chapter 6
Matrix Riccati Equations in Stability of Linear Stochastic Differential Equations with Delays

Here asymptotic mean-square stability conditions are obtained in terms of the existence of positive definite solutions of some matrix Riccati equations. Using the procedure of constructing Lyapunov functionals, we will obtain different matrix Riccati equations for one stochastic differential equation with delays. If a considered differential equation does not contain delays, then all these matrix Riccati equations coincide with a unique linear matrix equation.

Consider first the stochastic linear differential equation without delay

$$dx(t) = Ax(t)\,dt + Cx(t)\,dw(t). \tag{6.1}$$

Here A and C are constant $n \times n$-matrices, $x(t) \in \mathbf{R}^n$, and $w(t)$ is the scalar standard Wiener process.

Similarly to Theorem 1.3, the necessary and sufficient condition for asymptotic mean-square stability of the zero solution of (6.1) can be formulated in terms of the existence of a positive definite solution P of the linear matrix equation

$$A'P + PA + C'PC = -Q \tag{6.2}$$

for any positive definite matrix Q. But for differential equations with delays, the situation is more complicated.

6.1 Equations with Constant Delays

6.1.1 One Delay in Deterministic Part and One Delay in Stochastic Part of Equation

Below two different ways of Lyapunov functionals construction are considered.

L. Shaikhet, *Lyapunov Functionals and Stability of Stochastic Functional Differential Equations*, DOI 10.1007/978-3-319-00101-2_6,
© Springer International Publishing Switzerland 2013

6.1.1.1 The First Way of Constructing a Lyapunov Functional

Consider the stochastic linear differential equation

$$dx(t) = (Ax(t) + Bx(t - h)) dt + Cx(t - \tau) dw(t), \quad t \geq 0,$$
$$x(s) = \phi(s), \quad s \leq 0. \tag{6.3}$$

Formal application of the quadratic Lyapunov functional for (6.3) leads to a system of matrix ordinary and partial differential equations [134]. The approach, used in this chapter, gives Lyapunov functionals defined by the solutions of some nonlinear (Riccati-type) matrix equations.

Let L be the generator of (6.3). Consider the functional $V_1(x) = x'Px$. Calculating LV_1, we get

$$LV_1 = (Ax(t) + Bx(t - h))' Px(t) + x'(t)P(Ax(t) + Bx(t - h))$$
$$+ x'(t - \tau)C'PCx(t - \tau)$$
$$= x'(t)(A'P + PA)x(t) + x'(t - \tau)C'PCx(t - \tau)$$
$$+ x'(t - h)B'Px(t) + x'(t)PB(t - h).$$

Using Lemma 2.3 for $a = x(t - h)$, $b = B'Px(t)$, we have

$$LV_1 \leq x'(t)(A'P + PA + PBR^{-1}B'P)x(t)$$
$$+ x'(t - h)Rx(t - h) + x'(t - \tau)C'PCx(t - \tau), \tag{6.4}$$

which is the representation of type (2.35), where $D(t) = A'P + PA + PBR^{-1}B'P$, $k = 2$, $Q_1(t) = R$, $Q_2(t) = C'PC$, $\tau_1(t) = h$, $\tau_2(t) = \tau$, and all other parameters are zeros.

Thus, by Theorem 2.5 we obtain the following theorem.

Theorem 6.1 *Suppose that for some positive definite matrices Q and R, there exists a positive definite solution P of the matrix Riccati equation*

$$A'P + PA + C'PC + R + PBR^{-1}B'P = -Q. \tag{6.5}$$

Then the trivial solution of (6.3) is asymptotically mean-square stable.

Remark 6.1 Using Lemma 2.3 for $a = x(t)$ and $b = PBx(t - h)$, instead of (6.4), we obtain the inequality

$$LV_1 \leq x'(t)(A'P + PA + R)x(t)$$
$$+ x'(t - h)B'PR^{-1}PBx(t - h) + x'(t - \tau)C'PCx(t - \tau).$$

Thus, in Theorem 6.1, instead of (6.5), we can use the equation

$$A'P + PA + C'PC + R + B'PR^{-1}PB = -Q. \tag{6.6}$$

Example 6.1 In the scalar case a positive solution of (6.5) (or (6.6)) exists if and only if

$$A + |B| + \frac{1}{2}C^2 < 0.$$

Example 6.2 Let us show that using in the matrix Riccati equations (6.5) and (6.6) different positive definite matrices Q and R we can get positive definite solutions P by different conditions on the parameters of the considered equation. Consider, for instance, the two-dimensional equation (6.3) with

$$A = \begin{pmatrix} -a_1 & 0 \\ 0 & -a_2 \end{pmatrix}, \qquad B = \begin{pmatrix} b_1 & b_2 \\ -b_2 & b_1 \end{pmatrix}, \qquad C = \begin{pmatrix} c_1 & 0 \\ 0 & c_2 \end{pmatrix}.$$

Let I be the 2×2 identity matrix, $Q = qI$, $R = rI$, $q, r > 0$. It is easy to get that a solution P of the matrix Riccati equation (6.5) with the elements $p_{11} > 0$, $p_{22} > 0$, $p_{12} = 0$ is defined by the equation

$$\frac{1}{r}\left(b_1^2 + b_2^2\right)p_{ii}^2 - 2\left(a_i - \frac{1}{2}c_i^2\right)p_{ii} + r + q = 0, \quad i = 1, 2.$$

This equation has a positive root p_{ii} for arbitrary positive q and r if and only if

$$a_i > \sqrt{b_1^2 + b_2^2} + \frac{1}{2}c_i^2, \quad i = 1, 2.$$

Let us obtain from the matrix Riccati equation (6.6) a positive definite solution P if the obtained condition does not hold. Put, for instance, $Q = I$, $R = P$,

$$b_1 = 0, \quad a_1 = \frac{3}{2} + \frac{1}{2}c_1^2 > \sqrt{b_1^2 + b_2^2} + \frac{1}{2}c_1^2 = 1 + \frac{1}{2}c_1^2,$$

$$b_2 = 1, \quad a_2 = \frac{7}{8} + \frac{1}{2}c_2^2 < \sqrt{b_1^2 + b_2^2} + \frac{1}{2}c_2^2 = 1 + \frac{1}{2}c_2^2.$$

It is easy to check that in this case the matrix Riccati equation (6.6) has a positive definite solution P with the elements $p_{11} = 3.5$, $p_{22} = 6$, $p_{12} = 0$.

6.1.1.2 The Second Way of Constructing a Lyapunov Functional

Consider now another representation of the initial equation leading to other matrix Riccati equations.

Let $\|B\|$ be the operator norm of a matrix B. Rewrite now (6.3) as follows:

$$\dot{z}(t) = (A + B)x(t) + C x(t - \tau)\dot{w}(t),$$

$$z(t) = x(t) + B \int_{t-h}^{t} x(s)\,ds, \tag{6.7}$$

and following the procedure of constructing Lyapunov functionals, consider the functional $V_1 = z'(t) P z(t)$. Calculating $L V_1$, by (6.7) we get

$$
\begin{aligned}
L V_1 &= z'(t) P (A + B) x(t) + x'(t)(A + B)' P z(t) + x'(t - \tau) C' P C x(t - \tau) \\
&= x'(t) \big[P(A + B) + (A + B)' P \big] x(t) + x'(t - \tau) C' P C x(t - \tau) \\
&\quad + \int_{t-h}^{t} \big[x'(t)(A + B)' P B x(s) + x'(s) B' P(A + B) x(t) \big] ds.
\end{aligned}
$$

Using Lemma 2.3 for $a = x(s)$ and $b = B' P(A + B) x(t)$, we have

$$
\int_{t-h}^{t} \big[x'(t)(A + B)' P B x(s) + x'(s) B' P(A + B) x(t) \big] ds
$$

$$
\leq h x' t (A + B)' P B R^{-1} B' P (A + B) x(t) + \int_{t-h}^{t} x'(s) R x(s) \, ds. \quad (6.8)
$$

Thus,

$$
\begin{aligned}
L V_1 &\leq x'(t) \big[(A + B)' P + P(A + B) + h(A + B)' P B R^{-1} B' P(A + B) \big] x(t) \\
&\quad + \int_{t-h}^{t} x'(s) R x(s) \, ds + x'(t - \tau) C' P C x(t - \tau).
\end{aligned}
$$

So, we get the representation of type (2.35), where

$$
D(t) = (A + B)' P + P(A + B) + h(A + B)' P B R^{-1} B' P(A + B),
$$
$$
k = 1, \qquad Q_1(t) = C' P C, \qquad \tau_1(t) = \tau,
$$
$$
m = 0, \qquad d\mu_0(s) = \delta(s - h) \, ds, \qquad R_0 = R,
$$

and all other parameters are zeros. By Theorem 2.5 we obtain the following:

Theorem 6.2 *Suppose that the inequality $h \| B \| < 1$ holds and that, for some symmetric matrices $Q > 0$ and $R > 0$, there exists a positive definite solution P of the matrix Riccati equation*

$$
(A + B)' P + P(A + B) + C' P C
$$
$$
+ h \big[R + (A + B)' P B R^{-1} B' P(A + B) \big] = -Q. \quad (6.9)
$$

Then the trivial solution of (6.3) is asymptotically mean-square stable.

Remark 6.2 Using Lemma 2.3 for $a = (A + B) x(t)$ and $b = P B x(s)$, instead of (6.8), we obtain the inequality

$$\int_{t-h}^{t} \left[x'(t)(A+B)'PBx(s) + x'(s)B'P(A+B)x(t) \right] ds$$

$$\leq hx'(t)(A+B)'R(A+B)x(t) + \int_{t-h}^{t} x'(s)B'PR^{-1}PBx(s)\,ds.$$

So, in Theorem 6.2, in place of (6.9), we can use the equation

$$(A+B)'P + P(A+B) + C'PC$$

$$+ h\left[(A+B)'R(A+B) + B'PR^{-1}PB \right] = -Q. \qquad (6.10)$$

Example 6.3 In the scalar case a positive solution of (6.9) (or (6.10)) for arbitrary positive Q and R exists if and only if

$$(A+B)\left(1 - h|B|\right) + \frac{1}{2}c^2 < 0, \quad h|B| < 1.$$

6.1.2 Several Delays in Deterministic Part of Equation

Consider now the more general linear stochastic differential equation

$$\dot{x}(t) = Ax(t) + \sum_{i=1}^{m} B_i x(t - h_i) + Cx(t - \tau)\dot{w}(t). \qquad (6.11)$$

To construct a Lyapunov functional for (6.11), let us use both previous representations of the initial equation. Namely, represent (6.11) in the form

$$\dot{z}(t) = (A+B)x(t) + \sum_{i=m_0+1}^{m} B_i x(t - h_i) + Cx(t - \tau)\dot{w}(t),$$

$$z(t) = x(t) + \sum_{j=1}^{m_0} B_j \int_{t-h_j}^{t} x(s)\,ds, \qquad B = \sum_{j=1}^{m_0} B_j, \quad 0 \leq m_0 \leq m.$$

Following the procedure of constructing Lyapunov functionals, put $V_1 = z'(t)Pz(t)$. In this case,

$$LV_1 = \left((A+B)x(t) + \sum_{i=m_0+1}^{m} B_i x(t - h_i) \right)' Pz(t)$$

$$+ z'(t)P\left((A+B)x(t) + \sum_{i=m_0+1}^{m} B_i x(t - h_i) \right) + x'(t-\tau)C'PCx(t-\tau)$$

$$= x'(t)\left[(A+B)'P + P(A+B) \right]x(t) + x'(t-\tau)C'PCx(t-\tau)$$

$$+ \sum_{i=m_0+1}^{m} \left[x'(t-h_i)B_i'Px(t) + x'(t)PB_ix(t-h_i) \right]$$

$$+ \sum_{j=1}^{m_0} \int_{t-h_j}^{t} \left[x'(t)(A+B)'PB_jx(s) + x'(s)B_j'P(A+B)x(t) \right] ds$$

$$+ \sum_{i=m_0+1}^{m} \sum_{j=1}^{m_0} \int_{t-h_j}^{t} \left[x'(t-h_j)B_i'PB_jx(s) + x'(s)B_j'PB_ix(t-h_i) \right] ds.$$

Using Lemma 2.3, we obtain that for some $R_i > 0$, $i = m_0 + 1, \ldots, m$,

$$x'(t-h_i)B_i'Px(t) + x'(t)PB_ix(t-h_i)$$
$$\leq x'(t)PB_iR_i^{-1}B_i'Px(t) + x'(t-h_i)R_ix(t-h_i), \tag{6.12}$$

for some $G_j > 0$, $j = 1, \ldots, m_0$,

$$x'(t)(A+B)'PB_jx(s) + x'(s)B_j'P(A+B)x(t)$$
$$\leq x'(t)(A+B)'PB_jG_j^{-1}B_j'P(A+B)x(t) + x'(s)G_jx(s), \tag{6.13}$$

and, for some $S_{ij} > 0$, $i = m_0 + 1, \ldots, m$, $j = 1, \ldots, m_0$,

$$x'(t-h_i)B_i'PB_jx(s) + x'(s)B_j'PB_ix(t-h_i)$$
$$\leq x'(t-h_i)B_i'PB_jS_{ij}^{-1}B_j'PB_i'x(t-h_i) + x'(s)S_{ij}x(s). \tag{6.14}$$

By inequalities (6.12)–(6.14) we have

$$LV_1 \leq x'(t)\left[(A+B)'P + P(A+B) + \sum_{i=m_0+1}^{m} PB_iR_i^{-1}B_i'P \right.$$

$$+ \sum_{j=1}^{m} h_j(A+B)'PB_jG_j^{-1}B_j'P(A+B) \bigg] x(t) + x'(t-\tau)C'PCx(t-\tau)$$

$$+ \sum_{i=m_0+1}^{m} x'(t-h_i)R_ix(t-h_i) + \sum_{j=1}^{m_0} \int_{t-h_j}^{t} x'(s)G_jx(s)\,ds$$

$$+ \sum_{i=m_0+1}^{m} \sum_{j=1}^{m_0}\left[h_jx'(t-h_i)B_i'PB_jS_{ij}^{-1}B_j'PB_ix(t-h_i) \right.$$

$$+ \int_{t-h_j}^{t} x'(s)S_{ij}x(s)\,ds \bigg].$$

Using the representation (2.35), similarly to Theorem 6.2, we obtain the following theorem.

Theorem 6.3 *Suppose that, for some* $m_0 = 0, 1, \ldots, m$, *the inequality* $\sum_{j=1}^{m_0} h_j \|B_j\|$
< 1 *holds and that, for some matrices* $R_i > 0$, $G_j > 0$, $S_{ij} > 0$, *and* $Q > 0$, *there
exists a positive definite solution* P *of the matrix Riccati equation*

$$
(A + B)' P + P(A + B) + C' PC + \sum_{i=m_0+1}^{m} \left(R_i + P B_i R_i^{-1} B_i' P \right)
$$

$$
+ \sum_{j=1}^{m_0} h_j \left[G_j + (A + B)' P B_j G_j^{-1} B_j' P (A + B) \right]
$$

$$
+ \sum_{i=m_0+1}^{m} \sum_{j=1}^{m_0} h_j \left(S_{ij} + B_i' P B_j S_{ij}^{-1} B_j' P B_i \right) = -Q, \tag{6.15}
$$

where

$$
B = \sum_{j=1}^{m_0} B_j, \quad 0 \le m_0 \le m.
$$

Then the trivial solution of (6.11) *is asymptotically mean-square stable.*

Remark 6.3 Similarly to inequalities (6.12)–(6.14), we can use, for instance, the
inequalities

$$
x'(t - h_i) B_i' Px(t) + x'(t) P B_i x(t - h_i)
$$
$$
\le x'(t - h_i) B_i' P R_i^{-1} P B_i x(t - h_i) + x'(t) R_i x(t), \tag{6.16}
$$
$$
x'(t)(A + B)' P B_j x(s) + x'(s) B_j' P (A + B)x(t)
$$
$$
\le x'(s) B_j' P (A + B) G_j^{-1} (A + B)' P B_j x(s) + x'(t) G_j x(t), \tag{6.17}
$$
$$
x'(t - h_i) B_i' P B_j x(s) + x'(s) B_j' P B_i x(t - h_i)
$$
$$
\le x'(s) B_j' P B_i S_{ij}^{-1} B_i' P B_j x(s) + x'(t - h_i) S_{ij} x(t - h_i). \tag{6.18}
$$

Using different combinations of inequalities (6.12)–(6.14) and (6.16)–(6.18) and
different representations of type (2.35) in place of (6.15), we can use one from
other eight different matrix Riccati equations. For example, using in place of all
inequalities (6.12)–(6.14) inequalities (6.16)–(6.18) we obtain the equation

$$
(A + B)' P + P(A + B) + C' PC + \sum_{i=m_0+1}^{m} \left(R_i + B_i' P \tilde{R}_i^{-1} P B_i \right)
$$

$$
+ \sum_{j=1}^{m_0} h_j \left[G_j + B_j' P (A + B) G_j^{-1} (A + B)' P B_j \right]
$$

$$+ \sum_{i=m_0+1}^{m} \sum_{j=1}^{m_0} h_j \left(S_{ij} + B'_j P B_i S_{ij}^{-1} B'_i P B_j \right) = -Q.$$

Using by virtue of Lemma 2.3 other inequalities of type (6.12)–(6.14) or (6.16)–(6.18), we will obtain other matrix Riccati equations for the initial equation (6.11).

Remark 6.4 It is easy to see that representing (6.11) in the form

$$\dot{z}(t) = (A + B)x(t) + \sum_{i=1}^{m_0} B_i x(t - h_i) + Cx(t - \tau)\dot{w}(t),$$

$$z(t) = x(t) + \sum_{j=m_0+1}^{m} B_j \int_{t-h_j}^{t} x(s)ds, \qquad B = \sum_{j=m_0+1}^{m} B_j, \quad 0 \le m_0 \le m,$$

we obtain a new modification of Theorem 6.3 after replacement of all sums $\sum_{j=1}^{m_0}$ by $\sum_{j=m_0+1}^{m}$ and of all sums $\sum_{i=m_0+1}^{m}$ by $\sum_{i=1}^{m_0}$. Using other combinations of the summands for representations of (6.11), we will obtain other modifications of Theorem 6.3.

Example 6.4 Consider the scalar case of (6.11) with $m = 2$. Using Theorem 6.3 and Remark 6.4 for different values of $m_0 = 0, 1, 2$, we obtain four different sufficient conditions for asymptotic mean-square stability of the trivial solution of (6.11):

$$A + |B_1| + |B_2| + \frac{1}{2}C^2 < 0,$$

$$(A + B_1)(1 - h_1|B_1|) + |B_2|(1 + h_1|B_1|) + \frac{1}{2}C^2 < 0, \quad h_1|B_1| < 1,$$

$$(A + B_2)(1 - h_2|B_2|) + |B_1|(1 + h_2|B_2|) + \frac{1}{2}C^2 < 0, \quad h_2|B_2| < 1,$$

$$(A + B_1 + B_2)(1 - h_1|B_1| - h_2|B_2|) + \frac{1}{2}C^2 < 0, \quad h_1|B_1| + h_2|B_2| < 1.$$

6.2 Distributed Delay

Consider now the stochastic linear differential equation with distributed delay

$$\dot{x}(t) = Ax(t) + \sum_{i=1}^{m} B_i \int_{t-h_i}^{t} x(s)\,ds + C \int_{t-\tau}^{t} x(s)\,ds\,\dot{w}(t). \qquad (6.19)$$

Rewrite (6.19) in the form

$$\dot{z}(t) = \left(A + (B, h)\right)x(t) + \sum_{i=m_0+1}^{m} B_i \int_{t-h_i}^{t} x(s)\,ds + C \int_{t-\tau}^{t} x(s)\,ds\,\dot{w}(t),$$

$$z(t) = x(t) + \sum_{j=1}^{m_0} B_j \int_{t-h_j}^{t} (s - t + h_j) x(s) \, ds,$$

$$(B, h) = \sum_{j=1}^{m_0} B_j h_j, \quad 0 \le m_0 \le m.$$

In this case, for the functional $V_1 = z'(t) P z(t)$, we have

$$L V_1 = \left(\left(A + (B, h) \right) x(t) + \sum_{i=m_0+1}^{m} B_i \int_{t-h_i}^{t} x(s) \, ds \right)' P z(t)$$

$$+ z'(t) P \left(\left(A + (B, h) \right) x(t) + \sum_{i=m_0+1}^{m} B_i \int_{t-h_i}^{t} x(s) \, ds \right)$$

$$+ \int_{t-\tau}^{t} x'(s) \, ds \, C' P C \int_{t-\tau}^{t} x(\theta) \, d\theta$$

$$= x'(t) \left[\left(A + (B, h) \right)' P + P \left(A + (B, h) \right) \right] x(t)$$

$$+ \int_{t-\tau}^{t} x'(s) \, ds \, C' P C \int_{t-\tau}^{t} x(\theta) \, d\theta$$

$$+ \sum_{i=m_0+1}^{m} \int_{t-h_i}^{t} \left[x'(s) B_i' P x(t) + x'(t) P B_i x(s) \right] ds$$

$$+ \sum_{j=1}^{m_0} \int_{t-h_j}^{t} (s - t + h_j) \left[x'(t) \left(A + (B, h) \right)' P B_j x(s) \right.$$

$$+ x'(s) B_j' P \left(A + (B, h) \right) x(t) \right] ds$$

$$+ \sum_{i=m_0+1}^{m} \sum_{j=1}^{m_0} \int_{t-h_i}^{t} \int_{t-h_j}^{t} (s - t + h_j) \left[x'(\theta) B_i' P B_j x(s) \right.$$

$$+ x'(s) B_j' P B_i x(\theta) \right] ds \, d\theta.$$

Note that using Lemma 2.3 for $a = x(s)$, $b = C' P C x(\theta)$, and $R > 0$, we have

$$\int_{t-\tau}^{t} x'(s) \, ds \, C' P C \int_{t-\tau}^{t} x(\theta) \, d\theta$$

$$= \frac{1}{2} \int_{t-\tau}^{t} \int_{t-\tau}^{t} \left[x'(\theta) C' P C x(s) + x'(s) C' P C x(\theta) \right] ds \, d\theta$$

$$\le \frac{\tau}{2} \int_{t-\tau}^{t} x'(s) R x(s) \, ds + \frac{\tau}{2} \int_{t-\tau}^{t} x'(\theta) C' P C R^{-1} C' P C x(\theta) \, d\theta.$$

Using inequalities similar to (6.12)–(6.14), we obtain

$$
LV_1 \leq x'(t) \Bigg[\big(A + (B, h)\big)' P + P\big(A + (B, h)\big) + \sum_{i=m_0+1}^{m} h_i \, P B_i \, R_i^{-1} B_i' P
$$

$$
+ \frac{1}{2} \sum_{j=1}^{m_0} h_j^2 \big(A + (B, h)\big)' P B_j G_j^{-1} B_j' P \big(A + (B, h)\big) \Bigg] x(t)
$$

$$
+ \frac{\tau}{2} \int_{t-\tau}^{t} x'(s) \big(R + C' P C R^{-1} C' P C \big) x(s) \, ds
$$

$$
+ \sum_{i=m_0+1}^{m} \int_{t-h_i}^{t} x'(s) R_i x(s) \, ds + \sum_{j=1}^{m_0} \int_{t-h_j}^{t} (s - t + h_j) x'(s) G_j x(s) \, ds
$$

$$
+ \sum_{i=m_0+1}^{m} \sum_{j=1}^{m_0} \Bigg[\frac{1}{2} h_j^2 \int_{t-h_i}^{t} x'(s) B_i' P B_j S_{ij}^{-1} B_j' P B_i x(s) \, ds
$$

$$
+ \int_{t-h_j}^{t} (s - t + h_j) x'(s) S_{ij} x(s) \, ds \Bigg].
$$

By Theorem 2.5 we obtain the following:

Theorem 6.4 *Suppose that, for some $m_0 = 0, 1, \ldots, m$, the inequality $\sum_{j=1}^{m_0} h_j^2 \|B_j\|$ < 2 holds and that, for some matrices $R > 0$, $R_i > 0$, $G_j > 0$, $S_{ij} > 0$, and $Q > 0$, there exists a positive definite solution P of the matrix Riccati equation*

$$
\big(A + (B, h)\big)' P + P\big(A + (B, h)\big) + \frac{\tau^2}{2} \big(R + C' P C R^{-1} C' P C \big)
$$

$$
+ \sum_{i=m_0+1}^{m} h_i \big(R_i + P B_i R_i^{-1} B_i' P \big)
$$

$$
+ \frac{1}{2} \sum_{j=1}^{m_0} h_j^2 \big(G_j + \big(A + (B, h)\big)' P B_j G_j^{-1} B_j' P \big(A + (B, h)\big) \big)
$$

$$
+ \frac{1}{2} \sum_{i=m_0+1}^{m} h_i \sum_{j=1}^{m_0} h_j^2 \big(S_{ij} + B_i' P B_j S_{ij}^{-1} B_j' P B_i \big) = -Q. \tag{6.20}
$$

Then the trivial solution of (6.19) is asymptotically mean-square stable.

Remark 6.5 Using the arguments from Remark 6.3, one can show that in place of (6.20) in Theorem 6.4 other matrix Riccati equations can be used, for instance,

$$
\big(A + (B, h)\big)' P + P\big(A + (B, h)\big) + \frac{\tau^2}{2} \big(R + C' P C R^{-1} C' P C \big)
$$

$$+ \sum_{i=m_0+1}^{m} h_i \left(R_i + B_i' P R_i^{-1} P B_i \right)$$

$$+ \frac{1}{2} \sum_{j=1}^{m_0} h_j^2 \left(G_j + B_j' P \left(A + (B,h) \right) G_j^{-1} \left(A + (B,h) \right)' P B_j \right)$$

$$+ \frac{1}{2} \sum_{i=m_0+1}^{m} h_i \sum_{j=1}^{m_0} h_j^2 \left(S_{ij} + B_j' P B_i S_{ij}^{-1} B_i' P B_j \right) = -Q.$$

Remark 6.6 Using another representation of (6.19) in the form

$$\dot{z}(t) = \left(A + (B,h)_1 \right) x(t) + \sum_{i=1}^{m_0} B_i \int_{t-h_i}^{t} x(s)\,ds + C \int_{t-\tau}^{t} x(s)\,ds\,\dot{w}(t),$$

$$z(t) = x(t) + \sum_{j=m_0+1}^{m} B_j \int_{t-h_j}^{t} (s - t + h_j) x(s)\,ds,$$

$$(B,h)_1 = \sum_{j=m_0+1}^{m} B_j h_j, \quad 0 \le m_0 \le m,$$

we obtain a new modification of Theorem 6.4 after replacement of all sums $\sum_{j=1}^{m_0}$ by $\sum_{j=m_0+1}^{m}$ and all sums $\sum_{i=m_0+1}^{m}$ by $\sum_{i=1}^{m_0}$. Using other combinations of the summands for representations of (6.19), we will obtain other modifications of Theorem 6.4.

Example 6.5 Consider the scalar case of (6.19) for $m = 2$. Using Theorem 6.4 and Remark 6.6 for different values of $m_0 = 0, 1, 2$, we obtain four different sufficient conditions for asymptotic mean-square stability of the trivial solution of (6.19):

$$A + h_1 |B_1| + h_2 |B_2| + \frac{\tau^2}{2} C^2 < 0,$$

$$(A + h_1 B_1) \left(1 - \frac{1}{2} h_1^2 |B_1| \right) + h_2 |B_2| \left(1 + \frac{1}{2} h_1^2 |B_1| \right) + \frac{\tau^2}{2} C^2 < 0,$$

$$h_1^2 |B_1| < 2,$$

$$(A + h_2 B_2) \left(1 - \frac{1}{2} h_2^2 |B_2| \right) + h_1 |B_1| \left(1 + \frac{1}{2} h_2^2 |B_2| \right) + \frac{\tau^2}{2} C^2 < 0,$$

$$h_2^2 |B_2| < 2,$$

$$(A + h_1 B_1 + h_2 B_2) \left(1 - \frac{1}{2} h_1^2 |B_1| - \frac{1}{2} h_2^2 |B_2| \right) + \frac{\tau^2}{2} C^2 < 0,$$

$$h_1^2 |B_1| + h_2^2 |B_2| < 2.$$

6.3 Combination of Discrete and Distributed Delays

Similar stability conditions can be obtained for systems with different combinations of discrete and distributed delays. For example, consider the linear stochastic differential equation

$$\dot{x}(t) = Ax(t) + B_1 x(t - h_1) + B_2 \int_{t-h_2}^{t} x(s)\, ds$$

$$+ \left(C_1 x(t - \tau_1) + C_2 \int_{t-\tau_2}^{t} x(s)\, ds \right) \dot{w}(t). \qquad (6.21)$$

Using four different representations of this equation, we will construct four different Lyapunov functionals for it.

First, consider (6.21). Using the functional $V_1 = x'(t) P x(t)$, we have

$$LV_1 = \left(Ax(t) + B_1 x(t - h_1) + B_2 \int_{t-h_2}^{t} x(s)\, ds \right)' P x(t)$$

$$+ x'(t) P \left(Ax(t) + B_1 x(t - h_1) + B_2 \int_{t-h_2}^{t} x(s)\, ds \right)$$

$$+ \left(C_1 x(t - \tau_1) + C_2 \int_{t-\tau_2}^{t} x(s)\, ds \right)' P \left(C_1 x(t - \tau_1) + C_2 \int_{t-\tau_2}^{t} x(s)\, ds \right)$$

$$= x(t) \left(A'P + PA \right) x(t) + x'(t - \tau_1) C_1' P C_1 x(t - \tau_1)$$

$$+ x'(t) P B_1 x(t - h_1) + x'(t - h_1) B_1' P x(t)$$

$$+ \int_{t-h_2}^{t} \left(x'(t) P B_2 x(s) + x'(s) B_2' P x(t) \right) ds$$

$$+ \int_{t-\tau_2}^{t} \left(x'(t - \tau_1) C_1' P C_2 x(s) + x'(s) C_2' P C_1 x(t - \tau_1) \right) ds$$

$$+ \frac{1}{2} \int_{t-\tau_2}^{t} \int_{t-\tau_2}^{t} \left(x'(\theta) C_2' P C_2 x(s) + x'(s) C_2' P C_2 x(\theta) \right) ds\, d\theta.$$

Using Lemma 2.3 for some positive definite matrices R_i, G_i, $i = 1, 2$, we obtain

$$LV_1 \leq x(t) \left(A'P + PA + R_1 + h_2 R_2 \right) x(t) + x'(t - \tau_1) C_1' P C_1 x(t - \tau_1)$$

$$+ x'(t - h_1) B_1' P R_1^{-1} P B_1 x(t - h_1) + \tau_2 x'(t - \tau_1) G_1 x(t - \tau_1)$$

$$+ \int_{t-h_2}^{t} x'(s) B_2' P R_2^{-1} P B_2 x(s)\, ds + \int_{t-\tau_2}^{t} x'(s) C_2' P C_1 G_1^{-1} C_1' P C_2 x(s)\, ds$$

$$+ \frac{\tau_2}{2} \int_{t-\tau_2}^{t} x'(s) \left(G_2 + C_2' P C_2 G_2^{-1} C_2' P C_2 \right) x(s)\, ds.$$

By the representation of type (2.35) with $S_i(t) = 0$, $dK_l = 0$,

$$D = A'P + PA + R_1 + h_2 R_2, \qquad k = 2, \qquad m = 0,$$

$$Q_1 = C_1'PC_1 + \tau_2 G_1, \qquad \tau_1(t) = \tau_1,$$

$$Q_2 = B_1'PR_1^{-1}PB_1 + \tau_2 G_1, \qquad \tau_2(t) = h_1,$$

$$d\mu(s) = \delta(s - h_2)\,ds, \qquad R(\tau + h_2, t) = B_2'PR_2^{-1}PB_2,$$

$$d\mu_0(s) = \delta(s - \tau_2)\,ds,$$

$$R_0 = C_2'PC_1 G_1^{-1}C_1'PC_2 + \frac{1}{2}\tau_2\big(G_2 + C_2'PC_2 G_2^{-1}C_2'PC_2\big),$$

and Theorem 2.5, we obtain the following theorem.

Theorem 6.5 *Let for some matrices $R_i > 0$, $G_i > 0$, $i = 1, 2$, and $Q > 0$, there exist a positive definite solution P of the matrix Riccati equation*

$$A'P + PA + C_1'PC_1 + R_1 + B_1'PR_1^{-1}PB_1 + h_2\big(R_2 + B_2'PR_2^{-1}PB_2\big)$$

$$+ \tau_2\big(G_1 + C_2'PC_1 G_1^{-1}C_1'PC_2\big) + \frac{\tau_2^2}{2}\big(G_2 + C_2'PC_2 G_2^{-1}C_2'PC_2\big) = -Q.$$

$$(6.22)$$

Then the trivial solution of (6.21) is asymptotically mean-square stable.

Remark 6.7 Similarly to Remark 6.3, one can show that in place of (6.22) in Theorem 6.5 one from the eight different matrix Riccati equations can be used, for example,

$$A'P + PA + C_1'PC_1 + R_1 + PB_1 R_1^{-1}B_1'P + h_2\big(R_2 + PB_2 R_2^{-1}B_2'P\big)$$

$$+ \tau_2\big(G_1 + C_1'PC_2 G_1^{-1}C_2'PC_1\big) + \frac{\tau_2^2}{2}\big(G_2 + C_2'PC_2 G_2^{-1}C_2'PC_2\big) = -Q.$$

Rewrite now (6.21) in the form

$$\dot{z}(t) = (A + B_1)x(t) + B_2 \int_{t-h_2}^{t} x(s)\,ds$$

$$+ \left(C_1 x(t - \tau_1) + C_2 \int_{t-\tau_2}^{t} x(s)\,ds\right)\dot{w}(t),$$

$$z(t) = x(t) + B_1 \int_{t-h_1}^{t} x(s)\,ds.$$

Then, for the functional $V_1 = z'(t)Pz(t)$, we have

$$LV_1 = \left((A+B_1)x(t) + B_2 \int_{t-h_2}^{t} x(s)\,ds \right)' Pz(t)$$

$$+ z'(t)P\left((A+B_1)x(t) + B_2 \int_{t-h_2}^{t} x(s)\,ds \right)$$

$$+ \left(C_1 x(t-\tau_1) + C_2 \int_{t-\tau_2}^{t} x(s)\,ds \right)' P\left(C_1 x(t-\tau_1) + C_2 \int_{t-\tau_2}^{t} x(s)\,ds \right)$$

$$= x(t)\big((A+B_1)'P + P(A+B_1)\big)x(t)$$

$$+ \int_{t-h_1}^{t} \big(x'(t)(A+B_1)'PB_1 x(s) + x'(s)B_1' P(A+B_1)x(t) \big)\,ds$$

$$+ \int_{t-h_2}^{t} \big(x'(t)PB_2 x(s) + x'(s)B_2' Px(t) \big)\,ds$$

$$+ \int_{t-h_1}^{t}\int_{t-h_2}^{t} \big(x'(\theta)B_1' PB_2 x(s) + x'(s)B_2' PB_1 x(\theta) \big)\,ds\,d\theta$$

$$+ x'(t-\tau_1)C_1' PC_1 x(t-\tau_1)$$

$$+ \int_{t-\tau_2}^{t} \big(x'(t-\tau_1)C_1' PC_2 x(s) + x'(s)C_2' PC_1 x(t-\tau_1) \big)\,ds$$

$$+ \frac{1}{2}\int_{t-\tau_2}^{t}\int_{t-\tau_2}^{t} \big(x'(\theta)C_2' PC_2 x(s) + x'(s)C_2' PC_2 x(\theta) \big)\,ds\,d\theta.$$

Using Lemma 2.3 for some $R_i > 0$, $i = 1, 2, 3$, $G_j > 0$, $j = 1, 2$, we obtain

$$LV_1 \le x(t)\big[(A+B_1)'P + P(A+B_1) + h_1(A+B_1)'R_1(A+B_1) + h_2 PR_2 P \big]x(t)$$

$$+ \int_{t-h_1}^{t} x'(s)\big[B_1' PR_1^{-1} PB_1 + h_2 B_1' PR_3 PB_1 \big]x(s)\,ds$$

$$+ \int_{t-h_2}^{t} x'(s)\big[h_1 B_2' R_3^{-1} B_2 + B_2' R_2^{-1} B_2 \big]x(s)\,ds$$

$$+ x'(t-\tau_1)\big[C_1' PC_1 + \tau_2 C_1' PG_1 PC_1 \big]x(t-\tau_1)$$

$$+ \int_{t-\tau_2}^{t} x'(s)\left[C_2' G_1^{-1} C_2 + \frac{\tau_2}{2}\big(C_2' G_2 C_2 + C_2' PG_2^{-1} PC_2 \big) \right]x(s)\,ds.$$

By the representation of type (2.35) and Theorem 2.5 the following theorem is proved.

Theorem 6.6 *Suppose that the inequality $h_1\|B_1\| < 1$ holds and that, for some matrices $R_i > 0$, $i = 1, 2, 3$, $G_i > 0$, $i = 1, 2$, and $Q > 0$, there exists a positive* *definite solution P of the matrix Riccati equation*

$$(A + B_1)'P + P(A + B_1) + C_1'PC_1$$

$$+ h_1\big((A + B_1)'R_1(A + B_1) + B_1'PR_1^{-1}PB_1\big)$$

$$+ h_2\big(PR_2P + B_2'R_2^{-1}B_2\big) + h_1h_2\big(B_1'PR_3PB_1 + B_2'R_3^{-1}B_2\big)$$

$$+ \tau_2\big(C_1'PG_1PC_1 + C_2'G_1^{-1}C_2\big) + \frac{\tau_2^2}{2}\big(C_2'G_2C_2 + C_2'PG_2^{-1}PC_2\big) = -Q.$$

$$(6.23)$$

Then the trivial solution of (6.21) is asymptotically mean-square stable.

Remark 6.8 Similarly to Remark 6.3, one can show that in place of (6.23) in Theorem 6.6 other different matrix Riccati equations can be used, for instance,

$$(A + B_1)'P + P(A + B_1) + C_1'PC_1$$

$$+ h_1\big(R_1 + B_1'P(A + B_1)R_1^{-1}(A + B_1)'PB_1\big)$$

$$+ h_2\big(R_2 + B_2'PR_2^{-1}PB_2\big) + h_1h_2\big(R_3 + B_2'PB_1R_3^{-1}B_1'PB_2\big)$$

$$+ \tau_2\big(G_1 + C_2'PC_1G_1^{-1}C_1'PC_2\big) + \frac{\tau_2^2}{2}\big(G_2 + C_2'PC_2G_2^{-1}C_2'PC_2\big) = -Q.$$

Remark 6.9 Rewriting (6.21) in the form

$$\dot{z}(t) = (A + h_2B_2)x(t) + B_1x(t - h_1) + \left(C_1x(t - \tau_1) + C_2\int_{t-\tau_2}^{t} x(s)\,ds\right)\dot{w}(t),$$

$$z(t) = x(t) + B_2\int_{t-h_2}^{t} (s - t + h_2)x(s)\,ds,$$

one can obtain the following theorem.

Theorem 6.7 *Suppose that the inequality $h_2^2\|B_2\| < 2$ holds and that, for some matrices $R_i > 0$, $i = 1, 2, 3$, $G_i > 0$, $i = 1, 2$, and $Q > 0$, there exists a positive definite solution P of the matrix Riccati equation*

$$(A + h_2B_2)'P + P(A + h_2B_2) + C_1'PC_1 + R_1 + PB_1R_1^{-1}B_1'P$$

$$+ \frac{1}{2}h_2^2\big[R_2 + R_3 + (A + h_2B_2)'PB_2R_2^{-1}B_2'P(A + h_2B_2)$$

$$+ B_1'PB_2R_3^{-1}B_2'PB_1\big] + \tau_2\big(G_1 + C_1'PC_2G_1^{-1}C_2'PC_1\big)$$

$$+ \frac{\tau_2^2}{2}\big(G_2 + C_2'PC_2G_2^{-1}C_2'PC_2\big) = -Q \qquad\qquad (6.24)$$

or

$$(A + h_2 B_2)' P + P(A + h_2 B_2) + C_1' P C_1 + R_1 + B_1' P R_1^{-1} P B_1$$

$$+ \frac{1}{2} h_2^2 \big[R_2 + R_3 + B_2' P(A + h_2 B_2) R_2^{-1} (A + h_2 B_2)' P B_2 + B_2' P B_1 R_3^{-1} B_1' P B_2 \big]$$

$$+ \tau_2 \big(G_1 + C_2' P C_1 G_1^{-1} C_1' P C_2 \big) + \frac{\tau_2^2}{2} \big(G_2 + C_2' P C_2 G_2^{-1} C_2' P C_2 \big) = -Q.$$

Then the trivial solution of (6.21) *is asymptotically mean-square stable.*

Remark 6.10 Rewriting (6.21) in the form

$$\dot{z}(t) = A_1 x(t) + \Big(C_1 x(t - \tau_1) + C_2 \int_{t - \tau_2}^{t} x(s) \, ds \Big) \dot{w}(t),$$

$$z(t) = x(t) + \int_{t - h_1}^{t} B_1 x(s) \, ds + \int_{t - h_2}^{t} (s - t + h_2) B_2 x(s) \, ds,$$

$$A_1 = A + B_1 + h_2 B_2,$$

one can obtain the following theorem.

Theorem 6.8 *Suppose that the inequality* $h_1 \| B_1 \| + \frac{1}{2} h_2^2 \| B_2 \| < 1$ *holds and that, for some matrices* $R_i > 0$, $G_i > 0$, $i = 1, 2$, *and* $Q > 0$, *there exists a positive definite solution* P *of the matrix Riccati equation*

$$A_1' P + P A_1 + C_1' P C_1 + h_1 \big(R_1 + A_1' P B_1 R_1^{-1} B_1' P A_1 \big)$$

$$+ \frac{1}{2} h_2^2 \big(R_2 + A_1' P B_2 R_2^{-1} B_2' P A_1 \big) + \tau_2 \big(G_1 + C_1' P C_2 G_1^{-1} C_2' P C_1 \big)$$

$$+ \frac{\tau_2^2}{2} \big(G_2 + C_2' P C_2 G_2^{-1} C_2' P C_2 \big) = -Q \qquad (6.25)$$

or

$$A_1' P + P A_1 + C_1' P C_1 + h_1 \big(R_1 + B_1' P A_1 R_1^{-1} A_1' P B_1 \big)$$

$$+ \frac{1}{2} h_2^2 \big(R_2 + B_2' P A_1 R_2^{-1} A_1' P B_2 \big)$$

$$+ \tau_2 \big(G_1 + C_2' P C_1 G_1^{-1} C_1' P C_2 \big) + \frac{\tau_2^2}{2} \big(G_2 + C_2' P C_2 G_2^{-1} C_2' P C_2 \big) = -Q.$$

Then the trivial solution of (6.21) *is asymptotically mean-square stable.*

Remark 6.11 Note that in some cases a matrix Riccati equation can be transformed to a linear matrix equation. Suppose, for instance, that the matrix B in (6.3) has the inverse matrix. Then putting in (6.5) $R = B'PB$, we transform it to the linear matrix equation

$$A'P + PA + P + B'PB + C'PC = -Q.$$

Suppose that for some $P > 0$ the matrix B satisfies the condition $PB = B'P > 0$. Then putting in (6.5) $R = PB$, for this P, we obtain the linear matrix equation

$$(A + B)'P + P(A + B) + C'PC = -Q.$$

We obtain the same linear matrix equations by putting $R = P$ or $R = PB$ in (6.6).

If in (6.21) the matrices C_1, C_2 have the inverse matrices, then putting $R_i = P$, $G_i = C_i'PC_i$, $i = 1, 2$, in (6.22), we transform it to the linear matrix equation

$$A'P + PA + (1 + h_2)P + B_1'PB_1 + h_2 B_2'PB_2$$
$$+ (1 + \tau_2)\left(C_1'PC_1 + \tau_2 C_2'PC_2\right) = -Q.$$

If in (6.21) the matrices B_1, B_2, C_1, C_2 have the inverse matrices, then putting $R_1 = B_1'PB_1$, $R_2 = B_2'PB_2$, $G_1 = G_2 = C_2'PC_2$ in (6.25), we transform it to the linear matrix equation

$$A_1'P + PA_1 + \left(h_1 + \frac{1}{2}h_2^2\right)A_1'PA_1 + h_1 B_1'PB_1 + \frac{1}{2}h_2^2 B_2'PB_2$$
$$+ (1 + \tau_2)\left(C_1'PC_1 + \tau_2 C_2'PC_2\right)$$
$$= -Q.$$

Example 6.6 Consider the scalar case of (6.21). Putting $C = |C_1| + \tau_2|C_2|$, by Theorems 6.5–6.8 we obtain four different sufficient conditions for asymptotic mean-square stability of the trivial solution

$$A + |B_1| + h_2|B_2| + \frac{1}{2}C^2 < 0,$$

$$(A + B_1)\left(1 - h_1|B_1|\right) + h_2|B_2|\left(1 + h_1|B_1|\right) + \frac{1}{2}C^2 < 0, \quad h_1|B_1| < 1,$$

$$(A + h_2 B_2)\left(1 - \frac{1}{2}h_2^2|B_2|\right) + |B^2|\left(1 + \frac{1}{2}h_2^2|B_2|\right) + \frac{1}{2}C^2 < 0, \quad h_2^2|B_2| < 2,$$

$$(A + B_1 + h_2 B_2)\left(1 - h_1|B_1| - \frac{1}{2}h_2^2|B_2|\right) + \frac{1}{2}C^2 < 0, \quad h_1|B_1| + \frac{1}{2}h_2^2|B_2| < 1.$$

6.4 Equations with Nonincreasing Delays

6.4.1 One Delay in Deterministic and One Delay in Stochastic Parts of Equation

Consider the stochastic linear differential equation with delays depending on time

$$\dot{x}(t) = Ax(t) + Bx\big(t - h(t)\big) + Cx\big(t - \tau(t)\big)\dot{w}(t),$$

$$x_0(s) = \phi(s), \ s \le 0. \tag{6.26}$$

Here it is supposed that the delays $h(t)$ and $\tau(t)$ are nonnegative differentiable functions satisfying the conditions

$$\dot{h}(t) \le 0, \qquad \dot{\tau}(t) \le 0. \tag{6.27}$$

6.4.1.1 The First Way of Constructing a Lyapunov Functional

Consider the Lyapunov functional $V_1 = x'Px$. Calculating LV_1, we get

$$\begin{aligned}
LV_1 &= \big(Ax(t) + Bx\big(t - h(t)\big)\big)' Px(t) + x'(t)P\big(Ax(t) + Bx\big(t - h(t)\big)\big) \\
&\quad + x'\big(t - \tau(t)\big)C'PCx\big(t - \tau(t)\big) \\
&= x'(t)\big(A'P + PA\big)x(t) + x'\big(t - \tau(t)\big)C'PCx\big(t - \tau(t)\big) \\
&\quad + x'\big(t - h(t)\big)B'Px(t) + x'(t)PBx\big(t - h(t)\big).
\end{aligned}$$

Using Lemma 2.3 for $R > 0$, $a = x(t - h(t))$, and $b = B'Px(t)$, we have

$$\begin{aligned}
LV_1 &\le x'(t)\big(A'P + PA + PBR^{-1}B'P\big)x(t) \\
&\quad + x'\big(t - h(t)\big)Rx\big(t - h(t)\big) + x'\big(t - \tau(t)\big)C'PCx\big(t - \tau(t)\big). \tag{6.28}
\end{aligned}$$

By the representation (2.35) with

$$D(t) = A'P + PA + PBR^{-1}B'P, \qquad k = 2,$$

$$Q_1(t) = R, \qquad \tau_1(t) = h(t), \qquad Q_2(t) = C'PC, \qquad \tau_2(t) = \tau(t),$$

and Theorem 2.5 we obtain the following:

Theorem 6.9 *Suppose that conditions (6.27) hold and that, for some positive definite matrices Q and R, there exists a positive definite solution P of the matrix Riccati equation (6.5). Then the trivial solution of (6.26) is asymptotically mean-square stable.*

Remark 6.12 Using Lemma 2.3 for $a = Px(t)$ and $b = Bx(t - h(t))$, one can get in Theorem 6.9 in place of (6.5) the matrix Riccati equation

$$A'P + PA + C'PC + B'RB + PR^{-1}P = -Q. \tag{6.29}$$

6.4.1.2 The Second Way of Constructing a Lyapunov Functional

Consider another way of constructing a Lyapunov functional for (6.26) leading to
another Riccati matrix equation.

Let us rewrite now (6.26) in the form of a neutral type equation

$$\dot{z}(t) = (A + B)x(t) + \dot{h}(t)Bx\big(t - h(t)\big) + Cx\big(t - \tau(t)\big)\dot{w}(t),$$

$$z(t) = x(t) + B\int_{t-h(t)}^{t} x(s)\,ds$$

and suppose that

$$h(0)\|B\| < 1, \qquad \hat{h} = \sup_{t \geq 0}\big|\dot{h}(t)\big| < \infty. \tag{6.30}$$

Following the procedure of constructing Lyapunov functionals, we will construct
now a functional V in the form $V = V_1 + V_2$, where $V_1 = z'(t)Pz(t)$. Calculating
LV_1, we get

$$
\begin{aligned}
LV_1 &= z'(t)'P\big((A + B)x(t) + \dot{h}(t)Bx\big(t - h(t)\big)\big) \\
&\quad + \big(x'(t)(A + B)' + \dot{h}(t)x'\big(t - h(t)\big)B'\big)Pz(t) \\
&\quad + x'\big(t - \tau(t)\big)C'PCx\big(t - \tau(t)\big) \\
&= x'(t)\big(P(A + B) + (A + B)'P\big)x(t) + x'\big(t - \tau(t)\big)C'PCx\big(t - \tau(t)\big) \\
&\quad + \dot{h}(t)\big(x'(t)PBx\big(t - h(t)\big) + x'\big(t - h(t)\big)B'Px(t)\big) \\
&\quad + \int_{t-h(t)}^{t}\big(x'(t)(A + B)'PBx(s) + x'(s)B'P(A + B)x(t)\big)\,ds \\
&\quad + \dot{h}(t)\int_{t-h(t)}^{t}\big(x'(s)B'PBx\big(t - h(t)\big) + x'\big(t - h(t)\big)B'PBx(s)\big)\,ds.
\end{aligned}
$$

$$\tag{6.31}$$

Using Lemma 2.3 for $R_1 > 0$, $a = x(s)$, and $b = B'P(A + B)x(t)$, for $R_2 > 0$, $a = x(t)$, and $b = PBx(t - h(t))$, and for $R_3 > 0$, $a = x(s)$, and $b = B'PBx(t - h(t))$,
we obtain

$$\int_{t-h(t)}^{t}\big[x'(t)(A + B)'PBx(s) + x'(s)B'P(A + B)x(t)\big]\,ds$$

$$\leq h(t)x'(t)(A + B)'PBR_1^{-1}B'P(A + B)x(t) + \int_{t-h(t)}^{t} x'(s)R_1x(s)\,ds,$$

$$\dot{h}(t)\big(x'(t)PBx\big(t - h(t)\big) + x'\big(t - h(t)\big)B'Px(t)\big)$$

$$\leq \big|\dot{h}(t)\big|\big(x'(t)R_2x(t) + x'\big(t - h(t)\big)B'PR_2^{-1}PBx\big(t - h(t)\big)\big), \tag{6.32}$$

$$\dot{h}(t)\int_{t-h(t)}^{t}\left(x'(s)B'PBx\big(t-h(t)\big)+x'\big(t-h(t)\big)B'PBx(s)\right)ds$$

$$\leq |\dot{h}(t)|\Bigg(\int_{t-h(t)}^{t}x'(s)R_3x(s)\,ds$$

$$+h(t)x'\big(t-h(t)\big)B'PBR_3^{-1}B'PBx\big(t-h(t)\big)\Bigg).$$

Then by (6.31)–(6.32) we have

$$LV_1\leq x'(t)\Big[(A+B)'P+P(A+B)+h(0)(A+B)'PBR_1^{-1}B'P(A+B)$$

$$+\hat{h}R_2\Big]x(t)+\hat{h}x'\big(t-h(t)\big)B'P\big(R_2^{-1}+h(0)BR_3^{-1}B'\big)PBx\big(t-h(t)\big)$$

$$+x'\big(t-\tau(t)\big)C'PCx\big(t-\tau(t)\big)+\int_{t-h(t)}^{t}x'(s)(R_1+\hat{h}R_3)x(s)\,ds.$$

Choosing the functional V_2 in the form

$$V_2=\int_{t-h(t)}^{t}\big(s-t+h(t)\big)x'(s)(R_1+\hat{h}R_3)x(s)\,ds+\int_{t-\tau(t)}^{t}x'(s)C'PCx(s)\,ds$$

$$+\hat{h}\int_{t-h(t)}^{t}x'(s)B'P\big(R_2^{-1}+h(0)BR_3^{-1}B'\big)PBx(s)\,ds,$$

by (6.27) we obtain

$$LV_2=h(t)x'(t)(R_1+\hat{h}R_3)x(t)-\big(1-\dot{h}(t)\big)\int_{t-h(t)}^{t}x'(s)(R_1+\hat{h}R_3)x(s)\,ds$$

$$+x'(t)C'PCx(t)-\big(1-\dot{\tau}(t)\big)x'\big(t-\tau(t)\big)C'PCx\big(t-\tau(t)\big)$$

$$+\hat{h}x'(t)B'P\big(R_2^{-1}+h(0)BR_3^{-1}B'\big)PBx(t)$$

$$-\hat{h}\big(1-\dot{h}(t)\big)x'\big(t-h(t)\big)B'P\big(R_2^{-1}+h(0)BR_3^{-1}B'\big)PBx\big(t-h(t)\big)$$

$$\leq x'(t)\Big[h(0)(R_1+\hat{h}R_3)+C'PC+\hat{h}B'P\big(R_2^{-1}+h(0)BR_3^{-1}B'\big)PB+\Big]x(t)$$

$$-\hat{h}x'\big(t-h(t)\big)B'P\big(R_2^{-1}+h(0)BR_3^{-1}B'\big)PBx\big(t-h(t)\big)$$

$$-x'\big(t-\tau(t)\big)C'PCx\big(t-\tau(t)\big)-\int_{t-h(t)}^{t}x'(s)(R_1+\hat{h}R_3)x(s)\,ds.$$

As a result, for the functional $V=V_1+V_2$, we have

$$LV\leq x'(t)\Big[(A+B)'P+P(A+B)+C'PC$$

$$+h(0)\big(R_1+(A+B)'PBR_1^{-1}B'P(A+B)\big)$$

$$+\hat{h}\big(R_2+B'PR_2^{-1}PB\big)+\hat{h}h(0)\big(R_3+B'PBR_3^{-1}B'PB\big)\Big]x(t).$$

So, we obtain the following theorem.

Theorem 6.10 *Suppose that conditions (6.27), (6.30) hold and that, for some matrices $R_i > 0$, $i = 1, 2, 3$, and $Q > 0$, there exists a positive definite solution P of the matrix Riccati equation*

$$(A + B)'P + P(A + B) + C'PC$$
$$+ h(0)\big(R_1 + (A + B)'PBR_1^{-1}B'P(A + B)\big)$$
$$+ \hat{h}\big(R_2 + B'PR_2^{-1}PB + h(0)\big(R_3 + B'PBR_3^{-1}B'PB\big)\big) = -Q. \quad (6.33)$$

Then the trivial solution of (6.26) is asymptotically mean-square stable.

Remark 6.13 Using Lemma 2.3 for $R_1 > 0$, $a = x(t)$, and $b = (A + B)'PBx(s)$, for $R_2 > 0$, $a = x(t - h(t))$, and $b = B'Px(t)$, and for $R_3 > 0$, $a = x(t - h(t))$, and $b = B'PBx(s)$, in place of (6.32) we obtain the inequalities

$$\int_{t-h(t)}^{t} \big[x'(t)(A + B)'PBx(s) + x'(s)B'P(A + B)x(t)\big]\,ds$$

$$\le h(t)x'(t)R_1x(t) + \int_{t-h(t)}^{t} x'(s)B'P(A + B)R_1^{-1}(A + B)'PBx(s)\,ds,$$

$$\dot{h}(t)\big(x'(t)PBx(t - h(t)) + x'(t - h(t))B'Px(t)\big)$$

$$\le |\dot{h}(t)|\big(x'(t - h(t))R_2x(t - h(t)) + x'(t)PBR_2^{-1}B'Px(t)\big), \quad (6.34)$$

$$\dot{h}(t)\int_{t-h(t)}^{t} \big(x'(s)B'PBx(t - h(t)) + x'(t - h(t))B'PBx(s)\big)\,ds$$

$$\le |\dot{h}(t)|\bigg(h(t)x'(t - h(t))R_3x(t - h(t))$$

$$+ \int_{t-h(t)}^{t} x'(s)B'PBR_3^{-1}B'PBx(s)\bigg)\,ds.$$

Using different combinations of inequalities (6.32) and (6.34) and choosing an appropriate functional V_2 in place of (6.33) in Theorem 6.10, one can use different matrix Riccati equations. For example, using all inequalities (6.34) in place of inequalities (6.32) and choosing the functional V_2 in the form

$$V_2 = \int_{t-h(t)}^{t} (s - t + h(t))x'(s)B'P(A + B)R_1^{-1}(A + B)'PBx(s)\,ds$$

$$+ \hat{h}\int_{t-h(t)}^{t} x'(s)R_2x(s)\,ds + \int_{t-\tau(t)}^{t} x'(s)C'PCx(s)\,ds$$

$$+ \hat{h}\int_{t-h(t)}^{t} (s - t + h(t))x'(s)B'PBR_3^{-1}B'PBx(s)\,ds,$$

we obtain that in Theorem 6.10 in place of (6.33) the following equation can be used:

$$
\begin{aligned}
(A + B)'P &+ P(A + B) + C'PC \\
&+ h(0)\big(R_1 + B'P(A + B)R_1^{-1}(A + B)'PB\big) \\
&+ \hat{h}\big(R_2 + PBR_2^{-1}B'P + h(0)\big(R_3 + B'PBR_3^{-1}B'PB\big)\big) = -Q. \quad (6.35)
\end{aligned}
$$

Example 6.7 In the scalar case a positive solution of (6.33) (or (6.35)) exists if and only if

$$
(A + B)\big(1 - h(0)|B|\big) + \hat{h}|B|\big(1 + h(0)|B|\big) + \frac{1}{2}c^2 < 0, \quad h(0)|B| < 1.
$$

6.4.2 Several Delays in Deterministic Part of Equation

Consider now the stochastic linear differential equation with delays

$$
\dot{x}(t) = Ax(t) + \sum_{i=1}^{m} B_i x\big(t - h_i(t)\big) + Cx\big(t - \tau(t)\big)\dot{w}(t), \quad (6.36)
$$

which is a generalization of (6.26). Here m is a positive integer, and the delays $h_i(t)$ are nonnegative differentiable functions satisfying the conditions

$$
\dot{\tau}(t) \le 0, \quad \dot{h}_i(t) \le 0, \quad \hat{h}_i = \sup_{t \ge 0}\big|\dot{h}_i(t)\big| < \infty, \quad i = 1, \ldots, m. \quad (6.37)
$$

To construct Lyapunov functionals for (6.36), we will use both previous representations of the initial equation. Namely, rewrite (6.36) in the form

$$
\dot{z}(t) = (A + B)x(t) + \sum_{l=1}^{m_0} \dot{h}_l(t)B_l x\big(t - h_l(t)\big)
$$

$$
+ \sum_{i=m_0+1}^{m} B_i x\big(t - h_i(t)\big) + Cx\big(t - \tau(t)\big)\dot{w}(t),
$$

$$
z(t) = x(t) + \sum_{j=1}^{m_0} B_j \int_{t-h_j(t)}^{t} x(s)\,ds,
$$

$$
B = \sum_{j=1}^{m_0} B_j, \quad 0 \le m_0 \le m.
$$

We will construct a Lyapunov functional V in the form $V = V_1 + V_2$, where $V_1 = z'(t)Pz(t)$. In this case,

$$
LV_1 = \left[(A+B)x(t) + \sum_{i=m_0+1}^{m} B_i x\big(t - h_i(t)\big) + \sum_{l=1}^{m_0} \dot{h}_l(t) B_l x\big(t - h_l(t)\big) \right]' Pz(t)
$$

$$
+ z'(t)P\left[(A+B)x(t) + \sum_{i=m_0+1}^{m} B_i x\big(t - h_i(t)\big) + \sum_{l=1}^{m_0} \dot{h}_l(t) B_l x\big(t - h_l(t)\big) \right]
$$

$$
+ x'\big(t - \tau(t)\big) C' P C x\big(t - \tau(t)\big)
$$

$$
= x'(t)\big[(A+B)'P + P(A+B)\big]x(t) + x'\big(t - \tau(t)\big)C'PCx\big(t - \tau(t)\big)
$$

$$
+ \sum_{i=m_0+1}^{m} \left[x'\big(t - h_i(t)\big) B_i' P x(t) + x'(t) P B_i x\big(t - h_i(t)\big) \right]
$$

$$
+ \sum_{j=1}^{m_0} \int_{t-h_j(t)}^{t} \left[x'(t)(A+B)' P B_j x(s) + x'(s) B_j' P (A+B) x(t) \right] ds
$$

$$
+ \sum_{j=1}^{m_0} \sum_{i=m_0+1}^{m} \int_{t-h_j(t)}^{t} \left[x'\big(t - h_i(t)\big) B_i' P B_j x(s) \right.
$$

$$
\left. + x'(s) B_j' P B_i x\big(t - h_i(t)\big) \right] ds
$$

$$
+ \sum_{l=1}^{m_0} \dot{h}_l(t)\left[x'\big(t - h_l(t)\big) B_l' P x(t) + x'(t) P B_l x\big(t - h_l(t)\big) \right]
$$

$$
+ \sum_{l=1}^{m_0} \sum_{j=1}^{m_0} \dot{h}_l(t) \int_{t-h_j(t)}^{t} \left[x'\big(t - h_l(t)\big) B_l' P B_j x(s) \right.
$$

$$
\left. + x'(s) B_j' P B_l x\big(t - h_l(t)\big) \right] ds.
$$

Using Lemma 2.3 for $R_i > 0$, $i = m_0 + 1, \ldots, m$, $a = x(t - h_i(t))$, and $b = B_i' P x(t)$, for $G_j > 0$, $j = 1, \ldots, m_0$, $a = x(s)$, and $b = B_j' P(A+B)x(t)$, for $S_{ij} > 0$, $i = m_0 + 1, \ldots, m$, $j = 1, \ldots, m_0$, $a = x(s)$, and $b = B_j' P B_i x(t - h_i(t))$, for $U_l > 0$, $l = 1, \ldots, m_0$, $a = x(t - h_l(t))$, and $b = B_l' P x(t)$, and for $Z_{lj} > 0$, $l = 1, \ldots, m_0$, $j = 1, \ldots, m_0$, $a = x(s)$, and $b = B_j' P B_l x(t - h_l(t))$, we obtain

$$
x'\big(t - h_i(t)\big) B_i' P x(t) + x'(t) P B_i x\big(t - h_i(t)\big)
$$

$$
\leq x'\big(t - h_i(t)\big) R_i x\big(t - h_i(t)\big) + x'(t) P B_i R_i^{-1} B_i' P x(t),
$$

$$
x'(t)(A+B)' P B_j x(s) + x'(s) B_j' P(A+B)x(t)
$$

$$
\leq x'(s) G_j x(s) + x'(t)(A+B)' P B_j G_j^{-1} B_j' P(A+B)x(t),
$$

$$x'\bigl(t - h_i(t)\bigr)B_i'PB_jx(s) + x'(s)B_j'PB_ix\bigl(t - h_i(t)\bigr)$$

$$\leq x'(s)S_{ij}x(s) + x'\bigl(t - h_i(t)\bigr)B_i'PB_jS_{ij}^{-1}B_j'PB_ix\bigl(t - h_i(t)\bigr), \tag{6.38}$$

$$\dot{h}_l(t)\bigl[x'\bigl(t - h_l(t)\bigr)B_l'Px(t) + x'(t)PB_lx\bigl(t - h_l(t)\bigr)\bigr]$$

$$\leq \hat{h}_l\bigl[x'\bigl(t - h_l(t)\bigr)U_lx\bigl(t - h_l(t)\bigr) + x'(t)PB_lU_l^{-1}B_l'Px(t)\bigr],$$

$$\dot{h}_l(t)\bigl[x'\bigl(t - h_l(t)\bigr)B_l'PB_jx(s) + x'(s)B_j'PB_lx\bigl(t - h_l(t)\bigr)\bigr]$$

$$\leq \hat{h}_l\bigl[x'(s)Z_{lj}x(s) + x'\bigl(t - h_l(t)\bigr)B_l'PB_jZ_{lj}^{-1}B_j'PB_lx\bigl(t - h_l(t)\bigr)\bigr].$$

By inequalities (6.38) we obtain

$$LV_1 \leq x'(t)\Biggl[(A + B)'P + P(A + B) + \sum_{l=1}^{m_0}\hat{h}_lPB_lU_l^{-1}B_l'P$$

$$+ \sum_{i=m_0+1}^{m}PB_iR_i^{-1}B_i'P + \sum_{j=1}^{m_0}h_j(0)(A + B)'PB_jG_j^{-1}B_j'P(A + B)\Biggr]x(t)$$

$$+ x'\bigl(t - \tau(t)\bigr)C'PCx\bigl(t - \tau(t)\bigr) + \sum_{i=m_0+1}^{m}x'\bigl(t - h_i(t)\bigr)R_ix\bigl(t - h_i(t)\bigr)$$

$$+ \sum_{j=1}^{m_0}\int_{t-h_j(t)}^{t}x'(s)G_jx(s)\,ds + \sum_{j=1}^{m_0}\sum_{i=m_0+1}^{m}\int_{t-h_j(t)}^{t}x'(s)S_{ij}x(s)\,ds$$

$$+ \sum_{i=m_0+1}^{m}\sum_{j=1}^{m_0}h_j(0)x'\bigl(t - h_i(t)\bigr)B_i'PB_jS_{ij}^{-1}B_j'PB_ix\bigl(t - h_i(t)\bigr)$$

$$+ \sum_{l=1}^{m_0}\hat{h}_l\Biggl[x'\bigl(t - h_l(t)\bigr)U_lx\bigl(t - h_l(t)\bigr) + \sum_{j=1}^{m_0}\int_{t-h_j(t)}^{t}x'(s)Z_{lj}x(s)\,ds\Biggr]$$

$$+ \sum_{l=1}^{m_0}\sum_{j=1}^{m_0}\hat{h}_lh_j(0)x'\bigl(t - h_l(t)\bigr)B_l'PB_jZ_{lj}^{-1}B_j'PB_lx\bigl(t - h_l(t)\bigr).$$

Choosing the functional V_2 in the form

$$V_2 = \int_{t-\tau(t)}^{t}x'(s)C'PCx(s)\,ds + \sum_{i=m_0+1}^{m}\int_{t-h_i(t)}^{t}x'(s)R_ix(s)\,ds$$

$$+ \sum_{j=1}^{m_0}\int_{t-h_j(t)}^{t}\bigl(s - t + h_j(t)\bigr)x'(s)G_jx(s)\,ds$$

$$+ \sum_{j=1}^{m_0} \sum_{i=m_0+1}^{m} \left[h_j(0) \int_{t-h_i(t)}^{t} x'(s) B_i' P B_j S_{ij}^{-1} B_j' P B_i x(s)\, ds \right.$$

$$\left. + \int_{t-h_j(t)}^{t} \big(s - t + h_j(t)\big) x'(s) S_{ij} x(s)\, ds \right]$$

$$+ \sum_{l=1}^{m_0} \hat{h}_l \left[\int_{t-h_l(t)}^{t} x'(s) U_l x(s)\, ds \right.$$

$$+ \sum_{j=1}^{m_0} \int_{t-h_j(t)}^{t} \big(s - t + h_j(t)\big) x'(s) Z_{lj} x(s)\, ds \right]$$

$$+ \sum_{l=1}^{m_0} \sum_{j=1}^{m_0} \hat{h}_l h_j(0) \int_{t-h_j(t)}^{t} x'(s) B_l' P B_j Z_{lj}^{-1} B_j' P B_l x(s)\, ds,$$

for the functional $V = V_1 + V_2$, we obtain

$$LV \le x'(t) \left[(A+B)' P + P(A+B) + C' P C + \sum_{l=1}^{m_0} \hat{h}_l \big(U_l + P B_l U_l^{-1} B_l' P \big) \right.$$

$$+ \sum_{j=1}^{m_0} h_j(0) \big[G_j + (A+B)' P B_j G_j^{-1} B_j' P(A+B) \big]$$

$$+ \sum_{i=m_0+1}^{m} \big(R_i + P B_i R_i^{-1} B_i' P \big) + \sum_{j=1}^{m_0} h_j(0) \sum_{l=1}^{m_0} \hat{h}_l \big(Z_{lj} + B_l' P B_j Z_{lj}^{-1} B_j' P B_l \big)$$

$$+ \left. \sum_{j=1}^{m_0} h_j(0) \sum_{i=m_0+1}^{m} \big(S_{ij} + B_i' P B_j S_{ij}^{-1} B_j' P B_i \big) \right] x(t).$$

Thus, the following theorem is proved.

Theorem 6.11 *Suppose that conditions (6.37) and the inequality $\sum_{j=1}^{m_0} h_j(0) \|B_j\| < 1$ for some $m_0 = 1, \dots, m$ hold and that, for some positive definite matrices U_l, R_i, G_j, S_{ij}, Z_{lj}, and Q, there exists a positive definite solution P of the matrix Riccati equation*

$$(A+B)' P + P(A+B) + C' P C + \sum_{l=1}^{m_0} \hat{h}_l \big(U_l + P B_l U_l^{-1} B_l' P \big)$$

$$+ \sum_{j=1}^{m_0} h_j(0) \big[G_j + (A+B)' P B_j G_j^{-1} B_j' P(A+B) \big]$$

$$+ \sum_{i=m_0+1}^{m} \left(R_i + PB_i R_i^{-1} B_i' P\right) + \sum_{j=1}^{m_0} h_j(0) \sum_{l=1}^{m_0} \hat{h}_l\left(Z_{lj} + B_l' PB_j Z_{lj}^{-1} B_j' PB_l\right)$$

$$+ \sum_{j=1}^{m_0} h_j(0) \sum_{i=m_0+1}^{m} \left(S_{ij} + B_i' PB_j S_{ij}^{-1} B_j' PB_i\right) = -Q. \tag{6.39}$$

Then the trivial solution of (6.36) *is asymptotically mean-square stable.*

Remark 6.14 Using other variants of inequalities (6.38) and choosing an appropriate form of the functional V_2, in Theorem 6.11 one can use in place of (6.39) other matrix Riccati equations, for instance,

$$(A+B)'P + P(A+B) + C'PC + \sum_{l=1}^{m_0} \hat{h}_l\left(U_l + B_l' PU_l^{-1} PB_l\right)$$

$$+ \sum_{j=1}^{m_0} h_j(0)\left[G_j + B_j' P(A+B)G_j^{-1}(A+B)'PB_j\right]$$

$$+ \sum_{i=m_0+1}^{m} \left(R_i + B_i' PR_i^{-1} PB_i\right) + \sum_{j=1}^{m_0} h_j(0) \sum_{l=1}^{m_0} \hat{h}_l\left(Z_{lj} + B_j' PB_l Z_{lj}^{-1} B_l' PB_j\right)$$

$$+ \sum_{j=1}^{m_0} h_j(0) \sum_{i=m_0+1}^{m} \left(S_{ij} + B_j' PB_i S_{ij}^{-1} B_i' PB_j\right) = -Q.$$

Remark 6.15 It is easy to see that rewriting equation (6.36) in the form

$$\dot{z}(t) = (A+B)x(t) + \sum_{j=m_0+1}^{m} \dot{h}_l(t)B_l x\left(t - h_l(t)\right)$$

$$+ \sum_{i=1}^{m_0} B_i x\left(t - h_i(t)\right) + Cx\left(t - \tau(t)\right)\dot{w}(t),$$

$$z(t) = x(t) + \sum_{j=m_0+1}^{m} \int_{t-h_j(t)}^{t} x(s)\,ds,$$

$$B = \sum_{j=m_0+1}^{m} B_j, \quad 0 \le m_0 \le m,$$

we obtain a new modification of Theorem 6.11 after replacement in (6.39) of all sums $\sum_{j=1}^{m_0}$ by $\sum_{j=m_0+1}^{m}$ and all sums $\sum_{i=m_0+1}^{m}$ by $\sum_{i=1}^{m_0}$. Using other combinations of the summands for representations of (6.36), we will obtain other modifications of Theorem 6.11.

Example 6.8 Consider the scalar case of (6.36) for $m = 2$. Using Theorem 6.11 for different values of $m_0 = 0, 1, 2$, we obtain four different sufficient conditions for asymptotic mean-square stability of the trivial solution of (6.36):

$$A + |B_1| + |B_2| + \frac{1}{2}C^2 < 0,$$

$$(A + B_1)\big(1 - h_1(0)|B_1|\big) + \big(\hat{h}_1|B_1| + |B_2|\big)\big(1 + h_1(0)|B_1|\big) + \frac{1}{2}C^2 < 0,$$

$$h_1(0)|B_1| < 1,$$

$$(A + B_2)\big(1 - h_2(0)|B_2|\big) + \big(|B_1| + \hat{h}_2|B_2|\big)\big(1 + h_2(0)|B_2|\big) + \frac{1}{2}C^2 < 0,$$

$$h_2(0)|B_2| < 1,$$

$$(A + B_1 + B_2)\big(1 - h_1(0)|B_1| - h_2(0)|B_2|\big) + \big(\hat{h}_1|B_1| + \hat{h}_2|B_2|\big)$$

$$\times \big(1 + h_1(0)|B_1| + h_2(0)|B_2|\big) + \frac{1}{2}C^2 < 0, \quad h_1(0)|B_1| + h_2(0)|B_2| < 1.$$

6.5 Equations with Bounded Delays

Consider the linear stochastic differential equation with distributed delays depending on time

$$\dot{x}(t) = Ax(t) + \int_{t-h(t)}^{t} Bx(s)\,ds + \int_{t-\tau(t)}^{t} Cx(s)\,ds\,\dot{w}(t),$$

$$x_0(s) = \phi(s), \ s \le 0. \tag{6.40}$$

We suppose that the delays $h(t)$ and $\tau(t)$ satisfy the following conditions:

$$0 \le h_0 \le h(t) \le h_1, \quad \hat{h} = h_1 - h_0 \ge 0, \quad 0 \le \tau(t) \le \tau_1. \tag{6.41}$$

6.5.1 The First Way of Constructing a Lyapunov Functional

Calculating LV_1 for $V_1 = x'(t)Px(t)$, we get

$$LV_1 = x'(t)\big(A'P + PA\big)x(t) + I_0 + I_1,$$

where

$$I_0 = \int_{t-\tau(t)}^{t} \int_{t-\tau(t)}^{t} x(s)C'PCx(\theta)\,d\theta\,ds,$$

$$I_1 = \int_{t-h(t)}^{t} \big(x'(s)B'Px(t) + x'(t)PBx(s)\big)\,ds.$$

Using (6.41) and Lemma 2.3 for I_0 with $R_0 > 0$, $a = x(s)$, and $b = C'PCx(\theta)$ and for I_1 with $R_1 > 0$, $a = x(t)$, and $b = PBx(s)$, we obtain

$$I_0 = \frac{1}{2} \int_{t-\tau(t)}^{t} \int_{t-\tau(t)}^{t} \left(x'(\theta) C' PCx(s) + x'(s) C' PCx(\theta) \right) d\theta\, ds$$

$$\leq \frac{\tau_1}{2} \int_{t-\tau_1}^{t} x'(s) \left(R_0 + C' PCR_0^{-1} C' PC \right) x(s)\, ds, \tag{6.42}$$

$$I_1 \leq h_1 x'(t) R_1 x(t) + \int_{t-h_1}^{t} x'(s) B' PR_1^{-1} PBx(s)\, ds.$$

Then

$$LV_1 \leq x'(t) \left(A'P + PA + h_1 R_1 \right) x(t)$$

$$+ \int_{t-h_1}^{t} x'(s) B' PR_1^{-1} PBx(s)\, ds$$

$$+ \frac{\tau_1}{2} \int_{t-\tau_1}^{t} x'(s) \left(R_0 + C' PCR_0^{-1} C' PC \right) x(s)\, ds.$$

By the representation (2.35) and Theorem 2.5 we obtain the following theorem.

Theorem 6.12 *Let for some positive definite matrices R_0, R_1, and Q, there exist a positive definite solution P of the matrix Riccati equation*

$$A'P + PA + h_1 \left(R_1 + B' PR_1^{-1} PB \right) + \frac{\tau_1^2}{2} \left(R_0 + C' PCR_0^{-1} C' PC \right) = -Q. \tag{6.43}$$

Then the trivial solution of (6.40) is asymptotically mean-square stable.

Remark 6.16 Using Lemma 2.3 with other representations for a and b, it is possible to get other matrix Riccati equations in Theorem 6.12. For example, using $R_0 > 0$, $a = Cx(s)$, and $b = PCx(\theta)$, in place of (6.43) we obtain the equation

$$A'P + PA + h_1 \left(PR_1 P + B' R_1^{-1} B \right) + \frac{\tau_1^2}{2} C' \left(R_0 + PR_0^{-1} P \right) C = -Q. \tag{6.44}$$

Example 6.9 In the scalar case both equations (6.43) and (6.44) have a positive solution if and only if $A + h_1 |B| + \frac{1}{2} \tau_1^2 C^2 < 0$.

6.5.2 The Second Way of Constructing a Lyapunov Functional

Rewrite (6.40) in the form of a stochastic differential equation of neutral type

$$\dot{z}(t) = (A + h_1 B) x(t) - \int_{t-h_1}^{t-h(t)} Bx(s)\, ds + \int_{t-\tau(t)}^{t} Cx(s)\, ds\, \dot{w}(t),$$

$$z(t) = x(t) + \int_{t-h_1}^{t} (s - t + h_1) B x(s) \, ds.$$

Calculating LV_1, for the functional $V_1 = z'(t) P z(t)$, we get

$$LV_1 = x'(t) \big[(A + h_1 B)' P + P(A + h_1 B) \big] x(t) + I_0 + I_1 + I_2 + I_3,$$

where I_0 is defined in (6.42), and

$$I_1 = \int_{t-h_1}^{t} (s - t + h_1) \big[x'(s) B' P (A + h_1 B) x(t) + x'(t)(A + h_1 B)' P B x(s) \big] \, ds,$$

$$I_2 = -\int_{t-h_1}^{t-h(t)} \big[x'(t) P B x(s) + x'(s) B' P x(t) \big] \, ds,$$

$$I_3 = -\int_{t-h_1}^{t} \int_{t-h_1}^{t-h(t)} (s - t + h_1) \big[x'(s) B' P B x(\theta) + x'(\theta) B' P B x(s) \big] \, d\theta \, ds.$$

Using Lemma 2.3, we obtain

$$I_1 \le \frac{1}{2} h_1^2 x'(t) R_1 x(t)$$

$$+ \int_{t-h_1}^{t} (s - t + h_1) x'(s) B' P (A + h_1 B)' P B R_1^{-1} (A + h_1 B)' P B x(s) \, ds,$$

$$I_2 \le \int_{t-h_1}^{t-h(t)} \big[x'(t) R_2 x(t) + x'(s) B' P R_2^{-1} P B x'(s) \big] \, ds$$

$$\le \hat{h} x'(t) R_2 x(t) + \int_{t-h_1}^{t-h_0} x'(s) B' P R_2^{-1} P B x'(s) \, ds,$$

$$I_3 \le \int_{t-h_1}^{t} \int_{t-h_1}^{t-h(t)} (s - t + h_1) \big[x'(\theta) R_3 x(\theta) + x'(s) B' P B R_3^{-1} B' P B x(s) \big] \, d\theta \, ds$$

$$\le \frac{1}{2} h_1^2 \int_{t-h_1}^{t-h_0} x'(\theta) R_3 x(\theta) \, d\theta$$

$$+ \hat{h} \int_{t-h_1}^{t} (s - t + h_1) x'(s) B' P B R_3^{-1} B' P B x(s) \, ds.$$

Then

$$LV_1 \le x'(t) \Big[(A + h_1 B)' P + P(A + h_1 B) + \frac{1}{2} h_1^2 R_1 + \hat{h} R_2 \Big] x(t)$$

$$+ \int_{t-h_1}^{t} (s - t + h_1) x'(s) B' P (A + h_1 B) R_1^{-1} (A + h_1 B)' P B x(s) \, ds$$

$$+ \int_{t-h_1}^{t-h_0} x'(s) \left[B'PR_2^{-1}PB + \frac{1}{2}h_1^2 R_3 \right] x(s)\,ds$$

$$+ \hat{h} \int_{t-h_1}^{t} (s-t+h_1)x'(s)B'PBR_3^{-1}B'PBx(s)\,ds$$

$$+ \frac{\tau_1}{2} \int_{t-\tau_1}^{t} x'(s)\left(R_0 + C'PCR_0^{-1}C'PC\right)x(s)\,ds.$$

By Theorem 2.5 we obtain the following theorem.

Theorem 6.13 *Suppose that the inequality $h_1^2\|B\| < 2$ holds and that, for some positive definite matrices R_i, $i = 0, 1, 2, 3$, and Q, there exists a positive definite solution P of the matrix Riccati equation*

$$(A + h_1 B)'P + P(A + h_1 B) + \frac{\tau_1^2}{2}\left(R_0 + C'PCR_0^{-1}C'PC\right)$$

$$+ \frac{1}{2}h_1^2\left(R_1 + B'P(A + h_1 B)R_1^{-1}(A + h_1 B)'PB\right)$$

$$+ \hat{h}\left(R_2 + B'PR_2^{-1}PB\right) + \frac{1}{2}h_1^2\hat{h}\left(R_3 + B'PBR_3^{-1}B'PB\right) = -Q. \quad (6.45)$$

Then the trivial solution of (6.40) is asymptotically mean-square stable.

Remark 6.17 By analogy with the previous remarks in Theorem 6.13, instead of (6.45) other variants of matrix Riccati equations can be used.

Remark 6.18 Similar results can be obtained for more general equation

$$\dot{x}(t) = Ax(t) + \sum_{i=1}^{m} \int_{t-h_i(t)}^{t} B_i x(s)\,ds + \sum_{i=1}^{k} \int_{t-\tau_i(t)}^{t} C_i x(s)\,ds\,\dot{w}(t).$$

6.6 Equations with Unbounded Delays

Suppose now that the delays $h(t)$ and $\tau(t)$ in (6.40) are nonnegative differentiable functions that satisfy the conditions

$$\dot{h}(t) \leq \alpha < 1, \qquad \dot{\tau}(t) \leq \beta < 1. \qquad (6.46)$$

From (6.46) in particular it follows that

$$h(t) \leq h(0) + \alpha t, \qquad \tau(t) \leq \tau(0) + \beta t,$$

i.e., the delays can increase to infinity.

We will construct a Lyapunov functional V in the form $V = V_1 + V_2$, where $V_1 = x'Px$. Calculating LV_1, we obtain

$$
LV_1 \le x'(t)\big(A'P + PA + PBR^{-1}B'P\big)x(t)
$$
$$
+ x'\big(t - h(t)\big)Rx\big(t - h(t)\big) + x'\big(t - \tau(t)\big)C'PCx\big(t - \tau(t)\big).
$$

By the representation (2.35) with

$$
D(t) = A'P + PA + PBR^{-1}B'P, \qquad k = 2,
$$
$$
Q_1 = R, \qquad \tau_1(t) = h(t), \qquad Q_2 = C'PC, \qquad \tau_2(t) = \tau(t),
$$

from Theorem 2.5 we have following:

Theorem 6.14 *Suppose that conditions* (6.46) *hold and that, for some positive definite matrices R and QR, there exists a positive definite solution P of the matrix Riccati equation*

$$
A'P + PA + PBR^{-1}B'P + \frac{1}{1-\alpha}R + \frac{1}{1-\beta}C'PC = -Q. \tag{6.47}
$$

Then the trivial solution of (6.40) *is asymptotically mean-square stable.*

Remark 6.19 Instead of (6.47) in Theorem 6.14 other matrix Riccati equations can be used, for example,

$$
A'P + PA + B'PR^{-1}PB + \frac{1}{1-\alpha}R + \frac{1}{1-\beta}C'PC = -Q.
$$

Putting $R = P$ in this equation, one can reduce it to the linear matrix equation

$$
A'P + PA + B'PB + \frac{1}{1-\alpha}P + \frac{1}{1-\beta}C'PC = -Q.
$$

Example 6.10 In the scalar case a positive solution of (6.47) exists if and only if

$$
A + \frac{|B|}{\sqrt{1-\alpha}} + \frac{C^2}{2(1-\beta)} < 0.
$$

Consider the differential equation

$$
\dot{x}(t) = Ax(t) + \int_0^{h(t)} \beta(s)Bx(t-s)\,ds + \int_0^{\tau(t)} \sigma(s)Cx(t-s)\,ds\,\dot{w}(t). \tag{6.48}
$$

Here we suppose that the delays $h(t)$, $\tau(t)$ and the scalar functions $\beta(s)$, $\sigma(s)$ satisfy the conditions

$$\hat{h} = \sup_{t \geq 0} h(t) \leq \infty, \qquad \hat{\tau} = \sup_{t \geq 0} \tau(t) \leq \infty,$$

$$\beta = \int_0^{\hat{h}} |\beta(s)| \, ds < \infty, \qquad \sigma = \int_0^{\hat{\tau}} |\sigma(s)| \, ds < \infty. \tag{6.49}$$

Calculating LV_1 for $V_1 = x'Px$, $P > 0$, we obtain

$$LV_1 = \left(Ax(t) + \int_0^{h(t)} \beta(s)Bx(t-s)\,ds \right)' Px(t)$$

$$+ x'(t)P\left(Ax(t) + \int_0^{h(t)} \beta(s)Bx(t-s)\,ds \right)$$

$$+ \left(\int_0^{\tau(t)} \sigma(s)Cx(t-s)\,ds \right)' P\left(\int_0^{\tau(t)} \sigma(s)Cx(t-s)\,ds \right)$$

$$= x'(t)\big[A'P + PA\big]x(t)$$

$$+ \int_0^{h(t)} \beta(s)\big[x'(t-s)B'Px(t) + x'(t)PBx(t-s)\big]\,ds$$

$$+ \int_0^{\tau(t)}\int_0^{\tau(t)} \sigma(s)\sigma(\theta)x'(t-s)C'PCx(t-\theta)\,d\theta\,ds.$$

Using Lemma 2.3, we have

$$\int_0^{h(t)} \beta(s)\big[x'(t-s)B'Px(t) + x'(t)PBx(t-s)\big]\,ds$$

$$\leq \int_0^{h(t)} |\beta(s)|\big[x'(t)PR_1Px(t) + x'(t-s)B'R_1^{-1}Bx(t-s)\big]\,ds$$

$$\leq \beta x'(t)PR_1Px(t) + \int_0^{\hat{h}} |\beta(s)|x'(t-s)B'R_1^{-1}Bx(t-s)\,ds$$

and analogously

$$\int_0^{\tau(t)}\int_0^{\tau(t)} \sigma(\theta)\sigma(s)x'(t-s)C'PCx(t-\theta)\,ds\,d\theta$$

$$= \frac{1}{2}\int_0^{\tau(t)}\int_0^{\tau(t)} \sigma(\theta)\sigma(s)\big[x'(t-s)C'PCx(t-\theta)$$

$$+ x'(t-\theta)C'PCx(t-s)\big]\,ds\,d\theta$$

$$\leq \frac{1}{2}\int_0^{\tau(t)}\int_0^{\tau(t)} |\sigma(\theta)||\sigma(s)|\big[x'(t-s)C'R_2Cx(t-s)$$

$$+ x'(t - \theta) C' P R_2^{-1} P C x(t - \theta) \big] ds\, d\theta$$

$$= \frac{1}{2} \int_0^{\tau(t)} \big| \sigma(\theta) \big|\, d\theta \int_0^{\tau(t)} \big| \sigma(s) \big| x'(t - s) C' R_2 C x(t - s)\, ds$$

$$+ \frac{1}{2} \int_0^{\tau(t)} \big| \sigma(s) \big|\, ds \int_0^{\tau(t)} \big| \sigma(\theta) \big| x'(t - \theta) C' P R_2^{-1} P C x(t - \theta)\, d\theta$$

$$\leq \frac{\sigma}{2} \int_0^{\hat{t}} \big| \sigma(s) \big| x'(t - s) C' \big[R_2 + P R_2^{-1} P \big] C x(t - s)\, ds.$$

Therefore,

$$L V_1 \leq x'(t) \big[A' P + P A + \beta P R_1 P \big] x(t)$$

$$+ \int_0^{\hat{h}} \big| \beta(s) \big| x'(t - s) B' R_1^{-1} B x(t - s)\, ds,$$

$$+ \frac{\sigma}{2} \int_0^{\hat{t}} \big| \sigma(s) \big| x'(t - s) C' \big[R_2 + P R_2^{-1} P \big] C x(t - s)\, ds.$$

By the representation (2.35) with

$$D = A' P + P A + \beta P R_1 P, \qquad l = 2,$$

$$S_1 = B' R_1^{-1} B, \qquad d v_1(s) = \big| \beta(s) \big|\, ds, \quad s \in [0, \hat{h}],$$

$$S_2 = C' \big[R_2 + P R_2^{-1} P \big] C, \qquad d v_1(s) = \frac{\sigma}{2} \big| \sigma(s) \big|\, ds, \quad s \in [0, \hat{t}],$$

and Theorem 2.5 we obtain the following:

Theorem 6.15 *Suppose that conditions (6.49) hold and that, for some positive definite matrices R_1, R_2, and Q, there exists a positive definite solution P of the matrix Riccati equation*

$$A' P + P A + \beta P R_1 P + \beta B' R_1^{-1} B + \frac{\sigma^2}{2} C' \big(R_2 + P R_2^{-1} P \big) C = -Q. \qquad (6.50)$$

Then the trivial solution of (6.48) is asymptotically mean-square stable.

Remark 6.20 Similarly to the previous remarks in Theorem 6.15, other matrix Riccati equations can be used.

Example 6.11 In the scalar case, (6.50) has a positive solution if and only if

$$A + \beta |B| + \frac{\sigma^2}{2} C^2 < 0.$$

Chapter 7
Stochastic Systems with Markovian Switching

Investigation of systems with Markovian switching has a long history (see, for instance, [119, 120, 197, 198, 266, 268, 278] and references therein). In this chapter sufficient conditions for asymptotic mean-square stability of the solutions of stochastic differential equations with delay and Markovian switching are obtained. In particular, an application to Markov chain with two states and numerical simulation of systems with Markovian switching are considered.

7.1 The Statement of the Problem

Consider the stochastic differential equation

$$\dot{x}(t) = f\big(t, x_t, \eta(t)\big) + g\big(t, x_t, \eta(t)\big)\dot{w}(t),$$

$$x_0 = \phi \in H. \tag{7.1}$$

Here $w(t) \in \mathbf{R}^m$ is the standard Wiener process, $\eta(t)$ is a scalar Markov chain with finite set of states a_i, $i \in \mathfrak{N} = \{1, 2, \ldots, N\}$, and the probabilities of transition

$$p_{ij}(t) = \mathbf{P}\big\{\eta(\tau + t) = a_j/\eta(\tau) = a_i\big\}, \quad t, \tau \geq 0.$$

We suppose that the Markov chain $\eta(t)$ is independent on the Wiener process $w(t)$ and the probabilities of transition can be represented in the form

$$p_{ij}(\Delta) = \begin{cases} \lambda_{ij}\Delta + o(\Delta), & j \neq i, \\ 1 + \lambda_{ii}\Delta + o(\Delta), & j - i, \end{cases}$$

where

$$\lambda_{ij} \geq 0, \quad j \neq i, \qquad \lambda_{ii} \leq 0, \qquad \sum_{j=1}^{N} \lambda_{ij} = 0, \quad i \in \mathfrak{N}.$$

L. Shaikhet, *Lyapunov Functionals and Stability of Stochastic Functional Differential Equations*, DOI 10.1007/978-3-319-00101-2_7,
© Springer International Publishing Switzerland 2013

For arbitrary symmetric matrix $F(s)$, $s \geq 0$, the inequality $dF(s) \geq 0$ means that for each function $\varphi \in H$,

$$\int_0^\infty \varphi'(-s)\,dF(s)\,\varphi(-s) \geq 0.$$

Here and below, all integrals of such a type are understood as Stieltjes integrals. For arbitrary symmetric matrices $F(s)$ and $G(s)$, $s \geq 0$, $dF(s) \geq dG(s)$ means that $dF(s) - dG(s) \geq 0$, and $|A|$ denotes some matrix norm of a matrix A.

The set of negative definite matrices K_i is called uniformly negative definite with respect to $i \in \mathfrak{N}$ if there exists $c > 0$ such that for each $x \in \mathbf{R}^n$,

$$x'K_i x \leq -c|x|^2, \quad i \in \mathfrak{N}.$$

It is assumed that the functionals $f(t, \varphi, a_i) \in \mathbf{R}^n$ and $g(t, \varphi, a_i) \in \mathbf{R}^{n \times m}$ are defined for $t \geq 0$, $\varphi \in H$, $i \in \mathfrak{N}$ and satisfy the usual conditions [84, 87] of existence and uniqueness of the solution of (7.1). It is assumed also that the following conditions hold:

(H_1) For every positive definite matrix P_i, there exist symmetric matrices Q_i and $F_i(s)$ such that

$$\varphi'(0)P_i f(t, \varphi, a_i) \leq \varphi'(0)Q_i\varphi(0) + \int_0^\infty \varphi'(-s)\,dF_i(s)\,\varphi(-s),$$

$$dF_i(s) \geq 0, \ i \in \mathfrak{N}. \tag{7.2}$$

(H_1') The functional $f(t, \varphi, a_i)$ has the form

$$f(t, \varphi, a_i) = A_i\varphi(0) + \int_0^\infty dB_i(s)\,\varphi(-s), \quad i \in \mathfrak{N}. \tag{7.3}$$

(H_2) For every positive definite matrix P_i, there exists a matrix $G_i(s)$ such that

$$\mathrm{Tr}\left[g'(t, \varphi, a_i)P_i g(t, \varphi, a_i)\right] \leq \int_0^\infty \varphi'(-s)\,dG_i(s)\,\varphi(-s),$$

$$dG_i(s) \geq 0, \ i \in \mathfrak{N}. \tag{7.4}$$

(H_3) For the matrices $F_i(s)$ and $G_i(s)$ from (H_1) and (H_2), there exists a matrix $R(s)$ such that

$$dR(s) \geq 2dF_i(s) + dG_i(s), \quad i \in \mathfrak{N},$$

$$R = \int_0^\infty dR(s), \quad |R| < \infty, \quad \int_0^\infty s\big|dR(s)\big| < \infty. \tag{7.5}$$

Remark 7.1 Note that if the functional $f(t, \varphi, a_i)$ satisfies (H_1'), then it satisfies (H_1) too. For instance, for

$$Q_i = \frac{1}{2}\left(P_i A_i + A_i' P_i + \int_0^\infty \big|P_i\,dB_i(s)\big|I\right), \quad dF_i(s) = \frac{1}{2}\big|P_i\,dB_i(s)\big|I. \tag{7.6}$$

Let us represent an arbitrary functional $V(t, \varphi, a_i)$ defined for $t \geq 0$, $\varphi \in H$, $i \in \mathfrak{N}$, in the form $V(t, \varphi, a_i) = V(t, \varphi(0), \varphi(s), a_i)$, $s < 0$, and put

$$V_\varphi(t, x, a_i) = V(t, \varphi, a_i) = V\big(t, x, x(t+s), a_i\big), \quad s < 0,$$

$$\varphi = x_t, \quad x = \varphi(0) = x(t).$$

Let D be the class of functionals $V(t, \varphi, a_i)$ for which the function $V_\varphi(t, x, a_i)$ is twice continuously differentiable with respect to x and continuously differentiable with respect to t for almost all $t \geq 0$. For the functionals from D, the generator L of (7.1) is defined by the formula

$$LV(t, \varphi, a_i) = \frac{\partial}{\partial t} V_\varphi(t, x, a_i) + f'(t, \varphi, a_i) \frac{\partial}{\partial x} V_\varphi(t, x, a_i)$$

$$+ \frac{1}{2} \operatorname{Tr}\left[g'(t, \varphi, a_i) \frac{\partial^2}{\partial x^2} V_\varphi(t, x, a_i) g(t, \varphi, a_i) \right]$$

$$+ \sum_{j \neq i} \lambda_{ij}\big(V(t, \varphi, a_j) - V(t, \varphi, a_i)\big). \tag{7.7}$$

Note that by the condition on λ_{ij} the last summand in (7.7) can be written also in the form

$$\sum_{j=1}^{N} \lambda_{ij} V(t, \varphi, a_j).$$

Similarly to (2.8) for $0 \leq s \leq t < \infty$, we have

$$\mathbf{E}V\big(t, x_t, \eta(t)\big) = \mathbf{E}V\big(s, x_s, \eta(s)\big) + \int_s^t \mathbf{E}LV\big(\theta, x_\theta, \eta(\theta)\big) d\theta.$$

Similarly to Theorem 2.1, one can prove that asymptotic mean-square stability conditions can be obtained by construction of some positive definite (or positive semidefinite for differential equations of neutral type) Lyapunov functionals $V(t, \varphi, a_i)$ for which the inequality

$$LV(t, \varphi, a_i) \leq -c|\varphi(0)|^2, \quad c > 0, \tag{7.8}$$

holds. Below sufficient conditions for the asymptotic mean-square stability of the trivial solution of (7.1) are obtained via the procedure of constructing Lyapunov functionals.

7.2 Stability Theorems

Here two theorems about sufficient conditions for asymptotic mean-square stability of the trivial solution of (7.1) are obtained.

Theorem 7.1 *Let hypotheses* H_1, H_2, H_3 *hold and suppose that the matrices*

$$K_i = 2Q_i + \sum_{j \neq i} \lambda_{ij}(P_j - P_i) + R \tag{7.9}$$

are negative definite for each $i \in \mathfrak{N}$. *Then the trivial solution of* (7.1) *is asymptotically mean-square stable.*

Proof Let us construct a Markov chain $P(t)$ with the set of states $P_1, P_2, \ldots, P_N$, where P_i, $i \in \mathfrak{N}$, are positive definite matrices. We will suppose that $P(t) = P_i$ if $\eta(t) = a_i$, $i \in \mathfrak{N}$, and the probabilities of transition

$$p_{ij}(t) = \mathbf{P}\{P(\tau + t) = P_j / P(\tau) = P_i\}, \quad t, \tau \geq 0,$$

are the same as for the Markov chain $\eta(t)$.

We will construct a Lyapunov functional V for (7.1) in the form $V = V_1 + V_2$, where $V_1(t, \varphi, a_i) = \varphi'(0) P_i \varphi(0)$. Calculating LV_1, we obtain

$$LV_1(t, \varphi, a_i) = 2\varphi'(0) P_i f(t, \varphi, a_i) + \mathrm{Tr}\big[g'(t, \varphi, a_i) P_i g(t, \varphi, a_i)\big]$$

$$+ \sum_{j \neq i} \lambda_{ij} \varphi'(0)(P_j - P_i)\varphi(0).$$

Using (7.2), (7.4), and (7.5), we have

$$LV_1(t, x_t, a_i) \leq 2x'(t) Q_i x(t) + 2 \int_0^\infty x'(t - s)\, dF_i(s)\, x(t - s)$$

$$+ \int_0^\infty x'(t - s)\, dG_i(s)\, x(t - s) + \sum_{j \neq i} \lambda_{ij} x'(t)(P_j - P_i) x(t)$$

$$\leq x'(t) \Big[2Q_i + \sum_{j \neq i} \lambda_{ij}(P_j - P_i) \Big] x(t)$$

$$+ \int_0^\infty x'(t - s)\, dR(s)\, x(t - s).$$

Consider the functional

$$V_2(t, x_t) = \int_0^\infty \int_{t-s}^t x'(\tau)\, dR(s)\, x(\tau)\, d\tau.$$

Since

$$LV_2(t, x_t) = \int_0^\infty \big[x'(t)\, dR(s)\, x(t) - x'(t - s)\, dR(s)\, x(t - s) \big],$$

using (7.9) and the condition that the matrices K_i are uniformly negative definite with respect to $i \in \mathfrak{N}$, for the functional $V = V_1 + V_2$, we obtain

$$LV(t, \varphi, a_i) \leq \varphi'(0) K_i \varphi(0) \leq -c|\varphi(0)|^2.$$

Therefore, the trivial solution of (7.1) is asymptotically mean-square stable. The proof is completed. $\qquad\qquad\square$

Consider the equation

$$\dot{x}(t) = A(\eta(t))x(t) + Bx(t - h) + g(t, x_t, \eta(t))\dot{w}(t). \qquad (7.10)$$

We suppose that if $\eta(t) = a_i$, then $A(\eta(t)) = A_i$, $g(t, \varphi, a_i)$ satisfies (7.4). Using Remark 7.1, it is easy to get that in this case

$$f(t, \varphi, a_i) = A_i \varphi(0) + B\varphi(-h),$$

$$Q_i = \frac{1}{2}\left(P_i A_i + A'_i P_i + |P_i B| I\right), \qquad dF_i(s) = \frac{1}{2}|P_i B| I \delta(s - h)\, ds.$$

Put

$$P = \sup_{i \in \mathfrak{N}} P_i, \qquad dG(s) = \sup_{i \in \mathfrak{N}} dG_i(s), \qquad G = \int_0^\infty dG(s)$$

and suppose that

$$|G| < \infty, \qquad \int_0^\infty s|dG(s)| < \infty.$$

From Theorem 7.1 it follows that if the matrices

$$K_i = P_i A_i + A'_i P_i + \left(|P_i| + |P|\right)|B| I + \sum_{j \neq i} \lambda_{ij}(P_j - P_i) + G \qquad (7.11)$$

are negative definite for each $i \in \mathfrak{N}$, then the trivial solution of (7.10) is asymptotically mean-square stable.

Let us obtain another stability condition.

Theorem 7.2 *Let* $|B|h < 1$ *and let* R_i, $i \in \mathfrak{N}$, *be uniformly negative definite matrices, where*

$$R_i = (A_i + B)' P_i + P_i(A_i + B) + \Lambda_i + G + h(\rho_i + \beta)I, \qquad (7.12)$$

$$\Lambda_i = \sum_{j \neq i} \lambda_{ij}(P_j - P_i), \qquad \rho_i = \left|(A_i + B)' P_i B + \Lambda_i B\right|,$$

$$\qquad\qquad\qquad\qquad\qquad\qquad\qquad\qquad\qquad\qquad\qquad (7.13)$$

$$\beta_i = \left(h|B' \Lambda_i B| + \rho_i\right), \qquad \beta = \sup_{i \in \mathfrak{N}} \beta_i.$$

Then the trivial solution of (7.10) *is asymptotically mean-square stable.*

Proof Reduce (7.10) to the form of a stochastic differential neutral-type equation

$$\frac{d}{dt}\left(x(t) + \int_{t-h}^{t} Bx(s)\,ds\right) = \left(A(\eta(t)) + B\right)x(t) + g\left(t, x_t, \eta(t)\right)\dot{w}(t).$$

Consider the functional

$$V_1(t, x_t, a_i) = \left(x(t) + \int_{t-h}^{t} Bx(s)\,ds\right)' P_i \left(x(t) + \int_{t-h}^{t} Bx(s)\,ds\right).$$

Calculating LV_1, we obtain

$$LV_1(t, x_t, a_i) = 2x'(t)(A_i + B)' P_i \left(x(t) + \int_{t-h}^{t} Bx(s)\,ds\right)$$

$$+ \mathrm{Tr}\left[g'(t, x_t, a_i) P_i g(t, x_t, a_i)\right]$$

$$+ \left(x(t) + \int_{t-h}^{t} Bx(s)\,ds\right)' \Lambda_i \left(x(t) + \int_{t-h}^{t} Bx(s)\,ds\right)$$

$$= 2x'(t)(A_i + B)' P_i x(t) + 2x'(t)(A_i + B)' P_i \int_{t-h}^{t} Bx(s)\,ds$$

$$+ \mathrm{Tr}\left[g'(t, x_t, a_i) P_i g(t, x_t, a_i)\right] + x'(t)\Lambda_i x(t)$$

$$+ 2\left(\int_{t-h}^{t} Bx(s)\,ds\right)' \Lambda_i x(t)$$

$$+ \left(\int_{t-h}^{t} Bx(s)\,ds\right)' \Lambda_i \left(\int_{t-h}^{t} Bx(s)\,ds\right)$$

$$= x'(t)\left[(A_i + B)' P_i + P_i(A_i + B) + \Lambda_i\right]x(t)$$

$$+ \mathrm{Tr}\left[g'(t, x_t, a_i) P_i g(t, x_t, a_i)\right]$$

$$+ 2x'(t)\left[(A_i + B)' P_i B + \Lambda_i B\right]\int_{t-h}^{t} x(s)\,ds$$

$$+ \int_{t-h}^{t} x'(s)\,ds\, B'\Lambda_i B \int_{t-h}^{t} x(\tau)\,d\tau.$$

By (7.4) and (7.13) we have

$$LV_1(t, x_t, a_i) \leq x'(t)\left[(A_i + B)' P_i + P_i(A_i + B) + \Lambda_i\right]x(t)$$

$$+ \int_{0}^{\infty} x'(t - s)\,dG_i(s)\,x(t - s) + |B'\Lambda_i B|\left(\int_{t-h}^{t} |x(s)|\,ds\right)^2$$

$$+ \rho_i \int_{t-h}^{t} \left(|x(t)|^2 + |x(s)|^2\right)ds$$

$$\leq x'(t)\big[\big(A(i)+B\big)'P_i + P_i\big(A(i)+B\big) + \Lambda_i + h\rho_i I\big]x(t)$$
$$+ \int_0^\infty x'(t-s)\,dG_i(s)\,x(t-s) + \beta \int_{t-h}^t \big|x(s)\big|^2\,ds.$$

Putting

$$V_2(t,x_t) = \int_0^\infty \int_{t-s}^t x'(\tau)\,dG(s)\,x(\tau)\,d\tau + \beta \int_{t-h}^t (s-t+h)\big|x(s)\big|^2\,ds,$$

we obtain

$$LV_2(t,x_t) = x'(t)Gx(t) + \beta h\big|x(t)\big|^2$$
$$- \int_0^\infty x'(t-s)\,dG(s)\,x(t-s) - \beta \int_{t-h}^t \big|x(s)\big|^2\,ds.$$

Therefore, using (7.12) and the negative definiteness of the matrices R_i, $i \in \mathfrak{N}$, for the functional $V = V_1 + V_2$, we have

$$LV(t,x_t,a_i) \leq x'(t)R_i x(t) \leq -c\big|x(t)\big|^2. \qquad (7.14)$$

From this and from $|B|h < 1$ it follows that the trivial solution of (7.10) is asymptotically mean-square stable. The proof is completed.   □

7.3 Application to Markov Chain with Two States

Consider the scalar stochastic differential equation

$$\dot{x}(t) = \eta(t)x(t) + bx(t-h) + \sigma x(t-\tau)\dot{w}(t), \qquad (7.15)$$

where $b > 0$, $h \geq 0$, and $\eta(t)$ is a Markov chain with two states $\{a_1, a_2\}$ such that

$$a_2 < 0, \quad a_1 > |a_2| > b + \frac{\sigma^2}{2}. \qquad (7.16)$$

It is clear that if $\eta(t) = a_1$, then the trivial solution of (7.15) is unstable. But for $\eta(t) = a_2$, from (7.16) the inequality

$$a_2 + b + \frac{\sigma^2}{2} < 0$$

follows, which is (see (3.11)) a sufficient condition for the asymptotic mean-square stability of the trivial solution of (7.15).

7.3.1 The First Stability Condition

Let us obtain a sufficient condition for asymptotic mean-square stability of the trivial solution of (7.15), supposing that the Markov chain $\eta(t)$ has two states $\{a_1, a_2\}$ with condition (7.16) and the transition rates λ_{12} and λ_{21} such that $\lambda_{12} > \lambda_{21}$.

From Theorem 7.1 and (7.11) it follows that the sufficient condition for the asymptotic mean-square stability of the trivial solution of (7.15) has the form

$$2p_i a_i + (p_i + p)b + p\sigma^2 + \lambda_{ij}(p_j - p_i) < 0, \quad i, j = 1, 2, \ j \neq i, \qquad (7.17)$$

where $p_1 > 0$, $p_2 > 0$, $p = \max(p_1, p_2)$. It is easy to see that if $i = 1$ and $p_1 \leq p_2$, then condition (7.17) is impossible. So, let $p_1 > p_2$. Therefore, $p = p_1$ and $\gamma = p_2 p_1^{-1} \in (0, 1)$. From (7.17) we have

$$2a_1 + 2b + \sigma^2 + \lambda_{12}(\gamma - 1) < 0,$$

$$2a_2\gamma + (\gamma + 1)b + \sigma^2 + \lambda_{21}(1 - \gamma) < 0.$$

From this and from (7.16) it follows that

$$\frac{\lambda_{21} + b + \sigma^2}{\lambda_{21} + 2|a_2| - b} < \gamma < 1 - \frac{2(a_1 + b) + \sigma^2}{\lambda_{12}}. \qquad (7.18)$$

Thus, if the condition

$$\frac{\lambda_{21} + b + \sigma^2}{\lambda_{21} + 2|a_2| - b} < 1 - \frac{2(a_1 + b) + \sigma^2}{\lambda_{12}} \qquad (7.19)$$

holds, then there exists a positive number $\gamma \in (0, 1)$ such that condition (7.18) holds too. From (7.19) and (7.16) it follows that the sufficient condition for the asymptotic mean-square stability of the trivial solution of (7.15) has the form

$$|a_2| < a_1 < \frac{\lambda_{12}(|a_2| - b - \varepsilon)}{\lambda_{21} + 2|a_2| - b} - b - \varepsilon, \quad \varepsilon = \frac{\sigma^2}{2}. \qquad (7.20)$$

Note that condition (7.20) is possible if and only if $\lambda_{12} > \lambda_{21}$ only. Indeed, from (7.20) it follows that

$$\lambda_{12} > (\lambda_{21} + 2|a_2| - b)\frac{|a_2| + b + \varepsilon}{|a_2| - b - \varepsilon} > \lambda_{21} + 2|a_2| - b > \lambda_{21}.$$

7.3.2 The Second Stability Condition

Let us obtain another stability condition. From Theorem 7.2 for (7.15) we have

$$2(a_i + b)p_i + p\sigma^2 + \lambda_{ij}(p_j - p_i) + h(\rho_i + \beta) < 0,$$

$$i, j = 1, 2, \ j \neq i, \qquad (7.21)$$

where

$$\rho_i = b\big|(a_i + b)p_i + \lambda_{ij}(p_j - p_i)\big|, \qquad \beta = \max(\beta_1, \beta_2),$$
$$\beta_i = hb^2\lambda_{ij}|p_j - p_i| + \rho_i, \quad i,j = 1,2, \ j \neq i. \tag{7.22}$$

Let us transform condition (7.21)–(7.22) to a more visual form. From (7.16) it follows that for $i = 1$ and $p_1 \leq p_2$, condition (7.21) is impossible. So, assume that $p_1 > p_2$. Then from (7.22) it follows that

$$\rho_1 < b\big((a_1 + b)p_1 + \lambda_{12}(p_1 - p_2)\big),$$
$$\rho_2 < b\big(|a_2 + b|p_2 + \lambda_{21}(p_1 - p_2)\big),$$
$$\beta_1 < \overline{\beta}_1 = b\big[(a_1 + b)p_1 + \lambda_{12}(1 + bh)(p_1 - p_2)\big], \tag{7.23}$$
$$\beta_2 < \overline{\beta}_2 = b\big[|a_2 + b|p_2 + \lambda_{21}(1 + bh)(p_1 - p_2)\big].$$

By (7.16) we have $|a_2 + b| = -a_2 - b = |a_2| - b < a_1 + b$. Since, besides, $p_1 > p_2$ and $\lambda_{12} > \lambda_{21}$, we have $\overline{\beta}_1 > \overline{\beta}_2$. Thus, condition (7.21) follows from

$$2(a_i + b)p_i + p_1\sigma^2 + \lambda_{ij}(p_j - p_i) + h(\rho_i + \overline{\beta}_1) < 0. \tag{7.24}$$

In addition, using (7.23), we obtain that (7.24) follows from

$$2(a_1 + b)p_1 + p_1\sigma^2 + \lambda_{12}(p_2 - p_1)$$
$$+ hb\big[2(a_1 + b)p_1 + \lambda_{12}(2 + bh)(p_1 - p_2)\big] < 0,$$
$$2(a_2 + b)p_2 + p_1\sigma^2 + \lambda_{21}(p_1 - p_2)$$
$$+ hb\big[(|a_2| - b)p_2 + (a_1 + b)p_1 + \big(\lambda_{12}(1 + bh) + \lambda_{21}\big)(p_1 - p_2)\big] < 0.$$

Using $\gamma = p_2 p_1^{-1} \in (0, 1)$, we can rewrite these inequalities in the form

$$2(a_1 + b) + \sigma^2 + \lambda_{12}(\gamma - 1)$$
$$+ hb\big[2(a_1 + b) + \lambda_{12}(2 + bh)(1 - \gamma)\big] < 0,$$
$$2(a_2 + b)\gamma + \sigma^2 + \lambda_{21}(1 - \gamma) \tag{7.25}$$
$$+ hb\big[(|a_2| - b)\gamma + a_1 + b + \big(\lambda_{12}(1 + bh) + \lambda_{21}\big)(1 - \gamma)\big] < 0.$$

Suppose now that

$$bh < \sqrt{2} - 1. \tag{7.26}$$

Then from (7.25) we obtain

$$\frac{A + (a_1 + b)bh + \sigma^2}{A + C} < \gamma < 1 - \frac{2(a_1 + b)(1 + bh) + \sigma^2}{B}, \tag{7.27}$$

where

$$A = (\lambda_{21} + \lambda_{12}bh)(1 + bh),$$

$$B = \lambda_{12}\left(1 - 2bh - b^2h^2\right), \tag{7.28}$$

$$C = (2 - bh)\left(|a_2| - b\right).$$

By (7.26) and (7.16) we have that $B > 0$ and $C > 0$.

Thus, if the condition

$$\frac{A + (a_1 + b)bh + \sigma^2}{A + C} < 1 - \frac{2(a_1 + b)(1 + bh) + \sigma^2}{B}$$

or (that is the same in combination with (7.16))

$$|a_2| < a_1 < \frac{BC - \sigma^2(A + B + C)}{Bbh + 2(1 + bh)(A + C)} - b \tag{7.29}$$

holds, then there exists $\gamma \in (0, 1)$ such that condition (7.27) holds too, and therefore the conditions of Theorem 7.2 hold. This means that condition (7.29) or (7.28) is a sufficient condition for the asymptotic mean-square stability of the trivial solution of (7.15).

In Fig. 7.1 the stability regions for (7.15), given by condition (7.20), are shown for $\lambda_{12} = 28$, $\lambda_{21} = 0.05$, $b = 1$, $\varepsilon = 0.25$ (the bound 1), and the stability regions, given by condition (7.29), (7.28), are shown for the same values of the parameters λ_{12}, λ_{21}, b, p and the different values of h: (2) $h = 0$, (3) $h = 0.01$, (4) $h = 0.02$, (5) $h = 0.03$, (6) $h = 0.04$. It is easy to see that in spite of rough estimates that were used for getting inequalities (7.23), the stability condition (7.29), (7.28) for small enough h is better than (7.20).

In Fig. 7.2 the stability region for (7.15), given by condition (7.29), (7.28), is shown for $\lambda_{12} = 15$, $\lambda_{21} = 1$, $b = 0.2$, $h = 0.2$, $\sigma = 0$. Putting $\lambda_{12} = 5$ and using the same values of the other parameters, we obtain the stability region shown in Fig. 7.3.

We can see that in the case $\lambda_{12} = 15$ (Fig. 7.2) the point $A(a_1, a_2) = A(1, -0.5)$ belongs to the stability region, and therefore at this point the trivial solution of (7.15) is asymptotically mean-square stable. On the other hand, in the case $\lambda_{12} = 5$ (Fig. 7.3) the point $A(1, -0.5)$ does not belong to the stability region. Since condition (7.29), (7.28) is a sufficient condition only, for $\lambda_{12} = 5$, at the point $A(1, -0.5)$ the trivial solution of (7.15) can be either stable or unstable.

Remark 7.2 Let us show that the trivial solution of (7.15) can be asymptotically mean-square stable not by conditions (7.16) only. Consider (7.15) by the conditions

$$a_1 = -1, \qquad a_2 = 1, \qquad b = -1,$$
$$\sigma = 0.1, \qquad \lambda_{12} = 1, \qquad \lambda_{21} = 5, \tag{7.30}$$

and obtain the maximum value of the delay h for which the conditions of Theorem 7.2 hold.

By Theorem 7.2 we obtain the stability conditions in the form (7.21) again, i.e.,

$$2(a_i + b)p_i + \lambda_{ij}(p_j - p_i) + p\sigma^2 + h(\rho_i + \beta) < 0, \quad i \neq j, \tag{7.31}$$

Fig. 7.1 Stability region for (7.15), given by condition (7.20), is shown for $\lambda_{12} = 28$, $\lambda_{21} = 0.05$, $b = 1$, $\varepsilon = 0.25(1)$, and stability regions, given by condition (7.29), (7.28), are shown for the same values of the parameters λ_{12}, λ_{21}, b, ε and the different values of h: (2) $h = 0$, (3) $h = 0.01$, (4) $h = 0.02$, (5) $h = 0.03$, (6) $h = 0.04$

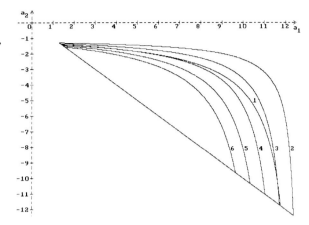

Fig. 7.2 Stability region for (7.15), given by the condition (7.29), (7.28), is shown for $\lambda_{12} = 15$, $\lambda_{21} = 1$, $b = 0.2$, $h = 0.2$, $\sigma = 0$

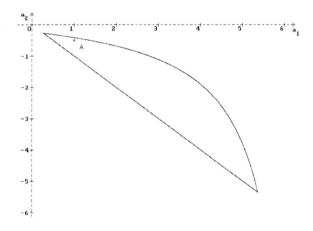

Fig. 7.3 Stability region for (7.15), given by the condition (7.29), (7.28), is shown for $\lambda_{12} = 5$, $\lambda_{21} = 1$, $b = 0.2$, $h = 0.2$, $\sigma = 0$

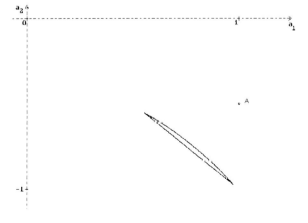

where $i = 1, 2$, $p = \max(p_1, p_2)$. But put now $p_2 = \gamma p_1$ with $1 < \gamma < 3$. Then from (7.13) and (7.30) we have

$$\rho_1 = (3 - \gamma)p_1, \qquad \rho_2 = 5(\gamma - 1)p_1,$$
$$\beta_1 = \big[h(\gamma - 1) + 3 - \gamma\big]p_1, \qquad \beta_2 = 5(\gamma - 1)(h + 1)p_1. \tag{7.32}$$

If, in addition, $\gamma \in (4/3, 3)$, then $\beta_2 > \beta_1$ for each $h > 0$, and therefore $\beta = \max(\beta_1, \beta_2) = \beta_2$. As a result, the stability conditions (7.31) take the form

$$5(\gamma - 1)h(h + 1) + h(3 - \gamma) + 1.01\gamma < 5,$$
$$5(\gamma - 1)h(h + 1) + 5h(\gamma - 1) + 0.01\gamma < 5(\gamma - 1). \tag{7.33}$$

From this it follows that

$$h < \sqrt{\left(\frac{2\gamma - 1}{5(\gamma - 1)}\right)^2 + \frac{5 - 1.01\gamma}{5(\gamma - 1)}} - \frac{2\gamma - 1}{5(\gamma - 1)},$$
$$h < \sqrt{\frac{9.99\gamma - 10}{5(\gamma - 1)}} - 1.$$

It is easy to check that for $\gamma = 1.900923$, the both inequalities take the form $h < 0.412721$. This means that by the values of the parameters (7.30) the inequality $h < 0.412721$ is a sufficient condition for the asymptotic mean-square stability of the trivial solution of (7.15).

However, it is necessary to remember that conditions (7.33) are sufficient conditions only. So, in reality, the trivial solution of (7.15) by the conditions (7.33) can be asymptotically mean-square stable also for $h \geq 0.412721$, as it will be shown below (in Sect. 7.4.2) by a numerical simulation.

7.4 Numerical Simulation of Systems with Markovian Switching

Taking into account that it is difficult enough in each case to get analytical conditions for stability, it is very important to have numerical methods for stability investigation. A numerical procedure for investigation of stability of stochastic systems with Markovian switching is considered here. This procedure can be used in the cases where analytical conditions of stability are absent. Some examples of using of the proposed numerical procedure are considered. the results of the calculations are presented by a lot of figures.

7.4.1 *System Without Stochastic Perturbations*

Consider the differential equation (7.15) again but with $\sigma = 0$, i.e., without stochastic perturbations of the type of white noise. In this case we have

$$\dot{x}(t) = \eta(t)x(t) + bx(t-h),$$
$$x(s) = \phi(s), \quad s \in [-h, 0].$$
(7.34)

Here, as before, $\eta(t)$ is a Markov chain with two states $\{a_1, a_2\}$, initial distribution

$$p_i = \mathbf{P}\{\eta(0) = a_i\}, \quad i = 1, 2,$$
(7.35)

and the probabilities of transition $p_{ij}(\Delta)$ of the form

$$p_{ij}(\Delta) = \mathbf{P}\{\eta(t+\Delta) = a_j / \eta(t) = a_i\} = \lambda_{ij}\Delta + o(\Delta),$$
$$i, j = 1, 2, \ i \neq j.$$
(7.36)

We suppose also that

$$a_2 < 0, \qquad a_1 > |a_2| > b > 0, \qquad \lambda_{12} > \lambda_{21} > 0.$$
(7.37)

Let us investigate the stability of the trivial solution of (7.34) at the point $A(a_1, a_2)$ with $a_1 = 1, a_2 = -0.5$ using a numerical simulation. Note that this point belongs to the stability region (Fig. 7.2). Consider the difference analogue of (7.34) in the form

$$x_{i+1} = (1 + \eta_i\Delta)x_i + bx_{i-m}\Delta,$$

where

$$x_i = x(t_i), \qquad \eta_i = \eta(t_i), \quad t_i = i\Delta, \qquad h = m\Delta, \quad \Delta > 0.$$

Simulation of the Markov chain η_i, $i = 0, 1, \ldots$, can be reduced to the simpler problem, simulation of a sequence of independent random variables ζ_i that are uniformly distributed on the interval $[0,1]$. Indeed, using (7.35), we have

$$p_1 = \mathbf{P}\{\eta_0 = a_1\} = \mathbf{P}\{\zeta_0 < p_1\},$$
$$p_2 = \mathbf{P}\{\eta_0 = a_2\} = \mathbf{P}\{\zeta_0 > p_1\}.$$
(7.38)

So, if after simulation of ζ_0, we obtain $\zeta_0 < p_1$, then we put $\eta_0 = a_1$, else we put $\eta_0 = a_2$.

Further, if $\eta_{i-1} = a_1$, $i > 0$, then from (7.36) for small enough $\Delta > 0$, we have

$$\mathbf{P}\{\eta_i = a_2 / \eta_{i-1} = a_1\} = \mathbf{P}\{\zeta_i < \lambda_{12}\Delta\} = \lambda_{12}\Delta,$$
$$\mathbf{P}\{\eta_i = a_1 / \eta_{i-1} = a_1\} = \mathbf{P}\{\zeta_i > \lambda_{12}\Delta\} = 1 - \lambda_{12}\Delta.$$
(7.39)

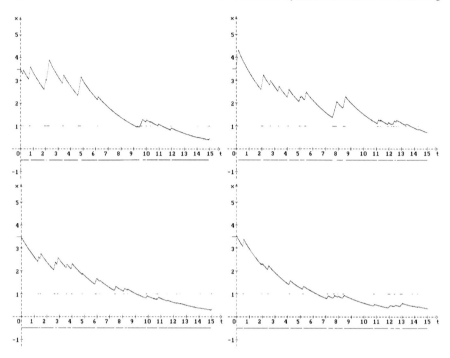

Fig. 7.4 Four possible trajectories of the Markov chain $\eta(t)$ and four appropriate trajectories of the solution of (7.34) are shown for the different values of the parameters: $a_1 = 1$, $a_2 = -0.5$, $b = 0.2$, $h = 0.2$, $x(s) = 3.5$, $s \in [-0.2, 0]$, $p_1 = p_2 = 0.5$, $\lambda_{12} = 15$, $\lambda_{21} = 1$, $\Delta = 0.01$

Therefore, we obtain the following algorithm: if after simulation of ζ_i, we have $\zeta_i > \lambda_{12}\Delta$, then we put $\eta_i = a_1$, else we put $\eta_i = a_2$. Analogously, if $\eta_{i-1} = a_2$, $i > 0$, then from (7.36) for small enough $\Delta > 0$, we have

$$
\begin{aligned}
\mathbf{P}\{\eta_i = a_1/\eta_{i-1} = a_2\} &= \mathbf{P}\{\zeta_i < \lambda_{21}\Delta\} = \lambda_{21}\Delta, \\
\mathbf{P}\{\eta_i = a_2/\eta_{i-1} = a_2\} &= \mathbf{P}\{\zeta_i > \lambda_{21}\Delta\} = 1 - \lambda_{21}\Delta.
\end{aligned}
\tag{7.40}
$$

By (7.38)–(7.40) we obtain the following algorithm: if after simulation of ζ_i, we have $\zeta_i < \lambda_{21}\Delta$, then we put $\eta_i = a_1$, else we put $\eta_i = a_2$.

In Fig. 7.4 four possible trajectories of the Markov chain $\eta(t)$ and four appropriate trajectories of the solution of (7.34) are shown for the following values of the parameters: $a_1 = 1$, $a_2 = -0.5$, $b = 0.2$, $h = 0.2$, $x(s) = 3.5$, $s \in [-0.2, 0]$, $p_1 = p_2 = 0.5$, $\lambda_{12} = 15$, $\lambda_{21} = 1$, $\Delta = 0.01$. In Fig. 7.5 hundred trajectories of the solution of (7.34) are shown for the same values of the parameters. We can see that all trajectories converge to zero in the whole accordance with the properties of stability.

Put now $\lambda_{12} = 5$ with the same values of the other parameters. In this case the point $A(1, -0.5)$ does not belong to the stability region (Fig. 7.3), and we have (Figs. 7.6 and 7.7) another situation: the trajectories of the solution fill by itself

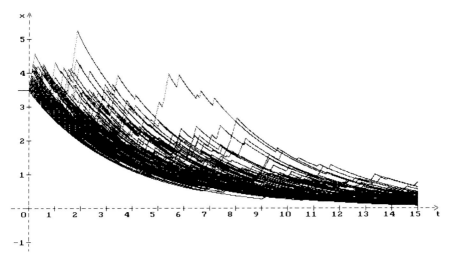

Fig. 7.5   Hundred trajectories of the solution of (7.34) are shown for the same as in Fig. 7.4 values of the parameters

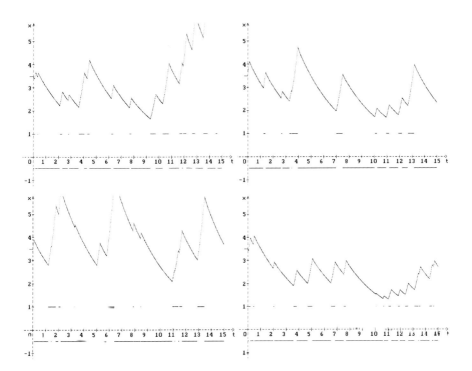

Fig. 7.6   Four possible trajectories of the Markov chain $\eta(t)$ and four appropriate trajectories of the solution of (7.34) are shown for the different values of the parameters: $a_1 = 1$, $a_2 = -0.5$, $b = 0.2$, $h = 0.2$, $x(s) = 3.5$, $s \in [-0.2, 0]$, $p_1 = p_2 = 0.5$, $\lambda_{12} = 5$, $\lambda_{21} = 1$, $\Delta = 0.01$

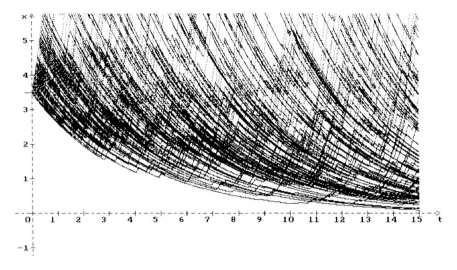

Fig. 7.7 Hundred trajectories of the solution of (7.34) are shown for the same as in Fig. 7.6 values of the parameters

the whole admissible space between the solutions of (7.34) with $\eta(t) \equiv a_1$ and $\eta(t) \equiv a_2$. This shows us the instability of the trivial solution of (7.34).

7.4.2 System with Stochastic Perturbations

Here we will investigate the stability of the trivial solution of stochastic differential equation (7.15) using a simultaneous numerical simulation of both stochastic processes: the Markov chain $\eta(t)$ with two states as it is described in the previous section and the Wiener process $w(t)$ as it is described in Sect. 2.1.1.

Consider the difference analogue of (7.15) of the form

$$x_{i+1} = (1 + \eta_i \Delta)x_i + bx_{i-m}\Delta + \sigma x_{i-l}w_{i+1},$$

where

$$x_i = x(t_i), \qquad \eta_i = \eta(t_i), \qquad w_{i+1} = W_n^{(c)}(t_{i+1}) - W_n^{(c)}(t_i),$$

$$t_i = i\Delta, \qquad h = m\Delta, \qquad \tau = l\Delta, \qquad \Delta > 0.$$

In Fig. 7.8 four realizations of one trajectory of the Wiener process $w(t)$, one trajectory of the Markov chain $\eta(t)$, and one trajectory of the solution $x(t)$ of (7.15) are shown for $a_1 = -1$, $a_2 = 1$, $b = -1$, $\sigma = 1$, $\lambda_{12} = 1$, $\lambda_{21} = 5$, $h = 0.99$, $\tau = 1.6$, $x_0 = 7.5$. In Fig. 7.9 we can see at the same time ten trajectories of the solution $x(t)$ of (7.15) for the referred above values of the parameters. All trajectories of the solution $x(t)$ go to zero. In Fig. 7.10 we can see the similar picture for $\sigma = 0.1$ and

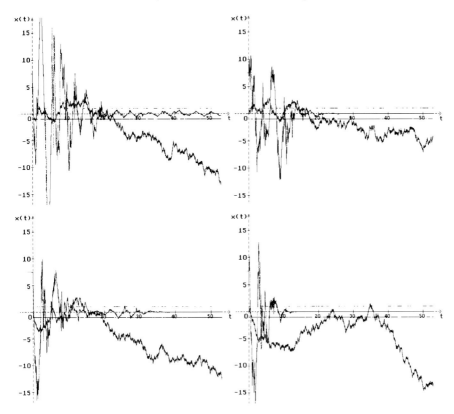

Fig. 7.8 Four realizations of one trajectory of the Wiener process $w(t)$, one trajectory of the Markov chain $\eta(t)$, and one appropriate trajectory of the solution $x(t)$ of (7.15) are shown for $a_1 = -1$, $a_2 = 1$, $b = -1$, $\sigma = 1$, $\lambda_{12} = 1$, $\lambda_{21} = 5$, $h = 0.99$, $\tau = 1.6$, $x_0 = 7.5$

the same values of the other parameters. We can see that in this case all trajectories of the solution $x(t)$ go to zero more quickly.

7.4.3 System with Random Delay

Consider the differential equation

$$\dot{x}(t) + bx\big(t - \eta(t)\big) = 0,$$
$$x(s) = \phi(s), \quad s \in [-h, 0], \tag{7.41}$$

with random delay. Here $b > 0$, $\eta(t)$ is a Markov chain with two states $\{a_1, a_2\}$ such that $0 < a_1 < a_2 = h$, the initial distribution (7.35), and the probabilities of transition (7.36).

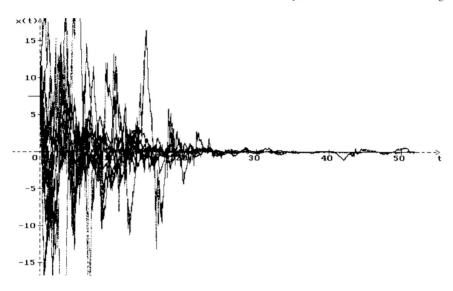

Fig. 7.9   Ten trajectories of the solution $x(t)$ of (7.15) for the values of the parameters in Fig. 7.8

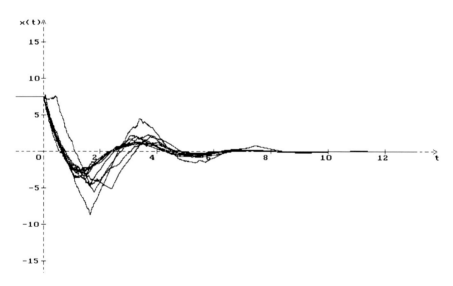

Fig. 7.10   Ten trajectories of the solution $x(t)$ of (7.15) for the values of the parameters in Fig. 7.8 except for $\sigma = 0.1$

As it was shown in Example 1.2, if $\eta(t) = h = $ const, then the inequality $bh < \frac{\pi}{2}$ is a necessary and sufficient condition for the asymptotic stability of the trivial solution of (7.41).

Let us investigate the stability of the trivial solution of (7.41) using the numerical simulation of the Markov chain $\eta(t)$ as in the previous examples. Consider the

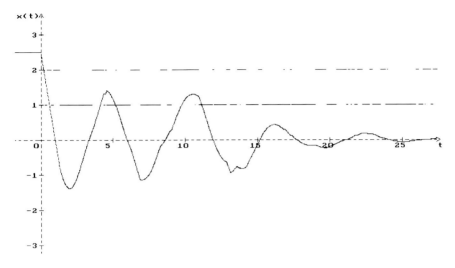

Fig. 7.11 One trajectory of the Markov chain $\eta(t)$ and one appropriate trajectory of the solution of (7.41) are shown for $b = 1$, $a_1 = 1$, $a_2 = 2$, $x(s) = 2.5$, $s \in [-2, 0]$, $p_1 = p_2 = 0.5$, $\lambda_{12} = 1$, $\lambda_{21} = 3$, $\Delta = 0.001$

difference analogue of (7.41) of the form

$$x_{i+1} = x_i - \Delta b x_{i-\eta_i},$$

$$x_i = x(t_i), \qquad \eta_i = \eta(t_i), \quad t_i = i\Delta, \ \Delta > 0.$$

Put $b = 1$, $a_1 = 1$, $a_2 = 2$, $x(s) = \text{const}$, $s \in [-h, 0]$, $\Delta = 0.001$. Several solutions of (7.41) with different values of the initial function are shown in Figs. 1.1, 1.2, and 1.3 for $\eta(t) \equiv a_1 = 1$, $\eta(t) \equiv \frac{1}{2}\pi$, $\eta(t) \equiv a_2 = 2$, respectively. We can see that all solutions converge to zero in the first case, are bounded only but do not converge to zero in the second case, and go to $\pm\infty$ in the last case.

In Fig. 7.11 one trajectory of the Markov chain $\eta(t)$ and one appropriate trajectory of the solution of (7.41) are shown for $p_1 = p_2 = 0.5$, $\lambda_{12} = 1$, $\lambda_{21} = 3$ and the same values of other parameters. In Fig. 7.12 hundred trajectories of the solution of (7.41) are shown for the same values of the parameters. We can see that all trajectories converge to zero. If $\lambda_{21} = 6$, then hundred trajectories of the solution of (7.41) converge to zero more quickly (Fig. 7.13).

In Fig. 7.14 one trajectory of the Markov chain $\eta(t)$ and one appropriate trajectory of the solution of (7.41) are shown for $p_1 = p_2 = 0.5$, $\lambda_{12} = 3$, $\lambda_{21} = 1$. In Fig. 7.15 hundred trajectories of the solution of (7.41) are shown for these values of λ_{12} and λ_{21}. We can see that in this case the solution of (7.41) is unstable.

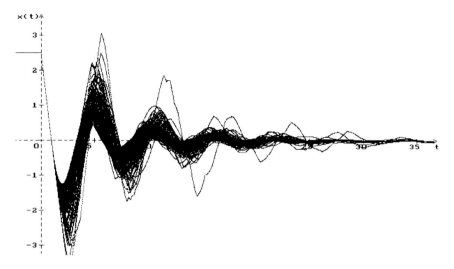

Fig. 7.12 Hundred trajectories of the solution of (7.41) are shown for the same as in Fig. 7.11 values of the parameters

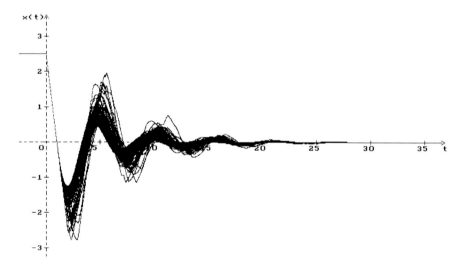

Fig. 7.13 Hundred trajectories of the solution of (7.41) are shown for the same values of the parameters as in Fig. 7.11 except for $\lambda_{21} = 6$

7.4.4 Some Generalization of Algorithm of Markov Chain Numerical Simulation

Let us show that the proposed numerical simulation (7.38)–(7.40) of the Markov chain with two states can be generalized to a Markov chain $\eta(t)$ with n states

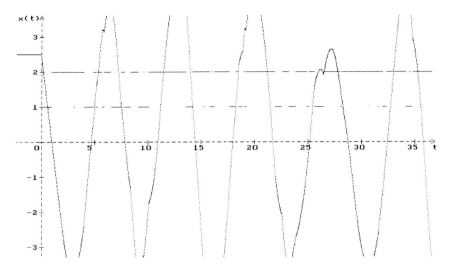

Fig. 7.14 One trajectory of the Markov chain $\eta(t)$ and one appropriate trajectory of the solution of (7.41) are shown for $b = 1$, $a_1 = 1$, $a_2 = 2$, $x(s) = 2.5$, $s \in [-2, 0]$, $p_1 = p_2 = 0.5$, $\lambda_{12} = 3$, $\lambda_{21} = 1$, $\Delta = 0.001$

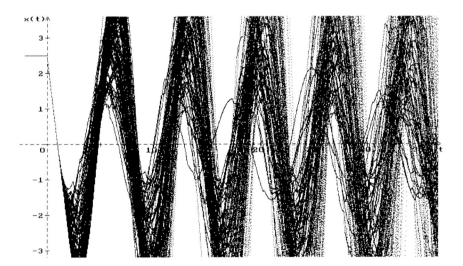

Fig. 7.15 Hundred trajectories of the solution of (7.41) are shown for the same values of the parameters as in Fig. 7.14

$\{a_1, \ldots, a_n\}$, initial distribution

$$p_i = \mathbf{P}\{\eta(0) = a_i\}, \quad i = 1, \ldots, n,$$

and the probabilities of transition

$$p_{ij}(\Delta) = \mathbf{P}\{\eta(t + \Delta) = a_j / \eta(t) = a_i\} = \lambda_{ij}\Delta + o(\Delta),$$

$$i, j = 1, \ldots, n, \ i \neq j.$$

Indeed, put

$$\eta_k = \eta(t_k), \quad t_k = k\Delta, \ \Delta > 0.$$

Reduce a simulation of the Markov chain η_k, $k = 0, 1, \ldots$, to a simulation of a sequence of independent random variables ζ_k that are uniformly distributed on $[0, 1]$. Note that

$$p_i = \mathbf{P}\{\eta_0 = a_i\} = \mathbf{P}\{S_{i-1} < \zeta_0 < S_i\}, \quad i = 1, \ldots, n,$$

where

$$S_0 = 0, \qquad S_i = \sum_{j=1}^{i} p_j, \quad i = 1, \ldots, n-1, \ S_n = 1.$$

It is easy to see that for each result of simulation ζ_0, there exists a number i such that $S_{i-1} < \zeta_0 < S_i$. So, we put $\eta_0 = a_i$.

Further, put $Q_{i0} = 0$,

$$Q_{ij} = \sum_{l=1}^{j} \lambda_{il}\Delta, \quad 1 \leq j < i,$$

$$Q_{ij} = \sum_{l=1}^{i-1} \lambda_{il}\Delta + \sum_{l=i+1}^{j} \lambda_{il}\Delta, \quad i < j \leq n.$$

Then

$$\mathbf{P}\{\eta_k = a_j / \eta_{k-1} = a_i\} = \mathbf{P}\{Q_{i,j-1} < \zeta_k < Q_{ij}\} = \lambda_{ij}\Delta$$

for $j < i$ or $j > i + 1$, and

$$\mathbf{P}\{\eta_k = a_{i+1} / \eta_{k-1} = a_i\} = \mathbf{P}\{Q_{i,i-1} < \zeta_k < Q_{i,i+1}\} = \lambda_{i,i+1}\Delta,$$

$$\mathbf{P}\{\eta_k = a_i / \eta_{k-1} = a_i\} = \mathbf{P}\{Q_{in} < \zeta_k\} = 1 - Q_{in}.$$

Thus, we obtain the following algorithm. Let $\eta_{k-1} = a_i$, $k > 0$. If after simulation of ζ_k, there exists a number $j \leq n$ such that $j \neq i$ and $Q_{i,j-1} < \zeta_k < Q_{ij}$ for $j < i$ or $j > i + 1$, $Q_{i,i-1} < \zeta_k < Q_{i,i+1}$ for $j = i + 1$, then we put $\eta_k = a_j$. If such a number j does not exist, i.e., $\zeta_k > Q_{in}$ for $i < n$ or $\zeta_k > Q_{n,n-1}$ for $i = n$, then we put $\eta_k = a_i$.

Chapter 8
Stabilization of the Controlled Inverted Pendulum by a Control with Delay

The problem of stabilization for the mathematical model of the controlled inverted pendulum (Fig. 8.1) during many years is very popular among the researchers (see, for instance, [1, 2, 33, 36, 37, 114, 121, 169, 170, 201, 209, 228, 256, 273, 276, 278, 282]). Unlike the classical way of stabilization in which the stabilized control is a linear combination of the state and velocity of the pendulum, here we propose another way of stabilization. We suppose that only the trajectory of the pendulum can be observed and stabilized control depends on the whole trajectory of the pendulum. We consider linear and nonlinear models of the controlled inverted pendulum by stochastic perturbations and investigate zero and steady-state nonzero solutions analytically and by numerical simulations.

8.1 Linear Model of the Controlled Inverted Pendulum

Here we consider the linearized mathematical model of the controlled inverted pendulum. We show that by some controls depending on a delay the trivial solution of the considered model can be asymptotically stable but by some other controls depending on a delay it is impossible. We also obtain sufficient conditions for stabilization of the trivial solution of the considered model by stochastic perturbations.

8.1.1 Stabilization by the Control Depending on Trajectory

The linearized mathematical model of the controlled inverted pendulum can be described by the linear second-order differential equation

$$\ddot{x}(t) - ax(t) = u(t), \quad a > 0, \ t \geq 0. \tag{8.1}$$

The classical way of stabilization [121] for (8.1) uses the control $u(t)$ in the form

$$u(t) = -b_1 x(t) - b_2 \dot{x}(t), \quad b_1 > a, \ b_2 > 0.$$

L. Shaikhet, *Lyapunov Functionals and Stability of Stochastic Functional Differential Equations*, DOI 10.1007/978-3-319-00101-2_8,
© Springer International Publishing Switzerland 2013

Fig. 8.1 Controlled inverted
pendulum

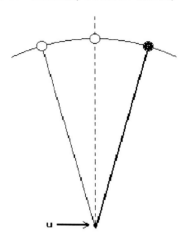

But this type of control, which represents instantaneous feedback, is quite difficult
to realize because usually it is necessary to have some finite time to make measurements of the coordinates and velocities, to treat the results of the measurements, and
to implement them in the control action.

Here we propose another way of stabilization. We suppose that only the trajectory of the pendulum is observed and the control $u(t)$ does not depend on the
velocity but it depends on the previous values of the trajectory $x(s)$, $s \le t$, and has
the form

$$u(t) = \int_0^\infty dK(\tau) x(t - \tau), \qquad (8.2)$$

where the kernel $K(\tau)$ is a right-continuous function of bounded variation on
$[0, \infty]$, and the integral is understood in the Stieltjes sense. This means in particular that both distributed and discrete delays can be used depending on the concrete
choice of the kernel $K(\tau)$.

The initial condition for system (8.1)–(8.2) has the form

$$x(s) = \phi(s), \qquad \dot{x}(s) = \dot{\phi}(s), \quad s \le 0, \qquad (8.3)$$

where $\phi(s)$ is a given continuously differentiable function.

Let us show that the inverted pendulum (8.1) can be stabilized by the control (8.2). Substituting (8.2) into (8.1) and putting $x_1(t) = x(t)$, $x_2(t) = \dot{x}(t)$, we
obtain the system of differential equations with delay

$$\dot{x}_1(t) = x_2(t),$$

$$\dot{x}_2(t) = ax_1(t) + \int_0^\infty dK(\tau) x_1(t - \tau). \qquad (8.4)$$

To prove the asymptotic stability of the trivial solution of (8.4), we will use the procedure of constructing Lyapunov functionals. Put

$$k_i = \int_0^\infty \tau^i \, dK(\tau), \quad i = 0, 1, \qquad k_j = \int_0^\infty \tau^j \, |dK(\tau)|, \quad j = 2, 3. \qquad (8.5)$$

Since

$$\int_{t-\tau}^t x_2(s) \, ds = \int_{t-\tau}^t \dot{x}_1(s) \, ds = x_1(t) - x_1(t - \tau),$$

then by (8.5) we have

$$\int_0^\infty dK(\tau) x_1(t - \tau) = k_0 x_1(t) - \int_0^\infty dK(\tau) \int_{t-\tau}^t x_2(s) \, ds.$$

Therefore, from the second equation of (8.4) it follows that

$$\dot{x}_2(t) = (a + k_0)x_1(t) - \int_0^\infty dK(\tau) \int_{t-\tau}^t x_2(s) \, ds. \qquad (8.6)$$

Put now

$$a_1 = -(a + k_0), \qquad z(t) = x_2(t) - G(t, x_{2t}),$$

$$G(t, x_{2t}) = \int_0^\infty dK(\tau) \int_{t-\tau}^t (s - t + \tau)x_2(s) \, ds. \qquad (8.7)$$

By (8.5) we have

$$\frac{d}{dt} \int_0^\infty dK(\tau) \int_{t-\tau}^t (s - t + \tau)x_2(s) \, ds = k_1 x_2(t) - \int_0^\infty dK(\tau) \int_{t-\tau}^t x_2(s) \, ds. \qquad (8.8)$$

Using (8.6), (8.7), and (8.8), we reduce (8.4) to the form

$$\dot{x}_1(t) = x_2(t),$$
$$\dot{z}(t) = -a_1 x_1(t) - k_1 x_2(t), \qquad (8.9)$$

which is the first step of the procedure.

Following the second step of the procedure of constructing Lyapunov functionals, we consider the auxiliary system of the ordinary differential equations

$$\dot{y}_1(t) = y_2(t),$$
$$\dot{y}_2(t) = -a_1 y_1(t) - k_1 y_2(t). \qquad (8.10)$$

Rewriting (8.10) in the matrix form

$$\dot{y}(t) = Ay(t), \quad y = \begin{pmatrix} y_1 \\ y_2 \end{pmatrix}, \quad A = \begin{pmatrix} 0 & 1 \\ -a_1 & -k_1 \end{pmatrix}, \qquad (8.11)$$

and using Corollary 1.1, we obtain that the inequalities

$$a_1 > 0, \qquad k_1 > 0, \tag{8.12}$$

are necessary and sufficient conditions for the asymptotic stability of the zero solution of (8.10).

From Theorem 1.3, Remark 1.1, and (1.29) it follows that by conditions (8.12) the matrix equation (1.27) with the matrix A defined in (8.11) and the symmetric matrix Q with the elements $q_{11} = q > 0$, $q_{22} = 1$, $q_{12} = 0$ has the matrix solution P with the elements

$$p_{11} = k_1 p_{12} + a_1 p_{22}, \quad p_{12} = \frac{q}{2a_1}, \quad p_{22} = \frac{p}{2a_1}, \quad p = \frac{q + a_1}{k_1}. \tag{8.13}$$

Following the third step of the procedure of constructing Lyapunov functionals, we will construct a Lyapunov functional for (8.9) in the form $V = V_1 + V_2$, where

$$V_1 = p_{11} x_1^2(t) + 2p_{12} x_1(t) z(t) + p_{22} z^2(t) \tag{8.14}$$

with p_{11}, p_{12}, p_{22}, and $z(t)$ are defined in (8.13) and (8.7).

Calculating $\dot{V}_1$, by (8.14), (8.9), and (8.7) we have

$$\dot{V}_1 = 2\big(p_{11} x_1(t) + p_{12} z(t)\big) x_2(t) - 2\big(p_{12} x_1(t) + p_{22} z(t)\big)\big(a_1 x_1(t) + k_1 x_2(t)\big)$$

$$= -2p_{12} a_1 x_1^2(t) - 2(k_1 p_{22} - p_{12}) x_2^2(t) + 2(p_{11} - k_1 p_{12} - a_1 p_{22}) x_1(t) x_2(t)$$

$$+ 2p_{22} a_1 x_1(t) G(t, x_{2t}) + 2(k_1 p_{22} - p_{12}) x_2(t) G(t, x_{2t}).$$

Using (8.7), (8.13), and arbitrary $\gamma > 0$, we obtain

$$\dot{V}_1 = -q x_1^2(t) - x_2^2(t) + p \int_0^\infty dK(\tau) \int_{t-\tau}^t (s - t + \tau) x_1(t) x_2(s)\, ds$$

$$+ \int_0^\infty dK(\tau) \int_{t-\tau}^t (s - t + \tau) x_2(t) x_2(s)\, ds$$

$$\leq -q x_1^2(t) - x_2^2(t) + \frac{p}{2} \int_0^\infty |dK(\tau)| \int_{t-\tau}^t (s - t + \tau)\left(\gamma x_1^2(t) + \frac{1}{\gamma} x_2^2(s)\right) ds$$

$$+ \frac{1}{2} \int_0^\infty |dK(\tau)| \int_{t-\tau}^t (s - t + \tau)\big(x_2^2(t) + x_2^2(s)\big)\, ds$$

$$= -\left(q - \frac{1}{4}\gamma p k_2\right) x_1^2(t) - \left(1 - \frac{1}{4} k_2\right) x_2^2(t)$$

$$+ \alpha \int_0^\infty |dK(\tau)| \int_{t-\tau}^t (s - t + \tau) x_2^2(s)\, ds, \tag{8.15}$$

where

$$\alpha = \frac{1}{2}\left(1 + \frac{p}{\gamma}\right). \tag{8.16}$$

Following the fourth step of the procedure of constructing Lyapunov functionals, let us choose the functional V_2 in the form

$$V_2 = \frac{\alpha}{2}\int_0^\infty |dK(\tau)| \int_{t-\tau}^t (s - t + \tau)^2 x_2^2(s)\,ds. \tag{8.17}$$

Then

$$\dot{V}_2 = \frac{\alpha k_2}{2}x_2^2(t) - \alpha \int_0^\infty |dK(\tau)| \int_{t-\tau}^t (s - t + \tau)x_2^2(s)\,ds, \tag{8.18}$$

and by (8.16), for the functional $V = V_1 + V_2$, we obtain

$$\dot{V} \le -\left(q - \frac{\gamma p k_2}{4}\right)x_1^2(t) - \left(1 - \frac{k_2}{2} - \frac{p k_2}{4\gamma}\right)x_2^2(t). \tag{8.19}$$

From the condition of positivity of the expressions in the brackets we have

$$\frac{4q}{pk_2} > \gamma > \frac{pk_2}{2(2 - k_2)} > 0.$$

So, if

$$\frac{4q}{pk_2} > \frac{pk_2}{2(2 - k_2)} > 0, \tag{8.20}$$

then there exists $\gamma > 0$ such that the Lyapunov functional V for some $c > 0$ satisfies the condition $\dot{V} \le -c(x_1^2(t) + x_2^2(t))$.

Using the representation (8.13) for p, rewrite (8.20) in the form

$$\frac{2(2 - k_2)k_1^2}{k_2^2} > \frac{(a_1 + q)^2}{4q} = \frac{a_1}{4}\left(2 + \frac{a_1}{q} + \frac{q}{a_1}\right)$$

and note that the right-hand part of the obtained inequality reaches its minimum for $q = a_1$. Using this q and (8.7), (8.12), we obtain the following theorem.

Theorem 8.1 *Let*

$$a_1 = -(a + k_0) > 0, \qquad k_1 > 0, \qquad k_2 < \frac{4}{1 + \sqrt{1 + 4a_1 k_1^{-2}}}. \tag{8.21}$$

Then the trivial solution of (8.4) is asymptotically stable.

Remark 8.1 Note that two first inequalities in (8.21) are necessary conditions for asymptotic stability of the trivial solution of (8.4), while the third inequality in (8.21) is a sufficient condition only.

Remark 8.2 Note that the third inequality in (8.21) can be represented in the form

$$k_1 > k_2 \sqrt{\frac{a_1}{2(2 - k_2)}}.$$

Remark 8.3 Let us show that the functional $G(t, x_{2t})$ in (8.7) satisfies condition (2.10) for a differential equation of neutral type. Indeed, transform $G(t, x_{2t})$ in the following way:

$$G(t, x_{2t}) = \int_0^\infty dK(\tau) \int_0^\tau (\tau - \theta) x_2(t - \theta) \, d\theta$$

$$= \int_0^\infty x_2(t - \theta) \int_\theta^\infty (\tau - \theta) \, dK(\tau) \, d\theta.$$

From (8.21) it follows that $k_2 < 2$. So, by (8.5),

$$\left| \int_0^\infty \int_\theta^\infty (\tau - \theta) \, dK(\tau) \, d\theta \right| \leq \int_0^\infty |dK(\tau)| \int_0^\tau (\tau - \theta) \, d\theta$$

$$= \frac{1}{2} \int_0^\infty \tau^2 |dK(\tau)| = \frac{k_2}{2} < 1.$$

8.1.2 Some Examples

Here we consider some different examples of the controls of type (8.2) and show that, for some of them, a stabilization is possible but, for some of them, a stabilization is impossible.

Example 8.1 Put in (8.2) $dK(\tau) = b\delta(\tau - h) \, d\tau$, where $h > 0$, and $\delta(\tau)$ is the Dirac delta-function. In this case system (8.1)–(8.2) has the form

$$\ddot{x}(t) - ax(t) = bx(t - h). \tag{8.22}$$

From (8.21) we obtain a contradiction for b: $k_0 = b < -a < 0$, $k_1 = bh > 0$. From Theorem 8.1 and Remark 8.1 it follows that the trivial solution of (8.22) cannot be asymptotically stable for any b and h, i.e., the inverted pendulum (8.1) cannot be stabilized by the control of the form $u(t) = bx(t - h)$.

Let us obtain the same statement using the characteristic equation of (8.22) $z^2 - a - be^{-hz} = 0$, $z = \alpha + i\beta$, that can be transformed to the following system of two equations for real α and β:

$$\alpha^2 - \beta^2 - a - be^{-h\alpha} \cos(h\beta) = 0, \qquad 2\alpha\beta + be^{-h\alpha} \sin(h\beta) = 0. \tag{8.23}$$

Let us show that system (8.23) for every b and h has at least one solution with $\alpha \geq 0$. Suppose first that $a + b > 0$. In this case, for $\beta = 0$, we have $\alpha^2 - a = be^{-h\alpha}$.

Consider the characteristic quasipolynomial $\Delta(\alpha) = \alpha^2 - a - be^{-h\alpha}$. Since $\Delta(0) = -(a+b) < 0$ and $\lim_{\alpha \to \infty} \Delta(\alpha) = +\infty$, then there exists $\alpha > 0$ such that $\Delta(\alpha) = 0$. Let now $a + b = 0$. In this case, $\alpha = \beta = 0$ is the solution of system (8.23). Let at last $a + b < 0$. Since $a > 0$, $b < -a < 0$. Choose β so that $0 < h\beta < \frac{\pi}{2}$. Then $\cos(h\beta) > 0$ and $\sin(h\beta) > 0$. From (8.23) it follows that

$$\frac{\alpha^2 - \beta^2 - a}{\cos(h\beta)} = -\frac{2\alpha\beta}{\sin(h\beta)} = be^{-h\alpha} < 0.$$

This means that α is a positive root of the equation $\alpha^2 + 2\alpha\gamma = a + \beta^2$, $\gamma = \beta \cot(h\beta)$, i.e., $\alpha = \sqrt{a + \beta^2 + \gamma^2} - \gamma > 0$.

Thus, for every b and h, system (8.23) has at least one solution with $\alpha \geq 0$, i.e., the trivial solution of (8.22) cannot be asymptotically stable.

Example 8.2 Put in (8.2) $dK(\tau) = (b_1\delta(\tau - h_1) + b_2\delta(\tau - h_2))d\tau$, where $h_1, h_2 > 0$, and $\delta(\tau)$ is the Dirac delta-function. In this case system (8.1)–(8.2) has the form

$$\ddot{x}(t) - ax(t) = b_1x(t - h_1) + b_2x(t - h_2), \tag{8.24}$$

and by (8.5) the stability conditions (8.21) are

$$a_1 = -(a + b_1 + b_2) > 0, \qquad k_1 = b_1h_1 + b_2h_2 > 0,$$

$$k_2 = |b_1|h_1^2 + |b_2|h_2^2 < k_m = \frac{4}{1 + \sqrt{1 + 4a_1k_1^{-2}}}. \tag{8.25}$$

Let us show that for an arbitrary $a > 0$, there exist b_1, b_2, h_1, h_2 such that conditions (8.25) hold and therefore the trivial solution of (8.24) is asymptotically stable.

First, note that (8.24) was already considered in Example 1.7, where it was shown that the inequality $a + b_1 + b_2 < 0$ is a necessary condition for asymptotic stability. This condition coincides with the first condition in (8.25).

Put now $b_1 = b$, $b_2 = -\alpha b$, $h_1 = h$, $h_2 = \beta h$. Here b is an arbitrary positive number, and positive numbers α, β, and h will be chosen below.

By that the first condition in (8.25) takes the form $a + b - \alpha b < 0$ and holds if $\alpha > 1 + ab^{-1}$. The second condition in (8.25) takes the form $bh - \alpha\beta bh = bh(1 - \alpha\beta) > 0$ and holds if $\beta < \alpha^{-1}$. The third condition in (8.25) in this case has the form

$$Ah^2 < \frac{4}{1 + \sqrt{1 + Bh^{-2}}}, \qquad A = b(1 + \alpha\beta^2), \quad B = \frac{4(\alpha - 1 - ab^{-1})}{b(1 - \alpha\beta)^2} > 0.$$

It is easy to show that it holds if $h < 4(\sqrt{A(AB + 8)})^{-1}$.

So, we have shown that for an arbitrary $a > 0$, the parameters b_1, b_2, h_1, and h_2 can be chosen such that all conditions (8.25) hold and therefore the trivial solution of (8.24) is asymptotically stable.

Example 8.3 Consider the control (8.2) in the form

$$u(t) = b \int_{h_2}^{h_1} x(t-s)\,ds, \quad h_1 > h_2 \ge 0. \tag{8.26}$$

In this case from (8.21) we obtain a contradiction for b:

$$k_0 = b(h_1 - h_2) < -a < 0, \qquad k_1 = \frac{b}{2}\left(h_1^2 - h_2^2\right) > 0.$$

Thus, there do not exist b, h_1, h_2 such that conditions (8.21) hold. This means that system (8.1) cannot be stabilized by the control (8.26).

On the other hand, the characteristic equation of system (8.1), (8.26) has the form $z^2 - a - b \int_{h_2}^{h_1} e^{-zs}\,ds = 0$, $z = \alpha + i\beta$ or, in the form of a system of two equations for real α and β,

$$\alpha^2 - \beta^2 - a - bI_c = 0, \qquad 2\alpha\beta + bI_s = 0,$$

$$I_c = \int_{h_2}^{h_1} e^{-\alpha s} \cos(\beta s)\,ds, \qquad I_s = \int_{h_2}^{h_1} e^{-\alpha s} \sin(\beta s)\,ds. \tag{8.27}$$

Let us show that for arbitrary b, h_1, h_2, there exists a solution of (8.27) with nonnegative α. Let first $a + b(h_1 - h_2) > 0$. Put $\beta = 0$ and consider the characteristic quasipolynomial

$$\Delta(\alpha) = \alpha^2 - a - b\int_{h_2}^{h_1} e^{-\alpha s}\,ds = \alpha^2 - a - \frac{b}{\alpha}\left(e^{-\alpha h_2} - e^{-\alpha h_1}\right).$$

It is easy to see that $\lim_{\alpha \to 0} \Delta(\alpha) = -[a + b(h_1 - h_2)] < 0$ but $\lim_{\alpha \to \infty} \Delta(\alpha) = \infty > 0$. Therefore, there exists at least one $\alpha_0 > 0$ such that $\Delta(\alpha_0) = 0$ and $(\alpha_0, 0)$ is a solution of (8.27). If $a + b(h_1 - h_2) = 0$, then $\alpha = \beta = 0$ is a solution of (8.27). Let now $a + b(h_1 - h_2) < 0$. Since $a > 0$, $b < 0$. Choose $\beta > 0$ such that $\beta h_1 < \frac{\pi}{2}$. Then $I_c > 0$ and $I_s > 0$. From (8.27) it follows that

$$\frac{\alpha^2 - \beta^2 - a}{I_c} = -\frac{2\alpha\beta}{I_s} = b < 0.$$

This means that there exists $\alpha > 0$ that is a positive root of the equation $f(\alpha) = 0$, where $f(\alpha) = \alpha^2 + 2\alpha\gamma - a - \beta^2$, $\gamma = \beta I_c I_s^{-1}$. Indeed, it is easy to see that $f(0) = -a - \beta^2 < 0$ and $\lim_{\alpha \to \infty} f(\alpha) = \infty > 0$. So, this equation has a positive root.

Example 8.4 Consider the control (8.2) in the form

$$u(t) = b_1 \int_{h_2}^{h_1} x(t-s)\,ds + b_2 \int_{h_4}^{h_3} x(t-s)\,ds,$$

$$0 < h_4 < h_3 < h_2 < h_1. \tag{8.28}$$

In this case from (8.21) we obtain

$$k_0 = b_1(h_1 - h_2) + b_2(h_3 - h_4) < -a, \tag{8.29}$$

$$k_1 = \frac{b_1}{2}\left(h_1^2 - h_2^2\right) + \frac{b_2}{2}\left(h_3^2 - h_4^2\right) > 0, \tag{8.30}$$

$$k_2 = \frac{|b_1|}{3}\left(h_1^3 - h_2^3\right) + \frac{|b_2|}{3}\left(h_3^3 - h_4^3\right) < \frac{4}{1 + \sqrt{1 + 4|a + k_0|k_1^{-2}}}. \tag{8.31}$$

Let us show that there exist $b_1, b_2, h_1, h_2, h_3, h_4$ such that conditions (8.29)–(8.31) hold. Put, for instance,

$$b_1 = \frac{a}{h}, \qquad b_2 = -\gamma b_1, \qquad h_1 = h, \qquad h_{i+1} = \alpha_i h, \quad i = 1, 2, 3, \tag{8.32}$$

where $h > 0$, $\gamma > 0$,

$$0 < \alpha_3 < \alpha_2 < \alpha_1 < 1. \tag{8.33}$$

In this case conditions (8.29)–(8.30) have the forms respectively

$$k_0 = a\left(1 - \alpha_1 - \gamma(\alpha_2 - \alpha_3)\right) < -a,$$

$$k_1 = \frac{ah}{2}\left(1 - \alpha_1^2 - \gamma(\alpha_2^2 - \alpha_3^2)\right) > 0.$$

These conditions hold if the parameters γ and α_i, $i = 1, 2, 3$, satisfy the inequalities

$$\frac{2 - \alpha_1}{\alpha_2 - \alpha_3} < \gamma < \frac{1 - \alpha_1^2}{\alpha_2^2 - \alpha_3^2}. \tag{8.34}$$

Substituting (8.32) into (8.31), we obtain

$$h < \sqrt{\frac{2}{\alpha a(1 + 2\alpha\beta)}}, \tag{8.35}$$

where

$$\alpha = \frac{1}{3}\left(1 - \alpha_1^3 + \gamma(\alpha_2^3 - \alpha_3^3)\right), \qquad \beta = \frac{\alpha_1 + \gamma(\alpha_2 - \alpha_3) - 2}{(1 - \alpha_1^2 - \gamma(\alpha_2^2 - \alpha_3^2))^2}. \tag{8.36}$$

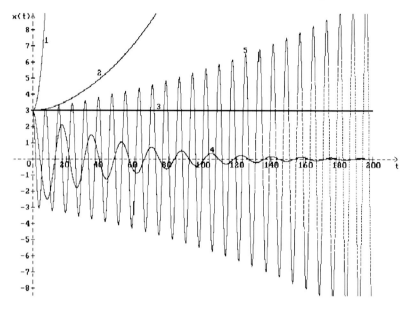

Fig. 8.2 Solutions of (8.1), (8.28) for $a = 1$, $h_1 = 0.48$, $h_2 = 0.24$, $h_3 = 0.12$, $h_4 = 0.06$, $b_1 = ah_1^{-1}$, $b_2 = -\gamma b_1$, $x(s) = 3$, $s \in [-0.48, 0]$, and the different values of γ: (1) $\gamma = 11.5$, (2) $\gamma = 11.99$, (3) $\gamma = 12$, (4) $\gamma = 13$, (5) $\gamma = 16.3$

Thus, if the parameters of (8.1) with the control (8.28) satisfy conditions (8.32)–(8.36), then the trivial solution of (8.1), (8.28) is asymptotically stable.

From (8.33)–(8.34) it follows that the parameters α_i, $i = 1, 2, 3$, should satisfy the conditions

$$0 < \alpha_3 < 2 - \sqrt{3},$$

$$\alpha_3 < \alpha_2 < 2(2 - \sqrt{3}) - \alpha_3, \tag{8.37}$$

$$\alpha_2 < \alpha_1 < \frac{1}{2}\left(\delta + \sqrt{\delta^2 - 8\delta + 4}\right), \quad \delta = \alpha_2 + \alpha_3.$$

Put, for instance, $a = 1$, $\alpha_i = 2^{-i}$, $i = 1, 2, 3$. Conditions (8.37) hold, and condition (8.34) takes the form $12 < \gamma < 16$. Putting $\gamma = 13$, from (8.35)–(8.36) we obtain $h < 1.02392$. Putting $\gamma = 12.1$, we have $h < 2.13984$.

In Fig. 8.2 the solutions of (8.1), (8.28) are shown for $a = 1$, $h_1 = 0.48$, $h_2 = 0.24$, $h_3 = 0.12$, $h_4 = 0.06$, $b_1 = ah_1^{-1}$, $b_2 = -\gamma b_1$, $x(s) = 3$, $s \leq [-0.48, 0]$, and different values of the parameter γ: (1) $\gamma = 11.5$, (2) $\gamma = 11.99$, (3) $\gamma = 12$, (4) $\gamma = 13$, (5) $\gamma = 16.3$. Only one of these solutions with $\gamma = 13$ is asymptotically stable.

8.1.3   About Stabilization by the Control Depending on Velocity or on Acceleration

Consider (8.1) with the control

$$u(t) = \int_0^\infty dK(\tau)\,\dot{x}(t - \tau), \tag{8.38}$$

where

$$\int_0^\infty |dK(\tau)| < \infty. \tag{8.39}$$

The characteristic equation of system (8.1), (8.38) has the form

$$z^2 - a - z \int_0^\infty dK(\tau)\,e^{-z\tau} = 0, \quad z = \alpha + \beta i. \tag{8.40}$$

Let us show that this equation has at least one root z with positive real part α. Indeed, put $\beta = 0$ and $\Delta(\alpha) = \alpha^2 - a - \alpha \int_0^\infty dK(\tau)\,e^{-\alpha\tau}$. Then (8.40) takes the form $\Delta(\alpha) = 0$. Note that $\Delta(0) = -a < 0$. Using (8.39), it is easy to get that

$$\lim_{\alpha \to \infty} \Delta(\alpha) = \lim_{\alpha \to \infty} \alpha^2 \left(1 - \frac{a}{\alpha^2} - \frac{1}{\alpha} \int_0^\infty dK(\tau)\,e^{-\alpha\tau} \right) = \infty.$$

Therefore, there exists at least one $\alpha > 0$ that is a root of the equation $\Delta(\alpha) = 0$. Thus, the inverted pendulum cannot be stabilized by the control of type (8.38) depending on the velocity only.

Consider (8.1) with the control

$$u(t) = \int_0^\infty dK(\tau)\,\ddot{x}(t - \tau), \tag{8.41}$$

where

$$|dK(0)| = |K(+0) - K(0)| < 1, \qquad \int_{+0}^\infty |dK(\tau)| < \infty. \tag{8.42}$$

The characteristic equation of system (8.1), (8.41) has the form

$$z^2 - a - z^2 \int_0^\infty dK(\tau)\,e^{-z\tau} = 0, \quad z = \alpha + \beta i. \tag{8.43}$$

Let us show that this equation has at least one root z with positive real part α. Indeed, put $\beta = 0$ and $\Delta(\alpha) = \alpha^2 - a - \alpha^2 \int_0^\infty dK(\tau)\,e^{-\alpha\tau}$. Then (8.43) takes the form $\Delta(\alpha) = 0$. Note that $\Delta(0) = -a < 0$. Using (8.42), it is easy to get that

$$\lim_{\alpha \to \infty} \Delta(\alpha) = \lim_{\alpha \to \infty} \alpha^2 \left(1 - \frac{a}{\alpha^2} - dK(0) - \int_{+0}^\infty dK(\tau)\,e^{-\alpha\tau} \right) = \infty.$$

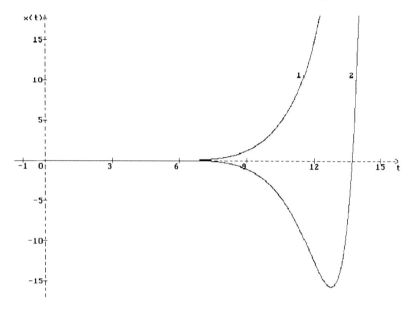

Fig. 8.3 Solutions of (8.44) for $a = 0.01$, $h = 5$, $x(s) = 0.01$, $s \in [-5, 0]$ and the different values of b: (1) $b = -0.01$, (2) $b = 0.01$

Therefore there exists at least one $\alpha > 0$ that is a root of the equation $\Delta(\alpha) = 0$. Thus, the inverted pendulum cannot be stabilized by the control of type (8.41) depending on the acceleration only.

In Fig. 8.3 the solution $x(t)$ of the equation

$$\ddot{x}(t) - ax(t) = b\ddot{x}(t - h) \tag{8.44}$$

is shown for $a = 0.01$, $h = 5$, $x(s) = 0.01$, $s \in [-5, 0]$, and the different values of b: (1) $b = -0.01$, (2) $b = 0.01$.

8.1.4 Stabilization by Stochastic Perturbations

Substituting (8.2) into (8.1) and supposing that the parameter a in (8.1) is under the influence of stochastic perturbations of the type of white noise, we obtain

$$\ddot{x}(t) - \left(a + \sigma \dot{w}(t)\right)x(t) = \int_0^\infty dK(\tau)x(t - \tau), \qquad a > 0, \ t \geq 0. \tag{8.45}$$

Here $w(t)$ is the standard Wiener process, and σ is a constant.

Theorem 8.2 *Let*

$$a_1 = -(a + k_0) > 0, \qquad k_1 > 0,$$

$$\sigma^2 < 2a_1 \left(k_1 - k_2 \sqrt{\frac{a_1}{2(2 - k_2)}} \right), \tag{8.46}$$

where k_i, $i = 0, 1, 2$, are defined in (8.5). Then the trivial solution of (8.45) is asymptotically mean-square stable.

Proof Put $x_1(t) = x(t)$ and $x_2(t) = \dot{x}(t)$. Then (8.45), similarly to (8.9), can be rewritten in the form of the system of stochastic differential equations of neutral type

$$\dot{x}_1(t) = x_2(t),$$
$$\dot{z}(t) = -a_1 x_1(t) - k_1 x_2(t) + \sigma x_1(t) \dot{w}(t), \tag{8.47}$$

with the initial conditions $x_1(s) = \phi(s)$, $x_2(s) = \dot{\phi}(s)$, $s \le 0$. Here k_1, a_1, and $z(t)$ are defined in (8.5), (8.7).

We will construct a Lyapunov functional for (8.47) in the form $V = V_1 + V_2$, where V_1 is defined by (8.14), (8.13). Calculating LV_1, where L is the generator of (8.47), and using (8.7), (8.13), we obtain

$$LV_1 = 2\big(p_{11}x_1(t) + p_{12}z(t)\big)x_2(t)$$

$$\quad - 2\big(p_{12}x_1(t) + p_{22}z(t)\big)\big(a_1x_1(t) + k_1x_2(t)\big) + \sigma^2 p_{22}x_1^2(t)$$

$$= -\big(2p_{12}a_1 - \sigma^2 p_{22}\big)x_1^2(t) - 2(k_1 p_{22} - p_{12})x_2^2(t)$$

$$\quad + 2(p_{11} - k_1 p_{12} - a_1 p_{22})x_1(t)x_2(t)$$

$$\quad + 2p_{22}a_1x_1(t)G(t, x_{2t}) + 2(k_1 p_{22} - p_{12})x_2(t)G(t, x_{2t})$$

$$= -\big(q - \sigma^2 p_{22}\big)x_1^2(t) - x_2^2(t) + p \int_0^\infty dK(\tau) \int_{t-\tau}^t (s - t + \tau)x_1(t)x_2(s)\,ds$$

$$\quad + \int_0^\infty dK(\tau) \int_{t-\tau}^t (s - t + \tau)x_2(t)x_2(s)\,ds.$$

Similarly to (8.15), for arbitrary $\gamma > 0$ and α defined by (8.16), we have

$$LV_1 \le -\left(q - \frac{1}{4}\gamma p k_2 - \sigma^2 p_{22} \right)x_1^2(t) - \left(1 - \frac{1}{4}k_2 \right)x_2^2(t)$$

$$\quad + \alpha \int_0^\infty |dK(\tau)| \int_{t-\tau}^t (s - t + \tau)x_2^2(s)\,ds.$$

Choosing the functional V_2 in the form (8.17) and using (8.16), for the functional $V = V_1 + V_2$, similarly to (8.19), we obtain

$$LV \leq -\left(q - \frac{\gamma pk_2}{4} - \sigma^2 p_{22}\right)x_1^2(t) - \left(1 - \frac{k_2}{2} - \frac{pk_2}{4\gamma}\right)x_2^2(t).$$

Supposing the positivity of the expressions in the brackets, we obtain

$$\frac{4(q - \sigma^2 p_{22})}{pk_2} > \gamma > \frac{pk_2}{2(2 - k_2)} > 0.$$

So, if

$$\frac{4(q - \sigma^2 p_{22})}{pk_2} > \frac{pk_2}{2(2 - k_2)} > 0, \qquad (8.48)$$

then there exists $\gamma > 0$ such that the Lyapunov functional V for some $c > 0$ satisfies the condition $LV \leq -c(x_1^2(t) + x_2^2(t))$.

By the representation (8.13) for p_{22} and p from (8.48) it follows that

$$\sigma^2 < 2a_1 k_1 \left(\frac{q}{q + a_1} - A(q + a_1)\right), \qquad A = \frac{k_2^2}{8k_1^2(2 - k_2)}. \qquad (8.49)$$

The right-hand part of inequality (8.49) reaches its maximum at $q = \sqrt{a_1 A^{-1}} - a_1$. Substituting this q into (8.49), we obtain the last inequality (8.46). The proof is completed. $\square$

Note that by Remark 8.2 conditions (8.46) by $\sigma = 0$ coincide with (8.21).

8.2 Nonlinear Model of the Controlled Inverted Pendulum

8.2.1 Stabilization of the Trivial Solution

Consider separately the deterministic case and the stochastic case.

8.2.1.1 The Deterministic Case

Consider the problem of stabilization for the nonlinear model of the controlled inverted pendulum

$$\ddot{x}(t) - a \sin x(t) = u(t), \qquad a > 0, \qquad (8.50)$$

via the control (8.2).

Let us suppose that conditions (8.21) hold. In Theorem 8.1 it is proved that by conditions (8.21) on the kernel $K(\tau)$ of the control (8.2) the appropriate linearized system (8.4) is asymptotically stable. So, by conditions (8.21) the trivial solution of the nonlinear system (8.50), (8.2) is asymptotically stable too if the initial function (8.3) belongs to some small enough neighborhood of the origin, called a region of attraction. Let us construct some estimate of the region of attraction for the trivial solution of system (8.50), (8.2).

Similarly to (8.9), let us represent (8.50), (8.2) in the form

$$\dot{x}_1(t) = x_2(t),$$
$$\dot{z}(t) = -a_1 x_1(t) - k_1 x_2(t) - af\big(x_1(t)\big),$$

(8.51)

where $f(x) = x - \sin x$, k_1, and a_1 and $z(t)$ are defined by (8.5), (8.7).

Following the procedure of constructing Lyapunov functionals, let us choose the auxiliary ordinary differential equations for system (8.51) in the form (8.10). We will construct a Lyapunov functional for (8.51) in the form $V = V_1 + V_2$, where the functional V_1 is defined again by (8.14), (8.13). Calculating $\dot{V}_1$ for system (8.51), similarly to (8.15), we obtain

$$
\dot{V}_1 \le -\left(q - \frac{1}{4}\gamma p k_2\right) x_1^2(t) - \left(1 - \frac{1}{4}k_2\right) x_2^2(t)
$$

$$
+ \alpha \int_0^\infty |dK(\tau)| \int_{t-\tau}^t (s - t + \tau) x_2^2(s)\, ds
$$

$$
- 2ap_{12} x_1(t) f\big(x_1(t)\big) - 2ap_{22} x_2(t) f\big(x_1(t)\big)
$$

$$
+ 2ap_{22} \int_0^\infty dK(\tau) \int_{t-\tau}^t (s - t + \tau) x_2(s) f\big(x_1(t)\big)\, ds,
$$

where p and α are defined by (8.13) and (8.16).

Suppose that for all $t \ge 0$ and some positive δ,

$$|x_1(t)| \le \delta.$$

(8.52)

Using (8.52), the inequality $|f(x)| = |x - \sin x| \le \frac{1}{6}|x|^3$, and some $\nu > 0$, we get

$$
\left|x_1(t) f\big(x_1(t)\big)\right| \le \frac{x_1^4(t)}{6} \le \frac{\delta^2}{6} x_1^2(t),
$$

$$
2\left|x_2(t) f\big(x_1(t)\big)\right| \le \frac{x_1^2(t)}{6}\left(\nu x_1^2(t) + \frac{1}{\nu} x_2^2(t)\right) \le \frac{\delta^2}{6}\left(\nu x_1^2(t) + \frac{1}{\nu} x_2^2(t)\right),
$$

and

$$2 \int_0^\infty |dK(\tau)| \int_{t-\tau}^t (s-t+\tau)|x_2(s)f(x_1(t))| \, ds$$

$$\leq \frac{\delta^2}{6} \int_0^\infty |dK(\tau)| \int_{t-\tau}^t (s-t+\tau)\left(vx_1^2(t)+\frac{1}{v}x_2^2(s)\right) ds$$

$$\leq \frac{\delta^2}{6}\left(\frac{vk_2}{2}x_1^2(t)+\frac{1}{v}\int_0^\infty |dK(\tau)| \int_{t-\tau}^t (s-t+\tau)x_2^2(s)\,ds\right).$$

As a result,

$$\dot{V}_1 \leq -\left[q-\frac{1}{4}\gamma pk_2 - \delta_1\left(2p_{12}+\left(1+\frac{k_2}{2}\right)p_{22}\right)v\right]x_1^2(t)$$

$$-\left(1-\frac{k_2}{4}-\frac{\delta_1 p_{22}}{v}\right)x_2^2(t)+\alpha_1\int_0^\infty |dK(\tau)| \int_{t-\tau}^t (s-t+\tau)x_2^2(s)\,ds,$$

where

$$\delta_1 = \frac{a\delta^2}{6}, \qquad \alpha_1 = \alpha + \frac{\delta_1 p_{22}}{v}. \tag{8.53}$$

Choosing V_2 in the form

$$V_2 = \frac{\alpha_1}{2}\int_0^\infty |dK(\tau)| \int_{t-\tau}^t (s-t+\tau)^2 x_2^2(s)\,ds \tag{8.54}$$

and using (8.53) and (8.16), for the functional $V = V_1 + V_2$, we obtain

$$\dot{V} \leq -\left(q-A-\frac{\gamma pk_2}{4}\right)x_1^2(t)-\left(1-B-\frac{pk_2}{4\gamma}\right)x_2^2(t), \tag{8.55}$$

where

$$A = A_0 + Dv, \qquad B = B_0 + D\frac{1}{v},$$

$$A_0 = 2p_{12}\delta_1, \; D = (1+B_0)p_{22}\delta_1, \; B_0 = \frac{k_2}{2}. \tag{8.56}$$

Supposing the positivity of the expressions in the brackets in (8.55), we obtain

$$\frac{4(q-A)}{pk_2} > \gamma > \frac{pk_2}{4(1-B)} > 0. \tag{8.57}$$

So, if

$$\frac{4(q-A)}{pk_2} > \frac{pk_2}{4(1-B)} > 0, \tag{8.58}$$

then there exists $\gamma > 0$ such that (8.57) holds.

Using (8.56), rewrite (8.58) in the form

$$(q - A_0 - Dv)\left(1 - B_0 - D\frac{1}{v}\right) > \left(\frac{pB_0}{2}\right)^2. \tag{8.59}$$

It is easy to show that the left-hand side of this inequality reaches its maximum at $v = \sqrt{(q - A_0)(1 - B_0)^{-1}}$. Substituting this v into (8.59), we obtain

$$\sqrt{(q - A_0)(1 - B_0)} > D + \frac{pB_0}{2}.$$

By (8.56) and (8.13) this inequality can be represented in the form

$$k_1\sqrt{(1 - \delta_0)(1 - B_0)} > \frac{1}{2}\left(\sqrt{q} + \frac{a_1}{\sqrt{q}}\right)\left((1 + B_0)\delta_0 + B_0\right), \qquad \delta_0 = \frac{\delta_1}{a_1}.$$

Note that the right-hand side of this inequality reaches its minimum at $q = a_1$. Using this q, we obtain

$$k_1\sqrt{(1 - \delta_0)(1 - B_0)} > \sqrt{a_1}\left((1 + B_0)\delta_0 + B_0\right).$$

This inequality can be rewritten in the form of the quadratic equation with respect to δ_0,

$$\delta_0^2 + 2\mu\delta_0 - \rho < 0, \tag{8.60}$$

where, by (8.56),

$$\mu = \frac{k_2}{2 + k_2} + \frac{k_1^2(2 - k_2)}{a_1(2 + k_2)^2}, \qquad \rho = \frac{2k_1^2 a_1^{-1}(2 - k_2) - k_2^2}{(2 + k_2)^2}.$$

Note that by Remark 8.2 $\rho > 0$. So, inequality (8.60) holds for $\delta_0 = 0$, and therefore it holds also for small enough $\delta_0 > 0$, namely $\delta_0 < \sqrt{\mu^2 + \rho} - \mu$, or by (8.53) and the definition of δ_0,

$$\delta^2 < \frac{6a_1}{a}\sqrt{\mu^2 + \rho} - \mu. \tag{8.61}$$

Let $\lambda_0 > 0$ and $\lambda_1 > 0$ be respectively the minimal and maximal eigenvalues of the positive definite matrix P with the elements p_{ij} defined by (8.13) with $q = a_1$. Then by (8.14), (8.55), (8.54), and (8.5) we have

$$\lambda_0 x_1^2(t) \leq V(t) \leq V(0) \leq \lambda_1\left(x_1^2(0) + z^2(0)\right) + \frac{\alpha_1 k_3}{6}\sup_{s \leq 0} x_2^2(s).$$

From (8.7) and (8.5) it follows that

$$|z(0)| \leq |x_2(0)| + \int_0^\infty |dK(\tau)|\int_{-\tau}^0 (s + \tau)|x_2(s)|\,ds \leq \left(1 + \frac{k_2}{2}\right)\sup_{s \leq 0}|x_2(s)|.$$

So,

$$\lambda_0 x_1^2(t) \le \lambda_1 x_1^2(0) + \left(\lambda_1 \left(1 + \frac{k_2}{2} \right)^2 + \frac{\alpha_1 k_3}{6} \right) \sup_{s \le 0} x_2^2(s),$$

and by (8.52) the domain of attraction for the trivial solution of system (8.50), (8.2) contains the set of initial functions satisfying the inequality

$$\frac{\lambda_1}{\lambda_0} \phi^2(0) + \left(\frac{\lambda_1}{\lambda_0} \left(1 + \frac{k_2}{2} \right)^2 + \frac{\alpha_1 k_3}{6 \lambda_0} \right) \sup_{s \le 0} \dot\phi^2(s) \le \delta^2.$$

By this condition on the initial function $\phi(s)$ with $\delta > 0$ that satisfies (8.61) the solution of system (8.50), (8.2) satisfies condition (8.52).

8.2.1.2 The Stochastic Case

Consider now the nonlinear model of the controlled inverted pendulum (8.50) under the influence of stochastic perturbations of the type of white noise

$$\ddot x(t) - \left(a + \sigma \dot w(t) \right) \sin x(t) = u(t), \quad a > 0, \ t \ge 0, \tag{8.62}$$

with the control (8.2). Similarly to (8.51), rewrite system (8.62), (8.2) in the form

$$\dot x_1(t) = x_2(t),$$
$$\dot z(t) = -a_1 x_1(t) - k_1 x_2(t) - af\left(x_1(t)\right) + \sigma\left(x_1(t) - f\left(x_1(t)\right)\right)\dot w(t), \tag{8.63}$$

where $f(x) = x - \sin x$, and k_1, a_1, and $z(t)$ are defined by (8.5) and (8.7).

Note that (8.47) is the linear part of (8.63). Since $|f(x)| \le \frac{x^3}{6}$, the order of non-linearity of (8.63) is 3. From Theorem 5.2 it follows that if the order of nonlinearity of the nonlinear system under consideration is higher than one, then the sufficient condition for asymptotic mean-square stability of the linear part of this system is at the same time a sufficient condition for stability in probability of the initial nonlinear system. Thus, we obtain the following theorem.

Theorem 8.3 *If conditions (8.46) hold, then the trivial solution of (8.62) by the control (8.2) is stable in probability.*

So, the nonlinear model of the controlled inverted pendulum under stochastic perturbations can be stabilized by a control that depends on the trajectory only.

8.2.2 Nonzero Steady-State Solutions

Here nonzero steady-state solutions of the nonlinear system (8.50), (8.2), (8.3) are studied. Substituting (8.2) into (8.50) and putting $x_1(t) = x(t)$ and $x_2(t) = \dot x(t)$, we represent this system as follows:

$$\dot{x}_1(t) = x_2(t),$$

$$\dot{x}_2(t) = a \sin x_1(t) + \int_0^\infty dK(\tau) x_1(t - \tau). \tag{8.64}$$

To get steady-state solutions of (8.64), let us suppose that $\dot{x}_1(t) \equiv 0$ and $\dot{x}_2(t) \equiv 0$. We obtain that $x_2(t) \equiv 0$ and $x_1(t) \equiv \hat{x}$ is a root of the equation

$$a \sin \hat{x} + k_0 \hat{x} = 0. \tag{8.65}$$

Suppose that $\hat{x} \neq 0$ and rewrite (8.65) in the form

$$S(\hat{x}) = 0, \tag{8.66}$$

where

$$S(x) = \frac{\sin x}{x} + \frac{k_0}{a}. \tag{8.67}$$

We will call the function $S(x)$ "the characteristic function of the system (8.64)."

Remark 8.4 The statements "$\hat{x}$ is a steady-state solution of the system (8.64)" and "$\hat{x}$ is a root of (8.66)" are equivalent.

Remark 8.5 For all $x \neq 0$,

$$-\alpha \leq \frac{\sin x}{x} < 1,$$

where $0.217233 < \alpha < 0.217234$. Therefore, if

$$-\alpha \leq -\frac{k_0}{a} < 1$$

or

$$0 < a + k_0 \leq (1 + \alpha)a, \tag{8.68}$$

then there exists at least one nonzero root of (8.66).

Remark 8.6 Since the function $S(x)$ is an even function, if $\hat{x}$ is a root of (8.66), then $-\hat{x}$ is a root of (8.66) too.

Remark 8.7 Condition (8.68) contradicts to the necessary condition

$$a_1 = -(a + k_0) > 0$$

for the asymptotic stability of the trivial solution of the corresponding linear system. Thus, by condition (8.68) the trivial solution of the corresponding linear system is unstable.

8.2.3 Stable, Unstable, and One-Sided Stable Points of Equilibrium

Here a stability of the steady-state solutions of system (8.64) is investigated. Let $\hat{x}$ be a root of (8.66). Put

$$x_1 = \hat{x} + y_1, \qquad x_2 = y_2. \tag{8.69}$$

Substituting (8.69) into (8.64) and using (8.65), we obtain

$$\dot{y}_1(t) = y_2(t),$$
$$\dot{y}_2(t) = a\left[\sin(\hat{x} + y_1(t)) - \sin\hat{x}\right] + \int_0^\infty dK(\tau)\, y_1(t - \tau),$$

or, after elementary trigonometric transformations,

$$\dot{y}_1(t) = y_2(t),$$
$$\dot{y}_2(t) = 2a \cos\left(\hat{x} + \frac{y_1(t)}{2}\right) \sin\left(\frac{y_1(t)}{2}\right) + \int_0^\infty dK(\tau)\, y_1(t - \tau). \tag{8.70}$$

It is easy to see that the linear approximation of system (8.70) has the form

$$\dot{y}_1(t) = y_2(t),$$
$$\dot{y}_2(t) = a \cos\hat{x}\, y_1(t) + \int_0^\infty dK(\tau)\, y_1(t - \tau). \tag{8.71}$$

Conditions (8.21) for system (8.71) have the form

$$a_2 = -(a \cos\hat{x} + k_0) > 0, \qquad k_1 > 0,$$
$$k_2 < k_m = \frac{4}{1 + \sqrt{1 + 4a_2 k_1^{-2}}}. \tag{8.72}$$

So, if for some root $\hat{x}$ of (8.66), conditions (8.72) hold, then the point $\hat{x}$ is a stable equilibrium of (8.64).

Theorem 8.4 *Let $\hat{x}$ be a positive root of (8.66). If $\hat{x}$ is a point of stable equilibrium of system (8.64), then*

$$\dot{S}(\hat{x}) < 0, \tag{8.73}$$

i.e., $\hat{x}$ is a point of decrease of the characteristic function $S(x)$.
If $\hat{x}$ is a point of increase of the characteristic function $S(x)$, i.e.,

$$\dot{S}(\hat{x}) > 0, \tag{8.74}$$

then $\hat{x}$ is a point of unstable equilibrium of system (8.64).

Proof Using the derivative of the function (8.67), i.e.,

$$\dot{S}(x) = \frac{1}{x}\left(\cos x - \frac{\sin x}{x}\right)$$

and (8.66), (8.67), (8.72), we obtain (8.73):

$$\dot{S}(\hat{x}) = \frac{1}{\hat{x}}\left(\cos\hat{x} - \frac{\sin\hat{x}}{\hat{x}}\right) = \frac{a\cos\hat{x} + k_0}{\hat{x}a} = -\frac{a_2}{\hat{x}a} < 0.$$

Let condition (8.74) hold. Then the first inequality of conditions (8.72) does not hold, and $\hat{x}$ cannot (see Remark 8.1) be a point of stable equilibrium. The proof is completed.   □

Remark 8.8 Let $\hat{x}$ be a point of extremum of the characteristic function $S(x)$. In this case, $\dot{S}(\hat{x}) = 0$, and $\hat{x}$ is a point of one-sided stable equilibrium of system (8.64). This means that if the system stays in a point x from a small enough neighborhood of $\hat{x}$ and $\dot{S}(x) < 0$, then the solution converges to $\hat{x}$. But if the system stays in a point x from a small enough neighborhood of $\hat{x}$ and $\dot{S}(x) > 0$, then the solution goes away from $\hat{x}$.

Remark 8.9 Since the function $S(x)$ is an even function, for negative roots of (8.66), the pictures are symmetrical.

8.3 Numerical Analysis of the Controlled Inverted Pendulum

Here we consider the results of numerical simulation of solutions of the linear and nonlinear mathematical models of the controlled inverted pendulum. For simplicity, the initial function from (8.3) is considered in the form $\phi(s) = x(0)$, $s \leq 0$. For greater visualization, the solutions are shown in the spaces (x_1, x_2) and $(t, x(t))$.

Note that the stability conditions for the difference analogue of the mathematical model of the controlled inverted pendulum were obtained in [273, 278].

8.3.1 Stability of the Trivial Solution and Limit Cycles

Consider now the linear model of the inverted pendulum in the form (8.24). Sufficient conditions for asymptotic stability of the trivial solution of (8.24) are (8.25). Recall that in accordance with Remark 8.1 two first inequalities of (8.25) are also necessary conditions for asymptotic stability, while the third inequality is a sufficient condition only.

Fig. 8.4 Solution of (8.24) for $x(0) = 6$, $a = 1$, $b_1 = 1$, $b_2 = -4$, $h_1 = 0.25$, $h_2 = 0.04$

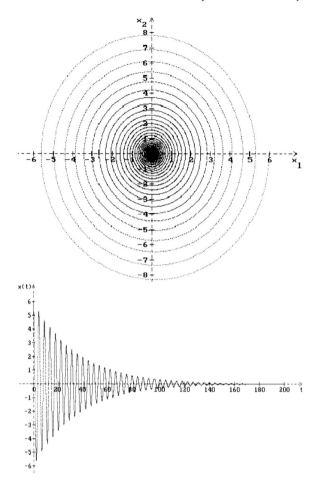

Put $x(0) = 6$, $a = 1$, $b_1 = 1$, $b_2 = -4$, $h_1 = 0.25$, $h_2 = 0.04$. In this case $k_0 = -3$, $k_1 = 0.09$, $k_2 = 0.0689$, $k_m = 0.123$. So, all conditions (8.25) hold, and the trivial solution of (8.24) is asymptotically stable. The appropriate solution of (8.24) is shown in Fig. 8.4.

Put $x(0) = 8$, $a = 3$, $b_1 = 1$, $b_2 = -4.5$, $h_1 = 0.6$, $h_2 = 0.1$. Then $k_0 = -3.5$, $k_1 = 0.15$, $k_2 = 0.405$, $k_m = 0.382$. So, two first conditions (8.25) hold, and the third condition (8.25) does not hold. But the trivial solution of the linear equation (8.24) is asymptotically stable (see Fig. 8.5), and the trivial solution of the corresponding nonlinear equation

$$\ddot{x}(t) - a \sin x(t) = b_1 x(t - h_1) + b_2 x(t - h_2) \tag{8.75}$$

is asymptotically stable too (see Fig. 8.6).

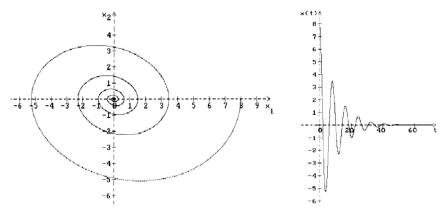

Fig. 8.5   Solution of (8.24) for $x(0) = 8$, $a = 3$, $b_1 = 1$, $b_2 = -4.5$, $h_1 = 0.6$, $h_2 = 0.1$

Fig. 8.6   Solution of (8.75) for $x(0) = 8$, $a = 3$, $b_1 = 1$, $b_2 = -4.5$, $h_1 = 0.6$, $h_2 = 0.1$

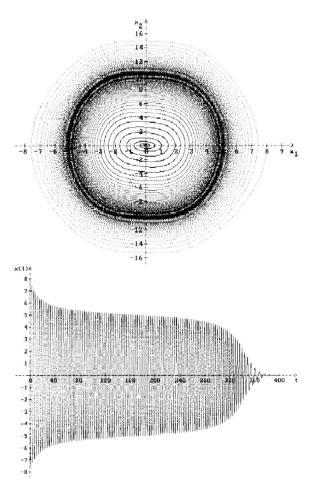

Fig. 8.7 Solution of (8.75) for $x(0) = 8$, $a = 3$, $b_1 = 1$, $b_2 = -4.5$, $h_1 = 0.59$, $h_2 = 0.1$

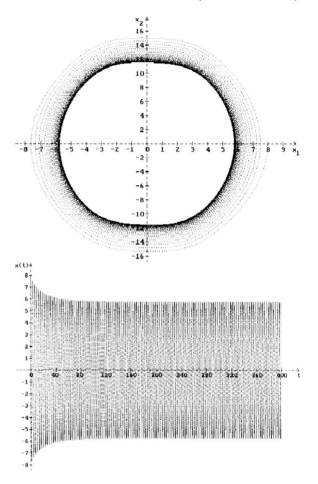

Let us investigate the influence of the delay h_1 on the behavior of the inverted pendulum. All other parameters have the same values as in Fig. 8.5. If $h_1 = 0.59$, then the trivial solution of the linear system (8.24) is asymptotically stable, but the nonlinear system (8.75) has a limit cycle (Fig. 8.7). The same phenomenon takes place for all values of the delay h_1 from $h_1 = 0.59$ to $h_1 = 0.54$ (Fig. 8.8). If $h_1 = 0.53$, then the limit cycle of the nonlinear system (8.75) is unstable (Fig. 8.9, the solution goes to infinity), but the trivial solution of the linear system (8.24) is asymptotically stable (Fig. 8.10, the solution goes to zero).

Consider the nonlinear model (8.75) for the following values of the parameters: $a = 10$, $b_1 = 1$, $b_2 = -2$, $h_1 = 0.8$, $h_2 = 0.36$. In this case the necessary condition for asymptotic stability $a + b_1 + b_2 < 0$ does not hold. So, the trivial solution of (8.75) is not stable. But for different initial conditions, there are different limit cycles. In Fig. 8.11 twelve different limit cycles are shown, which correspond to

Fig. 8.8 Solution of (8.75)
for $x(0) = 8$, $a = 3$, $b_1 = 1$,
$b_2 = -4.5$, $h_1 = 0.54$,
$h_2 = 0.1$

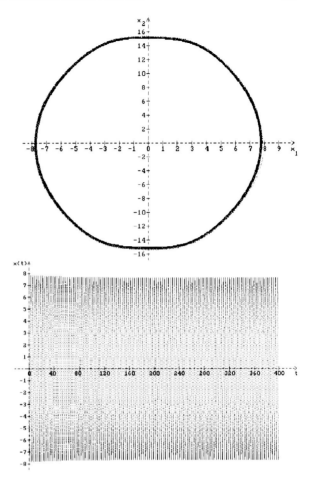

different initial conditions: $x(0) = 2$, $x(0) = 12$, $x(0) = 18$, $x(0) = 26$, $x(0) = 31$, $x(0) = 38$, $x(0) = 46$, $x(0) = 52$, $x(0) = 57$, $x(0) = 66$, $x(0) = 71$, $x(0) = 77$. Four from these limit cycles are shown separately in Fig. 8.12 by close-up.

Let us note that the considered limit cycles are stable. For instance, in Figs. 8.13 and 8.14 the first cycle from Fig. 8.12 is shown respectively for $x(0) = 2$ and $x(0) = 10$. In the first case the cycle is reached from the inside, and in the second case the cycle is reached from the outside. In both these cases the solution converges to the same limit cycle, which shows its asymptotic stability.

Note also that by small changing of the value of the parameter h_2 the asymptotic stability of all these limit cycles disappears. Changing $h_2 = 0.36$ to $h_2 = 0.37$, we can see (Fig. 8.15, $x(0) = 2$) that the trajectory of the solution goes to infinity. In Fig. 8.16 the same picture is shown for $x(0) = 0.1$.

Fig. 8.9 Solution of (8.75)
for $x(0) = 8$, $a = 3$, $b_1 = 1$,
$b_2 = -4.5$, $h_1 = 0.53$,
$h_2 = 0.1$

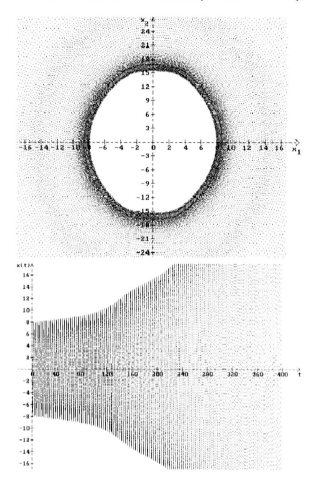

Put now $a = 3$, $b_1 = 1$, $b_2 = -4.5$, $h_1 = 0.59$, $h_2 = 0.1$. In this case $k_0 = -3.5$, $k_1 = 0.14$, $k_2 = 0.3931$, $k_m = 0.3587$, two first conditions (8.25) hold, but the third condition (8.25) does not hold. For $x(0) = 4.6$, the solution of (8.75) goes to zero (Fig. 8.17), but for $x(0) = 4.61$, the solution of (8.75) goes to a stable limit cycle. In Fig. 8.18 this cycle is reached from the inside, and in Fig. 8.19 it is reached from the outside for $x(0) = 100$. In Fig. 8.20 the appropriate solution of the linear equation (8.24) is shown, which goes to zero from the initial point $x(0) = 100$.

8.3.2 Nonzero Steady-State Solutions of the Nonlinear Model

Put in (8.75) $a = 1$, $b_1 = 1$, $b_2 = -1.08$, $h_1 = 0.8$, $h_2 = 0.3$. In this case, $k_0 = -0.08$, $k_1 = 0.476$, $k_2 = 0.7372$, and the first condition (8.25) does not hold. So,

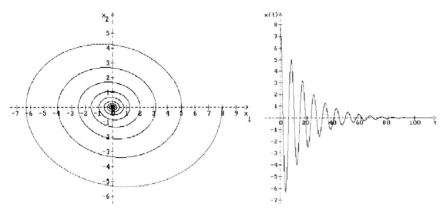

Fig. 8.10 Solution of (8.24) for $x(0) = 8$, $a = 3$, $b_1 = 1$, $b_2 = -4.5$, $h_1 = 0.53$, $h_2 = 0.1$

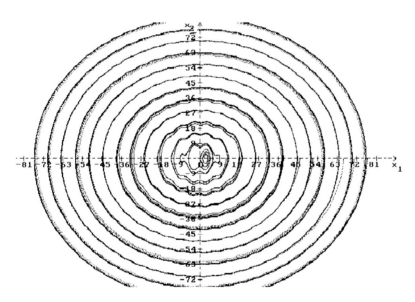

Fig. 8.11 Twelve limit cycles of (8.75) for $a = 10$, $b_1 = 1$, $b_2 = -2$, $h_1 = 0.8$, $h_2 = 0.36$, and different initial conditions: $x(0) = 2$, $x(0) = 12$, $x(0) = 18$, $x(0) = 26$, $x(0) = 31$, $x(0) = 38$, $x(0) = 46$, $x(0) = 52$, $x(0) = 57$, $x(0) = 66$, $x(0) = 71$, $x(0) = 77$

the trivial solution of (8.75) is unstable. On the other hand, the equation (8.66) has three positive roots $\hat{x}_1$, $\hat{x}_2$, $\hat{x}_3$ such that $2.906892 < \hat{x}_1 < 2.906893$, $6.864548 < \hat{x}_2 < 6.864549$, $8.659471 < \hat{x}_3 < 8.659472$. Therefore (Remark 8.4), these points are steady-state solutions of (8.75).

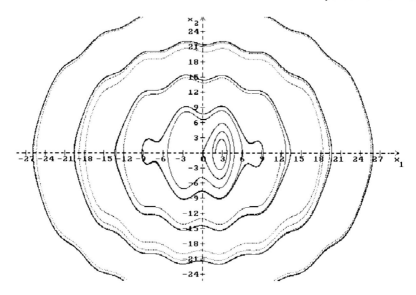

Fig. 8.12 Four limit cycles of (8.75) from Fig. 8.11 by close-up for different initial conditions: $x(0) = 2$, $x(0) = 12$, $x(0) = 18$, $x(0) = 26$

Fig. 8.13 The first cycle of (8.75) from Fig. 8.12 by $x(0) = 2$

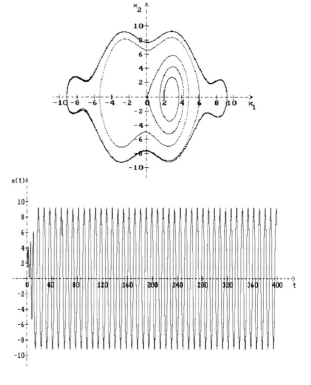

Fig. 8.14 The first cycle
of (8.75) from Fig. 8.12 by
$x(0) = 10$

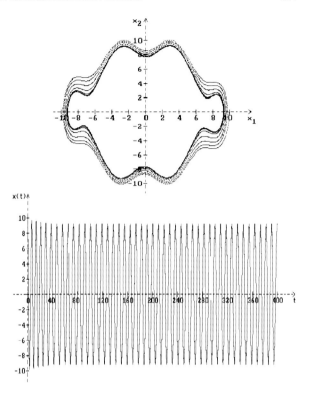

It is easy to check that

$$\dot{S}(\hat{x}_1) < 0, \qquad \dot{S}(\hat{x}_2) > 0, \qquad \dot{S}(\hat{x}_3) < 0.$$

By Theorem 8.4 the points $\hat{x}_1$ and $\hat{x}_3$ can be points of stable equilibrium of (8.75), and the point $\hat{x}_2$ is a point of unstable equilibrium of (8.75).

Note also that for the points $\hat{x}_1$ and $\hat{x}_3$, all conditions (8.72) hold, but for the point $\hat{x}_2$, the first condition (8.72) does not hold.

Let $x(0) = 6.864548$, i.e., the initial function is close enough to $\hat{x}_2$ and less than $\hat{x}_2$. In this case the solution of (8.75) goes away from the point of unstable equilibrium $\hat{x}_2$ and converges to the point of stable equilibrium $\hat{x}_1$. This situation is shown in Fig. 8.21.

Let $x(0) = 6.864549$, i.e., the initial function is close enough to $\hat{x}_2$ and greater than $\hat{x}_2$. In this case the solution of (8.75) goes away from the point of unstable equilibrium $\hat{x}_2$ and converges to the point of stable equilibrium $\hat{x}_3$. This situation is shown in Fig. 8.22.

Put now $a = 1$, $b_1 = 1$, $b_2 = -0.782766$, $h_1 = 0.8$, $h_2 = 0.3$. In this case $k_0 = 0.217233$, $k_1 = 0.565170$, $k_2 = 0.710449$, and (8.66) has one positive root

Fig. 8.15 Disappearance of limit cycles after changing $h_2 = 0.36$ to $h_2 = 0.37$, $x(0) = 2$

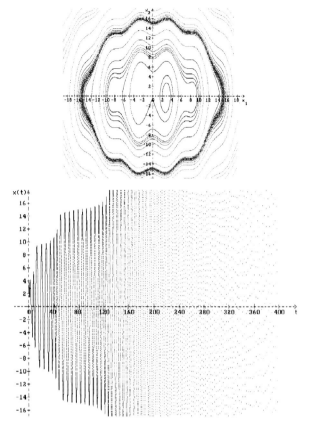

$\hat{x} = 4.493409$ only. This point is also a root of the equation $\dot{S}(\hat{x}) = 0$, i.e. (see Remark 8.8), it is a point of one-sided stable equilibrium of (8.75).

Let $x(0) = 4$, i.e., the initial function is close enough to $\hat{x}$ and less than $\hat{x}$. In this case, $\dot{S}(4) = -0.116111 < 0$, and the solution of (8.75) converges to the point $\hat{x}$. This situation is shown in Fig. 8.23.

Let $x(0) = 4.5$, i.e., the initial function is close enough to $\hat{x}$ and greater than $\hat{x}$. In this case, $\dot{S}(4.5) = 0.00142958 > 0$, and the solution of (8.75) goes away from the point $\hat{x}$ and goes to infinity. This situation is shown in Fig. 8.24.

Note that if the initial function is less than the point of one-sided stable equilibrium $\hat{x}$ and stays far enough from this point, then the system converges to infinity past by the point $\hat{x}$. This situation is shown in Fig. 8.25 for $x(0) = 3.7$.

Put $a = 7$, $b_1 = 1$, $b_2 = -2$, $h_1 = 0.4$, $h_2 = 0.02$. In this case, $a + k_0 = 6 > 0$, $k_1 = 0.476$, $k_2 = 0.7372$. So, the first condition (8.25) does not hold, and the trivial solution of (8.75) is unstable. But (8.66) has the positive root $\hat{x}_1 = 2.739489$ and the symmetric negative root $\hat{x}_2 = -2.739489$. For these roots, $a_2 = 7.441678 > 0$, so, two first conditions (8.72) hold, and $\dot{S}(\hat{x}_1) < 0$, but the third condition (8.72) does

Fig. 8.16 Disappearance of
limit cycles after changing
$h_2 = 0.36$ to $h_2 = 0.37$,
$x(0) = 0.1$

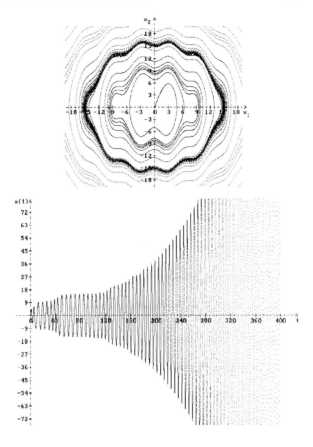

not hold since $k_2 > k_m = 0.3199$. In Fig. 8.26 the solution with the initial condition
$x(0) = 19$ converges to $\hat{x}_2$, and in Fig. 8.27 the solution with the initial condition
$x(0) = 12$ converges to $\hat{x}_1$.

Changing in the previous situation $a = 7$ on $a = 27$, we obtain nine posi-
tive points of equilibrium: $\hat{x}_1 = 3.029$, $\hat{x}_2 = 6.527$, $\hat{x}_3 = 9.082$, $\hat{x}_4 = 13.072$,
$\hat{x}_5 = 15.114$, $\hat{x}_6 = 19.665$, $\hat{x}_7 = 21.094$, $\hat{x}_8 = 26.513$, $\hat{x}_9 = 26.819$. It is easy to
check that $\dot{S}(\hat{x}_i) > 0$ for $i = 2, 4, 6, 8$ and $\dot{S}(\hat{x}_i) < 0$ for $i = 1, 3, 5, 7, 9$. So, by Theorem 8.4 the points $\hat{x}_i$ for $i = 1, 3, 5, 7, 9$ only can be points of stable equilibrium.
But for all these points, two first conditions (8.72) hold, and the third condition
(8.72) does not hold. So, these points can be points of unstable equilibrium. Indeed,
in Figs. 8.28 and 8.29 we can see that the solution of (8.75) has limit cycles around
the points $\hat{x}_1$ and $\hat{x}_1$.

Note that in the presence of an ample quantity of points of equilibrium we
can obtain different interesting limit cycles; see, for instance, Figs. 8.30, 8.31,
and 8.32.

Fig. 8.17 Solution of (8.75)
for $x(0) = 4.60$, $a = 3$,
$b_1 = 1$, $b_2 = -4.5$,
$h_1 = 0.59$, $h_2 = 0.10$

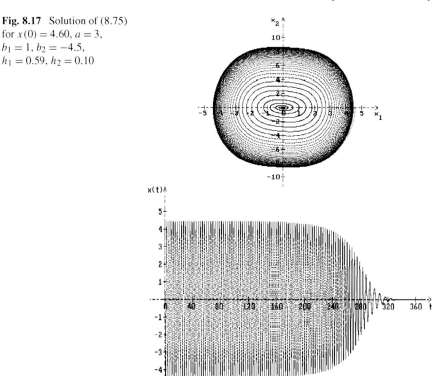

8.3.3 Stabilization of the Controlled Inverted Pendulum Under Influence of Markovian Stochastic Perturbations

Consider the linear model of the controlled pendulum in the form

$$\ddot{x}(t) + \eta(t)x(t) = u(t), \tag{8.76}$$

where $\eta(t)$ is a Markov chain with two states $\{a_1, a_2\}$ such that $a_1 > 0$ and $a_2 < 0$, the initial distribution

$$p_i = \mathbf{P}\{\eta(0) = a_i\}, \quad i = 1, 2, \tag{8.77}$$

and the probabilities of transition

$$p_{ij}(\Delta) = \mathbf{P}\{\eta(t + \Delta) = a_j / \eta(t) = a_i\} = \lambda_{ij}\Delta + o(\Delta),$$

$$i, j = 1, 2, \ i \neq j. \tag{8.78}$$

Fig. 8.18 Solution of (8.75)
for $x(0) = 4.61$, $a = 3$,
$b_1 = 1$, $b_2 = -4.5$,
$h_1 = 0.59$, $h_2 = 0.10$

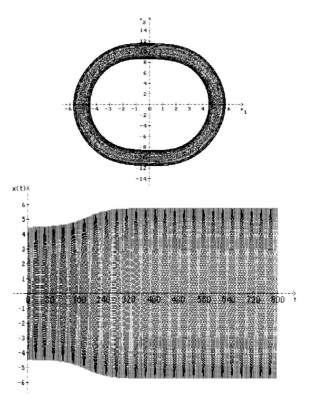

It is easy to see that by the control $u(t) \equiv 0$ the trivial solution of (8.76) is stable but not asymptotically stable in the case where $\eta(t) \equiv a_1 > 0$ (mathematical
pendulum) and unstable in the case where $\eta(t) \equiv a_2 < 0$ (inverted mathematical
pendulum).

Consider the problem of stabilization of the inverted mathematical pendulum by
the control

$$u(t) = b_1 x(t - h_1) + b_2 x(t - h_2). \tag{8.79}$$

By (8.25), if

$$k_0 = b_1 + b_2 < a_2, \qquad k_1 = b_1 h_1 + b_2 h_2 > 0,$$

$$k_2 = |b_1| h_1^2 + |b_2| h_2^2 < \frac{4}{1 + \sqrt{1 + 4(a_2 - k_0)k_1^{-2}}}, \tag{8.80}$$

then the trivial solution of (8.76), (8.79) (with $\eta(t) \equiv a_2 < 0$) is asymptotically
stable. It was shown also (Example 8.2) that for each $a_2 < 0$, there exist numbers

Fig. 8.19 Solution of (8.75)
for $x(0) = 100$, $a = 3$,
$b_1 = 1$, $b_2 = -4.5$,
$h_1 = 0.59$, $h_2 = 0.10$

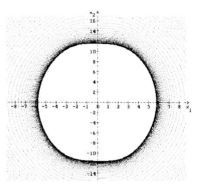

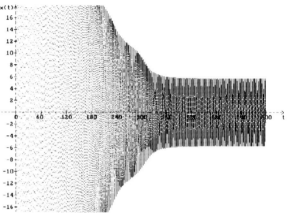

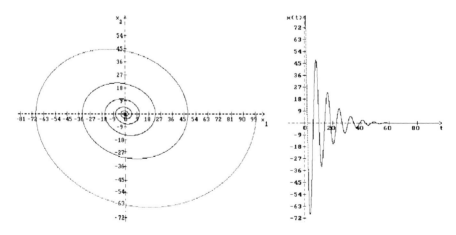

Fig. 8.20 Solution of (8.24) for $x(0) = 100$, $a = 3$, $b_1 = 1$, $b_2 = -4.5$, $h_1 = 0.59$, $h_2 = 0.10$

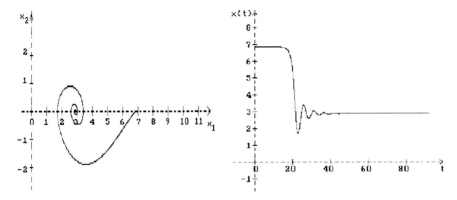

Fig. 8.21 The solution of (8.75) goes away from the left neighborhood of the point of unstable equilibrium $\widehat{x}_2$ and converges to the point of stable equilibrium $\widehat{x}_1$, $x(0) = 6.864548$

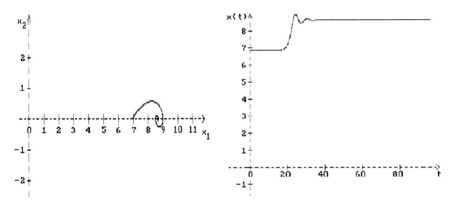

Fig. 8.22 The solution of (8.75) goes away from the right neighborhood the point of unstable equilibrium $\widehat{x}_2$ and converges to the point of stable equilibrium $\widehat{x}_3$, $x(0) = 6.864549$

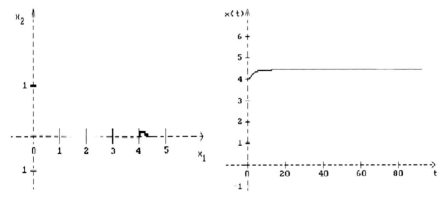

Fig. 8.23 The initial function of (8.75) is close enough to $\widehat{x}$ and is less than $\widehat{x}$, the solution converges to the point $\widehat{x}$, $x(0) = 4$

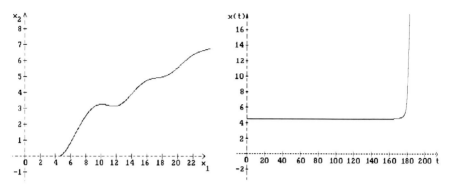

Fig. 8.24 The initial function of (8.75) is close enough to $\widehat{x}$ and is greater than $\widehat{x}$, the solution goes to infinity, $x(0) = 4.5$

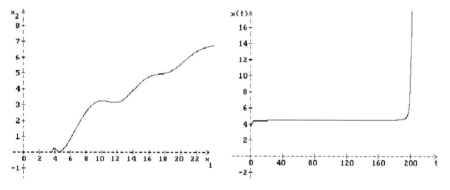

Fig. 8.25 The initial function of (8.75) is far enough from $\widehat{x}$ and is less than $\widehat{x}$, the solution goes to infinity, $x(0) = 3.7$

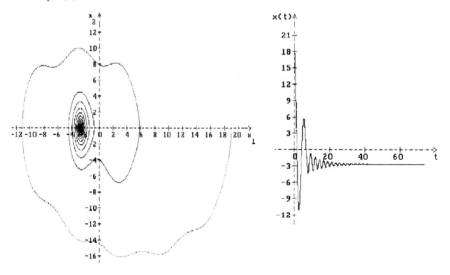

Fig. 8.26 Solution of (8.75) for $x(0) = 19$, $a = 7$, $b_1 = 1$, $b_2 = -2$, $h_1 = 0.4$, $h_2 = 0.02$

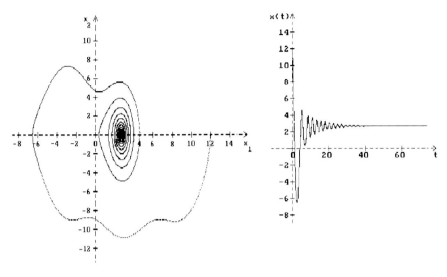

Fig. 8.27 Solution of (8.75) for $x(0) = 12$, $a = 7$, $b_1 = 1$, $b_2 = -2$, $h_1 = 0.4$, $h_2 = 0.02$

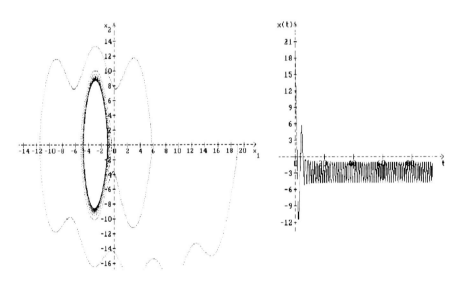

Fig. 8.28 Solution of (8.75) for $x(0) = 19$, $a = 27$, $b_1 = 1$, $b_2 = -2$, $h_1 = 0.4$, $h_2 = 0.02$

b_1, b_2, h_1, h_2 such that conditions (8.80) hold, and therefore the trivial solution of (8.76), (8.79) (with $\eta(t) \equiv a_2 < 0$) is asymptotically stable.

Let us investigate the stability of the trivial solution of (8.76), (8.79) by the numerical method and numerical simulation of the Markov chain $\eta(t)$ that was described in Sect. 7.4.1.

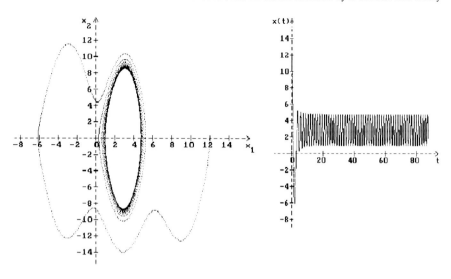

Fig. 8.29 Solution of (8.75) for $x(0) = 12$, $a = 27$, $b_1 = 1$, $b_2 = -2$, $h_1 = 0.4$, $h_2 = 0.02$

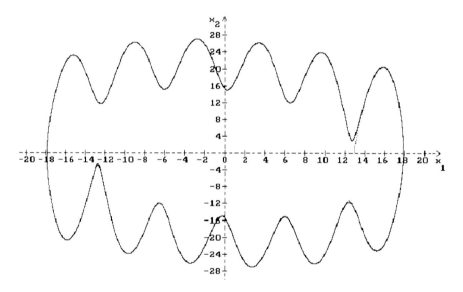

Fig. 8.30 Solution of (8.75) for $x(0) = 13$, $a = 123$, $b_1 = 1$, $b_2 = -2$, $h_1 = 0.4$, $h_2 = 0.02$

Put $a_1 = 1$, $a_2 = -1$, $b_1 = 1$, $b_2 = -2.1$, $h_1 = 0.8$, $h_2 = 0.3$, $x(s) = 3.5$, $s \leq 0$. If $\eta(t) \equiv a_1$, then the solution of (8.76), (8.79) goes to $\pm\infty$. The trajectories of $\eta(t)$ and $x(t)$ in this case are shown in Fig. 8.33. If $\eta(t) \equiv a_2$, then conditions (8.80) hold, and the solution of (8.76), (8.79) converges to zero (Fig. 8.34).

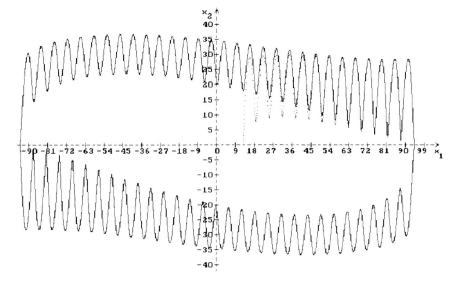

Fig. 8.31 Solution of (8.75) for $x(0) = 13$, $a = 200$, $b_1 = 1$, $b_2 = -1.1$, $h_1 = 0.4$, $h_2 = 0.02$

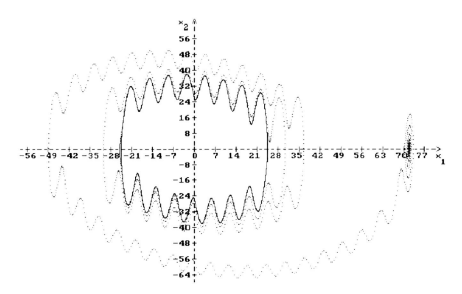

Fig. 8.32 Solution of (8.75) for $x(0) = 72$, $a = 200$, $b_1 = 1$, $b_2 = -2$, $h_1 = 0.4$, $h_2 = 0.02$

Put in (8.77), (8.78) $p_1 = p_2 = 0.5$, $\lambda_{21} = 15$, $\lambda_{12} = 1$. One of possible trajectories of the Markov chain $\eta(t)$ and the corresponding trajectory of the solution $x(t)$ of (8.76), (8.79) are shown in Fig. 8.35. Hundred trajectories of the solution $x(t)$ are

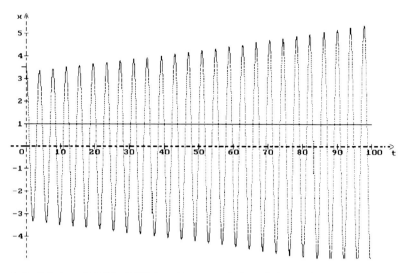

Fig. 8.33 Solution of (8.76), (8.79) by $\eta(t) \equiv a_1 = 1$, $b_1 = 1$, $b_2 = -2.1$, $h_1 = 0.8$, $h_2 = 0.3$, $x(s) = 3.5$, $s \leq 0$

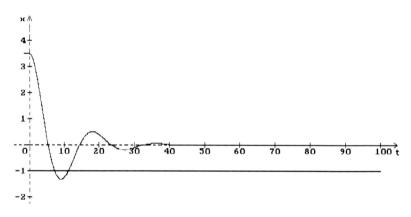

Fig. 8.34 Solution of (8.76), (8.79) by $\eta(t) \equiv a_2 = -1$, $b_1 = 1$, $b_2 = -2.1$, $h_1 = 0.8$, $h_2 = 0.3$, $x(s) = 3.5$, $s \leq 0$

shown in Fig. 8.36. One can see that in this case the trivial solution of (8.76), (8.79) is unstable.

Put now $\lambda_{21} = 1$, $\lambda_{12} = 15$. In this case the trivial solution of (8.76), (8.79) is asymptotically stable. One of the possible trajectories of the Markov chain $\eta(t)$ and the corresponding trajectory of the solution $x(t)$ of (8.76), (8.79) are shown in

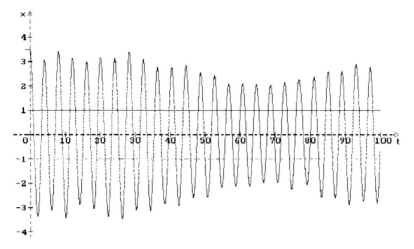

Fig. 8.35 Solution of (8.76), (8.79) by $a_1 = 1$, $a_2 = -1$, $p_1 = p_2 = 0.5$, $\lambda_{21} = 15$, $\lambda_{12} = 1$, $b_1 = 1$, $b_2 = -2.1$, $h_1 = 0.8$, $h_2 = 0.3$, $x(s) = 3.5$, $s \leq 0$

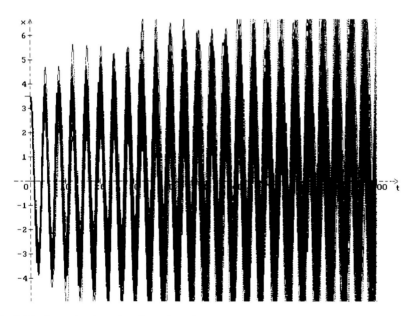

Fig. 8.36 Hundred trajectories of solution of (8.76), (8.79) by the values of the parameters as in Fig. 8.35

Fig. 8.37. Hundred trajectories of the solution $x(t)$ are shown in Fig. 8.38. All these trajectories converge to zero.

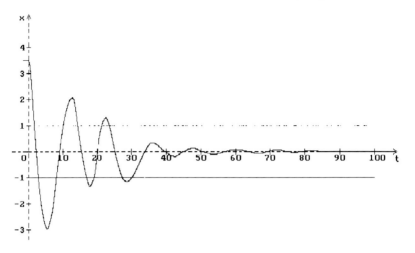

Fig. 8.37 Solution of (8.76), (8.79) by $a_1 = 1$, $a_2 = -1$, $p_1 = p_2 = 0.5$, $\lambda_{21} = 1$, $\lambda_{12} = 15$, $b_1 = 1$, $b_2 = -2.1$, $h_1 = 0.8$, $h_2 = 0.3$, $x(s) = 3.5$, $s \leq 0$

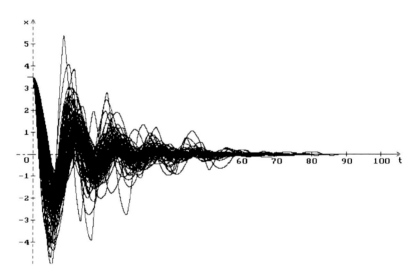

Fig. 8.38 Hundred trajectories of solution of (8.76), (8.79) by the values of the parameters as in Fig. 8.37

Chapter 9
Stability of Equilibrium Points of Nicholson's Blowflies Equation with Stochastic Perturbations

We consider the Nicholson blowflies equation (one of the most known models in ecology) with stochastic perturbations. We obtain sufficient conditions for stability in probability of the trivial and positive equilibrium points of this nonlinear differential equation with delay.

9.1 Introduction

Consider the nonlinear differential equation with exponential nonlinearity

$$\dot{x}(t) = ax(t-h)e^{-bx(t-h)} - cx(t). \tag{9.1}$$

It describes a population dynamics of the well-known Nicholson blowflies [224]. Here $x(t)$ is the size of the population at time t, a is the maximum per capita daily egg production rate, $1/b$ is the size at which the population reproduces at the maximum rate, c is the per capita daily adult death rate, and h is the generation time.

Nicholson's blowflies model is popular enough with researchers [7, 8, 30, 31, 38, 40, 61, 73, 99, 102, 129, 166, 173–176, 178, 184, 250–252, 278, 284–287, 297, 307, 312, 316, 320]. The majority of the results on (9.1) deal with the global attractiveness of the positive point of equilibrium and oscillatory behaviors of solutions [73, 102, 129, 166, 173, 174, 178, 250–252, 278, 285].

Below we will obtain sufficient conditions for stability in probability of the trivial and positive equilibrium points of (9.1) by stochastic perturbations. The basic stages of the proposed research are the following. It is assumed that the considered nonlinear differential equation has an equilibrium point and exposed to white-noise-type stochastic perturbations that are proportional to the deviation of the system current state from the considered equilibrium point. In this case the equilibrium point is a solution of the stochastic differential equation too. The constructed stochastic differential equation is centered around the considered equilibrium point and linearized in the neighborhood of this equilibrium point. Necessary and sufficient conditions for the asymptotic mean-square stability of the linear part of the considered equation are

L. Shaikhet, *Lyapunov Functionals and Stability of Stochastic Functional Differential Equations*, DOI 10.1007/978-3-319-00101-2_9,
© Springer International Publishing Switzerland 2013

obtained. Since the order of nonlinearity in (9.1) is higher than one (see Sect. 5.3), these conditions are sufficient for stability in probability of the equilibrium point of the initial nonlinear equation by stochastic perturbations.

9.2 Two Points of Equilibrium, Stochastic Perturbations, Centering, and Linearization

The points of equilibrium of (9.1) are defined by the condition $\dot{x}(t) = 0$ that can be represented in the form

$$ae^{-bx^*}x^* = cx^*. \tag{9.2}$$

From (9.2) it follows that (9.1) has two points of equilibrium

$$x_1^* = 0, \qquad x_2^* = \frac{1}{b}\ln\frac{a}{c}. \tag{9.3}$$

Similarly to Sect. 5.4, let us assume that (9.1) is exposed to stochastic perturbations that are of white noise type and are directly proportional to the deviation of $x(t)$ from the point of equilibrium x^* and influence $\dot{x}(t)$ immediately. In this way, (9.1) takes the form

$$\dot{x}(t) = ax(t-h)e^{-bx(t-h)} - cx(t) + \sigma\big(x(t) - x^*\big)\dot{w}(t). \tag{9.4}$$

Let us center (9.4) at the point of equilibrium x^* using the new variable $y(t) = x(t) - x^*$. By this way from (9.4) via (9.2) we obtain

$$\dot{y}(t) = -cy(t) + ae^{-bx^*}\big[y(t-h)e^{-by(t-h)} + x^*\big(e^{-by(t-h)} - 1\big)\big] + \sigma y(t)\dot{w}(t). \tag{9.5}$$

It is clear that the stability of an equilibrium point x^* of (9.4) is equivalent to the stability of the trivial solution of (9.5).

Along with (9.5), we will consider the linear part of this equation. Using the representation $e^y = 1 + y + o(y)$ (where $o(y)$ means that $\lim_{y\to 0}\frac{o(y)}{y} = 0$) and neglecting $o(y)$, we obtain the linear part (process $z(t)$) of (9.5) in the form

$$\dot{z}(t) = -cz(t) - ae^{-bx^*}\big(bx^* - 1\big)z(t-h) + \sigma z(t)\dot{w}(t). \tag{9.6}$$

As it follows from Remark 5.3, if the order of nonlinearity of the equation under consideration is higher than one, then a sufficient condition for the asymptotic mean-square stability of the linear part of the initial nonlinear equation is also a sufficient condition for the stability in probability of the initial equation. So, we will investigate sufficient conditions for the asymptotic mean-square stability of the linear part (9.6) of the nonlinear equation (9.5).

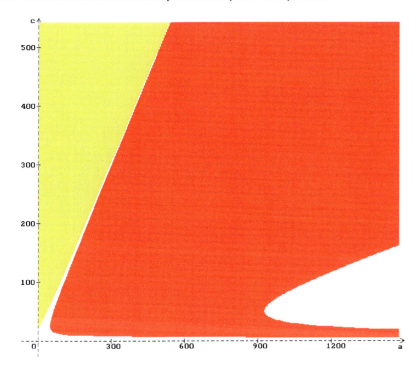

Fig. 9.1 Regions of stability in probability for zero equilibrium point (*yellow*) and positive equilibrium point (*red*) of (9.4) for $h = 0.02$, $p = 20$

9.3 Sufficient Conditions for Stability in Probability for Both Equilibrium Points

By (9.2)–(9.3) the nonlinear and linear equations (9.5)–(9.6) for the equilibrium points $x_1^* = 0$ respectively are

$$\dot{y}(t) = -cy(t) + ay(t - h)e^{-by(t-h)} + \sigma y(t)\dot{w}(t), \qquad (9.7)$$

$$\dot{z}(t) = -cz(t) + az(t - h) + \sigma z(t)\dot{w}(t). \qquad (9.8)$$

By Lemma 2.1 a necessary and sufficient condition for the asymptotic mean-square stability of the trivial solution of (9.8) is

$$G^{-1} > p, \quad p = \frac{1}{2}\sigma^2, \qquad (9.9)$$

where

$$G = \frac{1 - aq^{-1}\sinh(qh)}{c - a\cosh(qh)}, \quad c > a, \ q = \sqrt{c^2 - a^2}. \qquad (9.10)$$

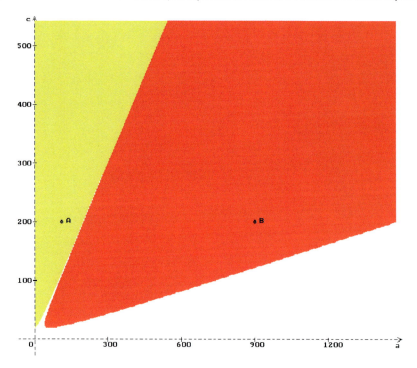

Fig. 9.2 Regions of stability in probability for zero equilibrium point (*yellow*) and positive equilibrium point (*red*) of (9.4) for $h = 0.1$, $p = 20$

In particular, if $p \geq 0$ and $h = 0$, then the stability condition (9.9)–(9.10) takes the form $c > a + p$.

Condition (9.9)–(9.10) is also a sufficient condition for the stability in probability of the zero equilibrium point $x^* = 0$ of (9.4).

Similarly to (9.9)–(9.10), for the equilibrium points $x_2^* = \frac{1}{b} \ln \frac{a}{c}$, the nonlinear and linear equations (9.5)–(9.6) respectively are

$$\dot{y}(t) = -cy(t) + cy(t-h)e^{-by(t-h)} + \frac{c}{b} \ln \frac{a}{c}\left(e^{-by(t-h)} - 1\right) + \sigma y(t)\dot{w}(t), \quad (9.11)$$

$$\dot{z}(t) = -cz(t) - c\left(\ln \frac{a}{c} - 1\right)z(t-h) + \sigma z(t)\dot{w}(t). \quad (9.12)$$

By Lemma 2.1 a necessary and sufficient condition for the asymptotic mean-square stability of the trivial solution of (9.12) is (9.9), where

$$G = \begin{cases} \frac{1+cq^{-1}(\ln(ac^{-1})-1)\sinh(qh)}{c[1+(\ln(ac^{-1})-1)\cosh(qh)]}, & c < a < ce^2, \ q = c\sqrt{\ln \frac{a}{c}\left(2 - \ln \frac{a}{c}\right)}, \\ \frac{1+ch}{2c}, & a = ce^2, \\ \frac{1+cq^{-1}(\ln(ac^{-1})-1)\sin(qh)}{c[1+(\ln(ac^{-1})-1)\cos(qh)]}, & a > ce^2, \ q = c\sqrt{\ln \frac{a}{c}\left(\ln \frac{a}{c} - 2\right)}. \end{cases} \quad (9.13)$$

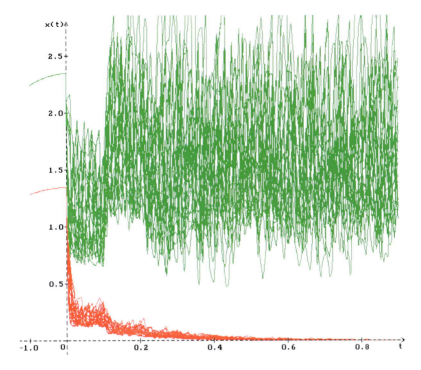

Fig. 9.3 Zero equilibrium point is stable at the point $A = (110, 200)$ (25 *red trajectories*) and is unstable in the point $B = (900, 200)$ (25 *green trajectories*)

In particular, if $p > 0$ and $h = 0$, then the stability condition takes the form $c \ln \frac{a}{c} > p$; if $p = 0$ and $h > 0$, then the region of stability is bounded by the lines $c = 0$, $c = a$, and $1 + (\ln \frac{a}{c} - 1) \cos(qh) = 0$ for $a > ce^2$.

Condition (9.9), (9.13) is also a sufficient condition for the stability in probability of the positive equilibrium point $x^* = \frac{1}{b} \ln \frac{a}{c}$ of (9.4).

Remark 9.1 Note that the stability conditions (9.9), (9.10) and (9.9), (9.13) have the following property: if the point (a, c) belongs to the stability region with some p and h, then for arbitrary positive α, the point $(a_0, c_0) = (\alpha a, \alpha c)$ belongs to the stability region with $p_0 = \alpha p$ and $h_0 = \alpha^{-1} h$.

9.4 Numerical Illustrations

In Fig. 9.1 the stability regions for (9.4) given by conditions (9.9), (9.10) for the zero equilibrium point (yellow) and (9.9), (9.13) for the positive equilibrium point (red) are shown in the space of the parameters (a, c) for $h = 0.02$ and $p = 20$. In Fig. 9.2 the similar regions of stability are shown for $h = 0.1$ and $p = 20$.

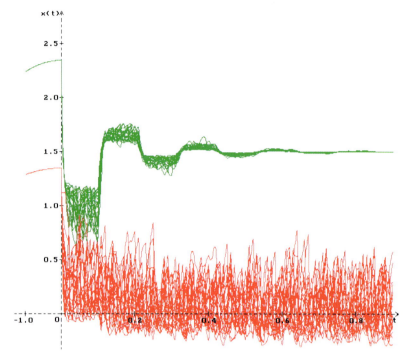

Fig. 9.4 Positive equilibrium point is unstable a the point $A = (110, 200)$ (25 *red trajectories*) and is stable at the point $B = (900, 200)$ (25 *green trajectories*)

For numerical simulation of the solution of (9.4), one uses the algorithm of numerical simulation of trajectories of the Wiener process (Chap. 2) and the Euler–Maruyama scheme [200]. Note that the stability of the difference analogue of (9.4) was investigated in detail in [38, 278].

Numerical simulation of the solution of (9.4) with $x^* = 0$ is shown in Fig. 9.3. At the point A with coordinates $a = 110$, $c = 200$ (see Fig. 9.2) the zero equilibrium point is stable in probability, so, all 25 trajectories (red) of the solution with the initial function $x(s) = 1.35 \cos(3s)$ converge to zero. At the point B with coordinates $a = 900$, $c = 200$ (see Fig. 9.2) the zero equilibrium point is unstable, so, 25 trajectories (green) of the solution with the initial function $x(s) = 2.35 \cos(3s)$ fill the whole space.

In Fig. 9.4 numerical simulation of the solution of (9.4) with the positive equilibrium point $x^* = \frac{1}{b} \ln \frac{a}{c}$ is shown by $b = 1$. At the point B with coordinates $a = 900$, $c = 200$ (see Fig. 9.2) the positive equilibrium point is stable in probability, so, all 25 trajectories (green) of the solution converge to $x^* = \ln(900/200) = 1.504$. At the point A with coordinates $a = 110$, $c = 200$ (see Fig. 9.2) the positive equilibrium point is unstable, and the trajectories (red) of the solution do not go to zero.

Chapter 10
Stability of Positive Equilibrium Point of Nonlinear System of Type of Predator–Prey with Aftereffect and Stochastic Perturbations

Here we consider a system of two nonlinear differential equations that is destined to unify different known mathematical models, in particular, very often investigated models of predator–prey type [47, 53, 60, 65, 72, 82, 83, 94, 107, 108, 112, 113, 127, 128, 153, 180, 235, 249, 267, 283, 288, 304, 305, 311, 314, 317, 321, 325]. The system under consideration is exposed to stochastic perturbations and is linearized in a neighborhood of the positive point of equilibrium. Asymptotic mean-square stability conditions for the trivial solution of the constructed linear system are at the same time sufficient conditions for the stability in probability of the positive equilibrium point of the initial nonlinear system by stochastic perturbations.

10.1 System Under Consideration

Consider the system of two nonlinear differential equations

$$\dot{x}_1(t) = x_1(t)\big(a - F_0(x_{1t}, x_{2t})\big) - F_1(x_{1t}, x_{2t}),$$

$$\dot{x}_2(t) = -x_2(t)\big(b + G_0(x_{1t}, x_{2t})\big) + G_1(x_{1t}, x_{2t}), \tag{10.1}$$

$$x_i(s) = \phi_i(s), \quad s \le 0, \ i = 1, 2.$$

Here $x_i(t)$, $i = 1, 2$, is the value of the process x_i at time t, and $x_{it} = x_i(t + s)$, $s \le 0$, is a trajectory of the process x_i to the point of time t.
   Put, for example,

$$F_0(x_{1t}, x_{2t}) = \int_0^\infty f_0\big(x_1(t - s)\big) \, dK_0(s),$$

$$F_1(x_{1t}, x_{2t}) = \prod_{i=1}^2 \int_0^\infty f_i\big(x_i(t - s)\big) \, dK_i(s), \tag{10.2}$$

L. Shaikhet, *Lyapunov Functionals and Stability of Stochastic Functional Differential Equations*, DOI 10.1007/978-3-319-00101-2_10,
© Springer International Publishing Switzerland 2013

$$G_0(x_{1t}, x_{2t}) = \int_0^\infty g_0\big(x_1(t-s)\big)\,dR_0(s),$$

$$G_1(x_{1t}, x_{2t}) = \prod_{i=1}^2 \int_0^\infty g_i\big(x_i(t-s)\big)\,dR_i(s),$$

where $K_i(s)$ and $R_i(s)$, $i = 0, 1, 2$, are nondecreasing functions such that

$$
\begin{aligned}
K_i &= \int_0^\infty dK_i(s) < \infty, & R_i &= \int_0^\infty dR_i(s) < \infty, \\
\hat{K}_i &= \int_0^\infty s\,dK_i(s) < \infty, & \hat{R}_i &= \int_0^\infty s\,dR_i(s) < \infty,
\end{aligned}
\tag{10.3}
$$

and all integrals are understood in the Stieltjes sense.

In the case (10.2)–(10.3) system (10.1) takes the form

$$\dot{x}_1(t) = x_1(t)\left(a - \int_0^\infty f_0\big(x_1(t-s)\big)\,dK_0(s)\right) - \prod_{i=1}^2 \int_0^\infty f_i\big(x_i(t-s)\big)\,dK_i(s),$$

$$\dot{x}_2(t) = -x_2(t)\left(b + \int_0^\infty g_0\big(x_1(t-s)\big)\,dR_0(s)\right) + \prod_{i=1}^2 \int_0^\infty g_i\big(x_i(t-s)\big)\,dR_i(s).$$

$$\tag{10.4}$$

Systems of type (10.4) are investigated in some biological problems. Put here, for example,

$$
\begin{aligned}
f_0(x) &= f_1(x) = f_2(x) = g_1(x) = g_2(x) = x, \\
g_0(x) &= 0, \qquad dK_1(s) = \delta(s)\,ds, \qquad dR_0(s) = 0
\end{aligned}
\tag{10.5}
$$

($\delta(s)$ is Dirac's function). If a and b are positive constants, $x_1(t)$ and $x_2(t)$ are respectively the densities of prey and predator populations, then (10.4) is transformed to the mathematical predator–prey model [267] with distributed delay

$$\dot{x}_1(t) = x_1(t)\left(a - \int_0^\infty x_1(t-s)\,dK_0(s) - \int_0^\infty x_2(t-s)\,dK_2(s)\right),$$

$$\dot{x}_2(t) = -bx_2(t) + \int_0^\infty x_1(t-s)\,dR_1(s)\int_0^\infty x_2(t-s)\,dR_2(s).$$

$$\tag{10.6}$$

Putting in (10.6)

$$
\begin{aligned}
dK_0(s) &= a_1\delta(s)\,ds, & dK_2(s) &= a_2\delta(s)\,ds, \\
dR_1(s) &= b_1\delta(s-h_1)\,ds, & dR_2(s) &= \delta(s-h_2)\,ds,
\end{aligned}
\tag{10.7}
$$

we obtain the known predator–prey mathematical model with fixed delays

$$\dot{x}_1(t) = x_1(t)\big(a - a_1 x_1(t) - a_2 x_2(t)\big),$$
$$\dot{x}_2(t) = -b x_2(t) + b_1 x_1(t - h_1) x_2(t - h_2). \tag{10.8}$$

If here $h_1 = h_2 = 0$, we have the classical Lotka–Volterra model

$$\dot{x}_1(t) = x_1(t)\big(a - a_1 x_1(t) - a_2 x_2(t)\big),$$
$$\dot{x}_2(t) = x_2(t)\big(-b + b_1 x_1(t)\big).$$

Many authors [15, 19, 23, 50, 69, 70, 116, 306, 309] consider the so-called ratio-dependent predator–prey models with delays of type

$$\dot{x}_1(t) = x_1(t)\left(a - \int_0^\infty x_1(t - s)\, dK_0(s)\right)$$
$$- \int_0^\infty \frac{x_1^k(t - s) x_2(t)}{x_1^k(t - s) + a_2 x_2^k(t - s)}\, dK_1(s), \tag{10.9}$$
$$\dot{x}_2(t) = -b x_2(t) + \int_0^\infty \frac{x_1^m(t - s) x_2(t)}{x_1^m(t - s) + b_2 x_2^m(t - s)}\, dR_1(s).$$

Here it is supposed that m and k are positive constants.

System (10.9) follows from (10.1) if

$$F_0(x_{1t}, x_{2t}) = \int_0^\infty x_1(t - s)\, dK_0(s), \qquad G_0(x_{1t}, x_{2t}) = 0,$$
$$F_1(x_{1t}, x_{2t}) = \int_0^\infty f\big(x_1(t - s), x_2(t - s)\big) x_2(t)\, dK_1(s),$$
$$G_1(x_{1t}, x_{2t}) = \int_0^\infty g\big(x_1(t - s), x_2(t - s)\big) x_2(t)\, dR_1(s), \tag{10.10}$$
$$f(x_1, x_2) = \frac{x_1^k}{x_1^k + a_2 x_2^k}, \qquad g(x_1, x_2) = \frac{x_1^m}{x_1^m + b_2 x_2^m}.$$

Putting in (10.9), for example,

$$dK_0(s) = a_0 \delta(s)\, ds, \qquad dK_1(s) = a_1 \delta(s)\, ds,$$
$$dR_1(s) = b_1 \delta(s - h)\, ds, \qquad k = m = 1, \tag{10.11}$$

we obtain the system

$$\dot{x}_1(t) = x_1(t)\left(a - a_0 x_1(t) - \frac{a_1 x_2(t)}{x_1(t) + a_2 x_2(t)}\right),$$
$$\dot{x}_2(t) = x_2(t)\left(-b + \frac{b_1 x_1(t - h)}{x_1(t - h) + b_2 x_2(t - h)}\right), \tag{10.12}$$

which was considered in [23, 50].

10.2 Equilibrium Points, Stochastic Perturbations, Centering, and Linearization

10.2.1 Equilibrium Points

Let in system (10.1) $F_i = F_i(\phi, \psi)$ and $G_i = G_i(\phi, \psi)$, $i = 0, 1$, be functionals defined on $H \times H$, where H is a set of functions $\phi = \phi(s)$, $s \leq 0$, with the norm $\|\phi\| = \sup_{s \leq 0} |\phi(s)|$, the functionals F_i and G_i are nonnegative for nonnegative functions ϕ and ψ. Let us suppose also that system (10.1) has a positive equilibrium point (x_1^*, x_2^*). This point is obtained from the conditions $\dot{x}_1(t) \equiv 0$, $\dot{x}_2(t) \equiv 0$ and is defined by the system of algebraic equations

$$x_1^*\left(a - F_0\left(x_1^*, x_2^*\right)\right) = F_1\left(x_1^*, x_2^*\right),$$
$$x_2^*\left(b + G_0\left(x_1^*, x_2^*\right)\right) = G_1\left(x_1^*, x_2^*\right). \tag{10.13}$$

From (10.13) it follows that system (10.1) has a positive solution by the condition

$$a > F_0\left(x_1^*, x_2^*\right) \tag{10.14}$$

only. For example, if $a > K_0 f_0(x_1^*)$, a positive equilibrium point of system (10.4) is defined by the system of algebraic equations

$$x_1^*\left(a - K_0 f_0\left(x_1^*\right)\right) = K_1 K_2 f_1\left(x_1^*\right) f_2\left(x_2^*\right),$$
$$x_2^*\left(b + R_0 g_0\left(x_1^*\right)\right) = R_1 R_2 g_1\left(x_1^*\right) g_2\left(x_2^*\right). \tag{10.15}$$

In particular, from (10.5), (10.14), (10.15) it follows that system (10.6) has a positive equilibrium point

$$x_1^* = \frac{b}{R_1 R_2}, \qquad x_2^* = \frac{a - K_0 x_1^*}{K_2} = \frac{a - (R_1 R_2)^{-1} K_0 b}{K_2}, \tag{10.16}$$

provided that $a > (R_1 R_2)^{-1} K_0 b$. For system (10.8), from (10.7), (10.16) we obtain

$$x_1^* = \frac{b}{b_1}, \qquad x_2^* = \frac{A}{a_2}, \qquad A = a - b\frac{a_1}{b_1} > 0. \tag{10.17}$$

From (10.13), (10.10) it follows that the positive equilibrium point for system (10.9) is

$$x_1^* = \frac{A}{K_0}, \qquad x_2^* = \frac{A}{B K_0}, \qquad A = a - \frac{K_1}{B + a_2 B^{1-k}} > 0, \quad B = \left(\frac{b b_2}{R_1 - b}\right)^{\frac{1}{m}} > 0.$$

In particular, by (10.11), for system (10.12), it is

$$x_1^* = \frac{A}{a_0}, \qquad x_2^* = \frac{A}{B a_0}, \qquad A = a - \frac{a_1}{B + a_2} > 0, \quad B = \frac{b b_2}{b_1 - b} > 0. \tag{10.18}$$

10.2.2 Stochastic Perturbations and Centering

Similarly to Sect. 9.2, we will assume that system (10.1) is exposed to stochastic perturbations that are of white noise type and are directly proportional to the deviations of the system state $(x_1(t), x_2(t))$ from the equilibrium point (x_1^*, x_2^*) and influence $\dot{x}_1(t)$, $\dot{x}_2(t)$, respectively. In this way system (10.1) is transformed to the form

$$\dot{x}_1(t) = x_1(t)\big(a - F_0(x_{1t}, x_{2t})\big) - F_1(x_{1t}, x_{2t}) + \sigma_1\big(x_1(t) - x_1^*\big)\dot{w}_1(t),$$
$$\dot{x}_2(t) = -x_2(t)\big(b + G_0(x_{1t}, x_{2t})\big) + G_1(x_{1t}, x_{2t}) + \sigma_2\big(x_2(t) - x_2^*\big)\dot{w}_2(t). \tag{10.19}$$

Here σ_1, σ_2 are constants, and $w_1(t)$, $w_2(t)$ are independent standard Wiener processes.

Centering system (10.19) at the positive point of equilibrium via the new variables $y_1 = x_1 - x_1^*$, $y_2 = x_2 - x_2^*$, we obtain

$$\dot{y}_1(t) = \big(y_1(t) + x_1^*\big)\big(a - F_0(y_{1t} + x_1^*, y_{2t} + x_2^*)\big)$$
$$- F_1\big(y_{1t} + x_1^*, y_{2t} + x_2^*\big) + \sigma_1 y_1(t)\dot{w}_1(t),$$
$$\dot{y}_2(t) = -\big(y_2(t) + x_2^*\big)\big(b + G_0(y_{1t} + x_1^*, y_{2t} + x_2^*)\big)$$
$$+ G_1\big(y_{1t} + x_1^*, y_{2t} + x_2^*\big) + \sigma_2 y_2(t)\dot{w}_2(t). \tag{10.20}$$

It is clear that the stability of equilibrium point (x_1^*, x_2^*) of system (10.19) is equivalent to the stability of the trivial solution of system (10.20).

For system (10.4), the representations (10.19) and (10.20) respectively take the forms

$$\dot{x}_1(t) = x_1(t)\left(a - \int_0^\infty f_0\big(x_1(t-s)\big)\,dK_0(s)\right)$$
$$- \prod_{i=1}^{2}\int_0^\infty f_i\big(x_i(t-s)\big)\,dK_i(s) + \sigma_1\big(x_1(t) - x_1^*\big)\dot{w}_1(t),$$
$$\dot{x}_2(t) = -x_2(t)\left(b + \int_0^\infty g_0\big(x_1(t-s)\big)\,dR_0(s)\right)$$
$$+ \prod_{i=1}^{2}\int_0^\infty g_i\big(x_i(t-s)\big)\,dR_i(s) + \sigma_2\big(x_2(t) - x_2^*\big)\dot{w}_2(t) \tag{10.21}$$

and

$$\dot{y}_1(t) = \big(y_1(t) + x_1^*\big)\left(a - \int_0^\infty f_0\big(y_1(t-s) + x_1^*\big)\,dK_0(s)\right)$$
$$- \prod_{i=1}^{2}\int_0^\infty f_i\big(y_i(t-s) + x_i^*\big)\,dK_i(s)$$

$$+ \sigma_1 y_1(t)\dot{w}_1(t), \tag{10.22}$$

$$\dot{y}_2(t) = -\left(y_2(t) + x_2^*\right)\left(b + \int_0^\infty g_0\left(x_1(t-s) + x_1^*\right)dR_0(s)\right)$$

$$+ \prod_{i=1}^{2}\int_0^\infty g_i\left(y_i(t-s) + x_i^*\right)dR_i(s) + \sigma_2 y_2(t)\dot{w}_2(t).$$

In particular, for system (10.6), from (10.21), (10.22) by (10.5), (10.16) we obtain

$$\dot{x}_1(t) = x_1(t)\left(a - \int_0^\infty x_1(t-s)\,dK_0(s) - \int_0^\infty x_2(t-s)\,dK_2(s)\right)$$

$$+ \sigma_1\left(x_1(t) - x_1^*\right)\dot{w}_1(t),$$

$$\dot{x}_2(t) = -bx_2(t) + \int_0^\infty x_1(t-s)\,dR_1(s)\int_0^\infty x_2(t-s)\,dR_2(s) \tag{10.23}$$

$$+ \sigma_2\left(x_2(t) - x_2^*\right)\dot{w}_2(t)$$

and

$$\dot{y}_1(t) = -\left(y_1(t) + x_1^*\right)\left(\int_0^\infty y_1(t-s)\,dK_0(s) + \int_0^\infty y_2(t-s)\,dK_2(s)\right)$$

$$+ \sigma_1 y_1(t)\dot{w}_1(t),$$

$$\dot{y}_2(t) = -by_2(t) + R_2 x_2^* \int_0^\infty y_1(t-s)\,dR_1(s)$$

$$+ R_1 x_1^* \int_0^\infty y_2(t-s)\,dR_2(s) \tag{10.24}$$

$$+ \prod_{i=1}^{2}\int_0^\infty y_i(t-s)\,dR_i(s) + \sigma_2 y_2(t)\dot{w}_2(t).$$

For (10.8), systems (10.23) and (10.24) take respectively the forms

$$\dot{x}_1(t) = x_1(t)\left(a - a_1 x_1(t) - a_2 x_2(t)\right) + \sigma_1\left(x_1(t) - x_1^*\right)\dot{w}_1(t),$$

$$\dot{x}_2(t) = -bx_2(t) + b_1 x_1(t-h_1)x_2(t-h_2) + \sigma_2\left(x_1(t) - x_2^*\right)\dot{w}_2(t) \tag{10.25}$$

and

$$\dot{y}_1(t) = -\left(y_1(t) + x_1^*\right)\left(a_1 y_1(t) + a_2 y_2(t)\right) + \sigma_1 y_1(t)\dot{w}_1(t),$$

$$\dot{y}_2(t) = -by_2(t) + b_1\left(x_2^* y_1(t-h_1) + x_1^* y_2(t-h_2)\right)$$

$$+ b_1 y_1(t-h_1)y_2(t-h_2) + \sigma_2 y_2(t)\dot{w}_2(t). \tag{10.26}$$

10.2.3 Linearization

Along with the considered nonlinear system, we will use the linear part of this system. Let us suppose that the functionals in (10.19) have the representations (10.2) with differentiable functions $f_i(x)$, $g_i(x)$, $i = 0, 1, 2$. Using for all these functions the representation

$$f(z + x^*) = f_0 + f_1 z + o(z), \quad f_0 = f(x^*), \quad f_1 = \frac{df}{dx}(x^*),$$

and neglecting $o(z)$, we obtain the linear part (process $(z_1(t), z_2(t))$) of system (10.22)

$$
\begin{aligned}
\dot{z}_1(t) &= (a - K_0 f_{00}) z_1(t) - \int_0^\infty z_1(t - s) \, dK(s) \\
&\quad - K_1 f_{10} f_{21} \int_0^\infty z_2(t - s) \, dK_2(s) + \sigma_1 z_1(t) \dot{w}_1(t), \\
\dot{z}_2(t) &= -(b + R_0 g_{00}) z_2(t) + \int_0^\infty z_1(t - s) \, dR(s) \\
&\quad + R_1 g_{10} g_{21} \int_0^\infty z_2(t - s) \, dR_2(s) + \sigma_2 z_2(t) \dot{w}_2(t),
\end{aligned}
\tag{10.27}
$$

where

$$
\begin{aligned}
dK(s) &= K_2 f_{20} f_{11} \, dK_1(s) + f_{01} x_1^* \, dK_0(s), \\
dR(s) &= R_2 g_{20} g_{11} \, dR_1(s) - g_{01} x_2^* \, dR_0(s).
\end{aligned}
\tag{10.28}
$$

Below we will speak about system (10.27) as about the linear part corresponding to system (10.22) or, for brevity, as about the linear part of system (10.22).

In particular, by conditions (10.5), (10.16), and (10.28) from (10.27) we obtain the linear part of system (10.24)

$$
\begin{aligned}
\dot{z}_1(t) &= -x_1^* \left(\int_0^\infty z_1(t - s) \, dK_0(s) + \int_0^\infty z_2(t - s) \, dK_2(s) \right) + \sigma_1 z_1(t) \dot{w}_1(t), \\
\dot{z}_2(t) &= -b z_2(t) + R_2 x_2^* \int_0^\infty z_1(t - s) \, dR_1(s) + R_1 x_1^* \int_0^\infty z_2(t - s) \, dR_2(s) \\
&\quad + \sigma_2 z_2(t) \dot{w}_2(t).
\end{aligned}
\tag{10.29}
$$

From (10.26) or, via (10.7), from (10.29) we have the linear part of system (10.26)

$$
\begin{aligned}
\dot{z}_1(t) &= -x_1^* \left(a_1 z_1(t) + a_2 z_2(t) \right) + \sigma_1 z_1(t) \dot{w}_1(t), \\
\dot{z}_2(t) &= -b z_2(t) + b_1 \left(x_2^* z_1(t - h_1) + x_1^* z_2(t - h_2) \right) + \sigma_2 z_2(t) \dot{w}_2(t).
\end{aligned}
\tag{10.30}
$$

As it is shown in Sect. 5.3, if the order of nonlinearity of the system under consideration is higher than one, then a sufficient condition for the asymptotic mean-square stability of the linear part of the considered nonlinear system is also a sufficient condition for the stability in probability of the initial system. So, below we will obtain sufficient conditions for the asymptotic mean-square stability of the linear part of considered nonlinear systems.

10.3 Stability of Equilibrium Point

Obtain now sufficient conditions for the asymptotic mean-square stability of the trivial solution of system (10.27) as the linear part of (10.22). The obtained conditions will be at the same time sufficient conditions for the stability in probability of the equilibrium point of (10.21).

Following the procedure of constructing Lyapunov functionals (Sect. 2.2.2), rewrite (10.27) in the form

$$\dot{Z}_1(t) = a_{11}z_1(t) + a_{12}z_2(t) + \sigma_1 z_1(t)\dot{w}_1(t),$$
$$\dot{Z}_2(t) = a_{21}z_1(t) + a_{22}z_2(t) + \sigma_2 z_2(t)\dot{w}_2(t), \tag{10.31}$$

where

$$Z_1(t) = z_1(t) - \int_0^\infty \int_{t-s}^t z_1(\theta)\,d\theta\,dK(s) - K_1 f_{10}f_{21}\int_0^\infty \int_{t-s}^t z_2(\theta)\,d\theta\,dK_2(s),$$

$$Z_2(t) = z_2(t) + \int_0^\infty \int_{t-s}^t z_1(\theta)\,d\theta\,dR(s) + R_1 g_{10}g_{21}\int_0^\infty \int_{t-s}^t z_2(\theta)\,d\theta\,dR_2(s), \tag{10.32}$$

and, by (10.15), (10.28),

$$a_{11} = a - K - K_0 f_{00} = K_1 K_2 f_{20}\left(\frac{f_{10}}{x_1^*} - f_{11}\right) - K_0 f_{01}x_1^*,$$

$$a_{12} = -K_1 K_2 f_{10}f_{21}, \qquad a_{21} = R = R_1 R_2 g_{20}g_{11} - R_0 g_{01}x_2^*, \tag{10.33}$$

$$a_{22} = R_1 R_2 g_{10}g_{21} - b - R_0 g_{00} = -R_1 R_2 g_{10}\left(\frac{g_{20}}{x_2^*} - g_{21}\right).$$

System (10.31), (10.32) is a system of stochastic differential equations of neutral type, so, following (2.10), we have to suppose that

$$\int_0^\infty s\,dK(s) + K_1|f_{10}f_{21}|\int_0^\infty s\,dK_2(s) < 1,$$

$$\int_0^\infty s\,dR(s) + R_1|g_{10}g_{21}|\int_0^\infty s\,dR_2(s) < 1,$$

or, by (10.3), (10.28), that

$$|f_{01}|x_1^*\hat{K}_0 + K_2|f_{20}f_{11}|\hat{K}_1 + K_1|f_{10}f_{21}|\hat{K}_2 < 1,$$
$$|g_{01}|x_2^*\hat{R}_0 + R_2|g_{20}g_{11}|\hat{R}_1 + R_1|g_{10}g_{21}|\hat{R}_2 < 1. \tag{10.34}$$

10.3.1 First Way of Constructing a Lyapunov Functional

Let $\hat{A} = \|a_{ij}\|$ be the matrix with the elements defined by (10.33), and $P = \|p_{ij}\|$ be the matrix with the elements defined by (1.29) for some $q > 0$. Represent p_{11}, p_{22} in the form

$$p_{ii} = \frac{1}{2}\left(qp_{ii}^{(0)} + p_{ii}^{(1)}\right), \quad i = 1, 2, \tag{10.35}$$

where

$$p_{11}^{(0)} = \frac{a_{22}^2 + \det(\hat{A})}{|\operatorname{Tr}(\hat{A})|\det(\hat{A})}, \qquad p_{11}^{(1)} = \frac{a_{21}^2}{|\operatorname{Tr}(\hat{A})|\det(\hat{A})},$$
$$p_{22}^{(0)} = \frac{a_{12}^2}{|\operatorname{Tr}(\hat{A})|\det(\hat{A})}, \qquad p_{22}^{(1)} = \frac{a_{11}^2 + \det(\hat{A})}{|\operatorname{Tr}(\hat{A})|\det(\hat{A})}, \tag{10.36}$$

and put

$$d\mu_{ij}(s) = q d\mu_{ij}^{(0)}(s) + d\mu_{ij}^{(1)}(s), \quad i, j = 1, 2, \tag{10.37}$$

where

$$d\mu_{11}^{(0)} = dK(s) - \frac{a_{12}}{|\operatorname{Tr}(\hat{A})|}dR(s), \qquad d\mu_{11}^{(1)} = \frac{a_{21}}{|\operatorname{Tr}(\hat{A})|}dR(s),$$

$$d\mu_{12}^{(0)} = K_1 f_{10} f_{21} dK_2(s) - \frac{a_{12}}{|\operatorname{Tr}(\hat{A})|}R_1 g_{10} g_{21} dR_2(s),$$

$$d\mu_{12}^{(1)} = \frac{a_{21}}{|\operatorname{Tr}(\hat{A})|}R_1 g_{10} g_{21} dR_2(s),$$

$$d\mu_{21}^{(0)} = -\frac{a_{12}}{|\operatorname{Tr}(\hat{A})|}dK(s), \qquad d\mu_{21}^{(1)} = \frac{a_{21}}{|\operatorname{Tr}(\hat{A})|}dK(s) - dR(s), \tag{10.38}$$

$$d\mu_{22}^{(0)} = -\frac{a_{12}}{|\operatorname{Tr}(\hat{A})|}K_1 f_{10} f_{21} dK_2(s),$$

$$d\mu_{22}^{(1)} = \frac{a_{21}}{|\operatorname{Tr}(\hat{A})|}K_1 f_{10} f_{21} dK_2(s) - R_1 g_{10} g_{21} dR_2(s),$$

and $dK(s)$, $dR(s)$ are defined by (10.28).

Put also

$$\delta_i = \frac{1}{2}\sigma_i^2, \quad i = 1, 2,$$

$$v_{ij}^{(m)} = \int_0^\infty s \left| d\mu_{ij}^{(m)}(s) \right|, \quad i, j = 1, 2, \ m = 0, 1,$$

(10.39)

and

$$A_1 = 1 - v_{11}^{(0)} - p_{11}^{(0)}\delta_1, \qquad A_2 = 1 - v_{22}^{(1)} - p_{22}^{(1)}\delta_2,$$

$$B_1 = v_{11}^{(1)} + p_{11}^{(1)}\delta_1, \qquad B_2 = v_{22}^{(0)} + p_{22}^{(0)}\delta_2,$$

$$C_1 = v_{12}^{(1)} + v_{21}^{(1)}, \qquad C_2 = v_{12}^{(0)} + v_{21}^{(0)}.$$

(10.40)

Theorem 10.1 *If $A_1 > 0$, $A_2 > 0$, and conditions (10.34) and*

$$\sqrt{(A_1C_1 + B_1C_2)(A_2C_2 + B_2C_1)} + B_1B_2 < A_1A_2 \qquad (10.41)$$

hold, then the trivial solution of system (10.27) is asymptotically mean-square stable and the equilibrium point of system (10.21) is stable in probability.

Proof We will consider now system (10.31)–(10.33) and suppose that the trivial solution of the appropriate auxiliary system without delays of type (2.60) with a_{ij}, $i, j = 1, 2$, defined by (10.33) is asymptotically mean-square stable, and so conditions (2.62) hold.

Consider the functional

$$V_1(t) = p_{11}Z_1^2(t) + 2p_{12}Z_1(t)Z_2(t) + p_{22}Z_2^2(t) \qquad (10.42)$$

with p_{ij}, $i, j = 1, 2$, defined by (1.29). Let L be the generator of system (10.31). Then, by (10.31), (10.42),

$$\begin{aligned}
LV_1(t) &= 2\big(p_{11}Z_1(t) + p_{12}Z_2(t)\big)\big(a_{11}z_1(t) + a_{12}z_2(t)\big) + p_{11}\sigma_1^2 z_1^2(t) \\
&\quad + 2\big(p_{12}Z_1(t) + p_{22}Z_2(t)\big)\big(a_{21}z_1(t) + a_{22}z_2(t)\big) + p_{22}\sigma_2^2 z_2^2(t) \\
&= 2(p_{11}a_{11} + p_{12}a_{21})Z_1(t)z_1(t) + 2(p_{12}a_{11} + p_{22}a_{21})Z_2(t)z_1(t) \\
&\quad + 2(p_{11}a_{12} + p_{12}a_{22})Z_1(t)z_2(t) + p_{11}\sigma_1^2 z_1^2(t) \\
&\quad + 2(p_{12}a_{12} + p_{22}a_{22})Z_2(t)z_2(t) + p_{22}\sigma_2^2 z_2^2(t).
\end{aligned}$$

(10.43)

Putting

$$\rho = \frac{a_{21} - a_{12}q}{|\mathrm{Tr}(\hat{A})|} \qquad (10.44)$$

and using (1.29), (2.62), we obtain

$$2(p_{11}a_{11} + p_{12}a_{21}) = -q, \qquad 2(p_{12}a_{12} + p_{22}a_{22}) = -1,$$

$$2(p_{12}a_{11} + p_{22}a_{21}) = \frac{-(a_{12}a_{22}q + a_{21}a_{11})a_{11} + (a_{11}^2 + \det(\hat{A}) + a_{12}^2 q)a_{21}}{|\operatorname{Tr}(\hat{A})| \det(\hat{A})}$$

$$= \frac{\det(\hat{A})a_{21} - (a_{11}a_{22} - a_{12}a_{21})a_{12}q}{|\operatorname{Tr}(\hat{A})| \det(\hat{A})} = \rho,$$

$$2(p_{11}a_{12} + p_{12}a_{22}) = \frac{((a_{22}^2 + \det(\hat{A}))q + a_{21}^2)a_{12} - (a_{12}a_{22}q + a_{21}a_{11})a_{22}}{|\operatorname{Tr}(\hat{A})| \det(\hat{A})}$$

$$= \frac{\det(A)a_{12}q - a_{21}(a_{11}a_{22} - a_{21}a_{12})}{|\operatorname{Tr}(\hat{A})| \det(\hat{A})} = -\rho.$$

So, (10.43) takes the form

$$LV_1(t) = -q Z_1(t)z_1(t) + \rho Z_2(t)z_1(t) + p_{11}\sigma_1^2 z_1^2(t)$$
$$- \rho Z_1(t)z_2(t) - Z_2(t)z_2(t) + p_{22}\sigma_2^2 z_2^2(t). \qquad (10.45)$$

Substituting (10.32) into (10.45), we have

$$LV_1 = \left(-q + p_{11}\sigma_1^2\right)z_1^2(t) + \left(-1 + p_{22}\sigma_2^2\right)z_2^2(t)$$

$$+ q \int_0^\infty \int_{t-s}^t z_1(t)z_1(\theta)\,d\theta\,dK(s)$$

$$+ q K_1 f_{10} f_{21} \int_0^\infty \int_{t-s}^t z_1(t)z_2(\theta)\,d\theta\,dK_2(s)$$

$$+ \rho \int_0^\infty \int_{t-s}^t z_1(t)z_1(\theta)\,d\theta\,dR(s)$$

$$+ \rho R_1 g_{10} g_{21} \int_0^\infty \int_{t-s}^t z_1(t)z_2(\theta)\,d\theta\,dR_2(s)$$

$$+ \rho \int_0^\infty \int_{t-s}^t z_2(t)z_1(\theta)\,d\theta\,dK(s)$$

$$+ \rho K_1 f_{10} f_{21} \int_0^\infty \int_{t-s}^t z_2(t)z_2(\theta)\,d\theta\,dK_2(s)$$

$$- \int_0^\infty \int_{t-s}^t z_2(t)z_1(\theta)\,d\theta\,dR(s)$$

$$- R_1 g_{10} g_{21} \int_0^\infty \int_{t-s}^t z_2(t)z_2(\theta)\,d\theta\,dR_2(s).$$

By (10.37), (10.38), (10.44), it can be written in the form

$$LV_1 = \left(-q + p_{11}\sigma_1^2\right)z_1^2(t) + \left(-1 + p_{22}\sigma_2^2\right)z_2^2(t)$$

$$+ \int_0^\infty \int_{t-s}^t z_1(t)z_1(\theta)\,d\theta\,d\mu_{11}(s) + \int_0^\infty \int_{t-s}^t z_1(t)z_2(\theta)\,d\theta\,d\mu_{12}(s)$$

$$+ \int_0^\infty \int_{t-s}^t z_2(t)z_1(\theta)\,d\theta\,d\mu_{21}(s) + \int_0^\infty \int_{t-s}^t z_2(t)z_2(\theta)\,d\theta\,d\mu_{22}(s).$$

$$(10.46)$$

Using (10.35), (10.37), (10.39) and some positive number γ from (10.46), we obtain

$$LV_1 \le \left(-q + qp_{11}^{(0)}\delta_1 + p_{11}^{(1)}\delta_1\right)z_1^2(t) + \left(-1 + qp_{22}^{(0)}\delta_2 + p_{22}^{(1)}\delta_2\right)z_2^2(t)$$

$$+ \frac{1}{2}\int_0^\infty \int_{t-s}^t \left(z_1^2(t) + z_1^2(\theta)\right)d\theta\left(q\left|d\mu_{11}^{(0)}(s)\right| + \left|d\mu_{11}^{(1)}(s)\right|\right)$$

$$+ \frac{1}{2}\int_0^\infty \int_{t-s}^t \left(\gamma^{-1}z_1^2(t) + \gamma z_2^2(\theta)\right)d\theta\left(q\left|d\mu_{12}^{(0)}(s)\right| + \left|d\mu_{12}^{(1)}(s)\right|\right)$$

$$+ \frac{1}{2}\int_0^\infty \int_{t-s}^t \left(\gamma z_2^2(t) + \gamma^{-1}z_1^2(\theta)\right)d\theta\left(q\left|d\mu_{21}^{(0)}(s)\right| + \left|d\mu_{21}^{(1)}(s)\right|\right)$$

$$+ \frac{1}{2}\int_0^\infty \int_{t-s}^t \left(z_2^2(t) + z_2^2(\theta)\right)d\theta\left(q\left|d\mu_{22}^{(0)}(s)\right| + \left|d\mu_{22}^{(1)}(s)\right|\right).$$

From this by (10.39) we have

$$LV_1 \le \left(-q + qp_{11}^{(0)}\delta_1 + p_{11}^{(1)}\delta_1\right)z_1^2(t) + \left(-1 + qp_{22}^{(0)}\delta_2 + p_{22}^{(1)}\delta_2\right)z_2^2(t)$$

$$+ \frac{1}{2}\left(qv_{11}^{(0)} + v_{11}^{(1)}\right)z_1^2(t) + \frac{1}{2}\int_0^\infty \int_{t-s}^t z_1^2(\theta)\,d\theta\left(q\left|d\mu_{11}^{(0)}(s)\right| + \left|d\mu_{11}^{(1)}(s)\right|\right)$$

$$+ \frac{\gamma^{-1}}{2}\left(qv_{12}^{(0)} + v_{12}^{(1)}\right)z_1^2(t)$$

$$+ \frac{\gamma}{2}\int_0^\infty \int_{t-s}^t z_2^2(\theta)\,d\theta\left(q\left|d\mu_{12}^{(0)}(s)\right| + \left|d\mu_{12}^{(1)}(s)\right|\right)$$

$$+ \frac{\gamma}{2}\left(qv_{21}^{(0)} + v_{21}^{(1)}\right)z_2^2(t)$$

$$+ \frac{\gamma^{-1}}{2}\int_0^\infty \int_{t-s}^t z_1^2(\theta)\,d\theta\left(q\left|d\mu_{21}^{(0)}(s)\right| + \left|d\mu_{21}^{(1)}(s)\right|\right)$$

$$+ \frac{1}{2}\left(qv_{22}^{(0)} + v_{22}^{(1)}\right)z_2^2(t) + \frac{1}{2}\int_0^\infty \int_{t-s}^t z_2^2(\theta)\,d\theta\left(q\left|d\mu_{22}^{(0)}(s)\right| + \left|d\mu_{22}^{(1)}(s)\right|\right)$$

$$= \left[q\left(-1 + \frac{1}{2}\left(v_{11}^{(0)} + \gamma^{-1}v_{12}^{(0)}\right) + p_{11}^{(0)}\delta_1 \right) \right.$$

$$\left. + \frac{1}{2}\left(v_{11}^{(1)} + \gamma^{-1}v_{12}^{(1)}\right) + p_{11}^{(1)}\delta_1 \right] z_1^2(t)$$

$$+ \left[-1 + \frac{1}{2}\left(\gamma v_{21}^{(1)} + v_{22}^{(1)}\right) + p_{22}^{(1)}\delta_2 + q\left(\frac{1}{2}\left(\gamma v_{21}^{(0)} + v_{22}^{(0)}\right) + p_{22}^{(0)}\delta_2 \right) \right] z_2^2(t)$$

$$+ \sum_{i=1}^{2} \int_0^\infty \int_{t-s}^t z_i^2(\theta)\, d\theta\, dF_i(s), \tag{10.47}$$

where

$$dF_i(s) = \frac{1}{2}\left(q\, dF_i^{(0)}(s) + dF_i^{(1)}(s)\right), \quad i = 1, 2,$$

$$dF_1^{(0)}(s) = \left|d\mu_{11}^{(0)}(s)\right| + \gamma^{-1}\left|d\mu_{21}^{(0)}(s)\right|,$$

$$dF_1^{(1)}(s) = \left|d\mu_{11}^{(1)}(s)\right| + \gamma^{-1}\left|d\mu_{21}^{(1)}(s)\right|,$$

$$dF_2^{(0)}(s) = \left|\gamma\, d\mu_{12}^{(0)}(s)\right| + \left|d\mu_{22}^{(0)}(s)\right|,$$

$$dF_2^{(1)}(s) = \left|\gamma\, d\mu_{12}^{(1)}(s)\right| + \left|d\mu_{22}^{(1)}(s)\right|.$$

Note that for the functional

$$V_2(t) = \sum_{i=1}^{2} \int_0^\infty \int_{t-s}^t (\theta - t + s) z_i^2(\theta)\, d\theta\, dF_i(s),$$

we have

$$LV_2(t) = \hat{F}_1 z_1^2(t) + \hat{F}_2 z_2^2(t) - \sum_{i=1}^{2} \int_0^\infty \int_{t-s}^t z_i^2(\theta)\, d\theta\, dF_i(s), \tag{10.48}$$

where

$$\hat{F}_1 = \frac{1}{2}\left[q\left(v_{11}^{(0)} + \gamma^{-1}v_{21}^{(0)}\right) + v_{11}^{(1)} + \gamma^{-1}v_{21}^{(1)}\right],$$

$$\hat{F}_2 = \frac{1}{2}\left[q\left(\gamma v_{12}^{(0)} + v_{22}^{(0)}\right) + \gamma v_{12}^{(1)} + v_{22}^{(1)}\right].$$

From (10.47), (10.48), for the functional $V = V_1 + V_2$, by (10.40) we obtain

$$LV(t) \le \left[q\left(-A_1 + \frac{\gamma^{-1}}{2}C_2 \right) + B_1 + \frac{\gamma^{-1}}{2}C_1 \right] z_1^2(t)$$

$$+ \left[-A_2 + \frac{\gamma}{2}C_1 + q\left(B_2 + \frac{\gamma}{2}C_2 \right) \right] z_2^2(t). \tag{10.49}$$

By Theorem 2.1, if there exist positive numbers q and γ such that

$$q\left(-A_1 + \frac{\gamma^{-1}}{2}C_2\right) + B_1 + \frac{\gamma^{-1}}{2}C_1 < 0,$$

$$-A_2 + \frac{\gamma}{2}C_1 + q\left(B_2 + \frac{\gamma}{2}C_2\right) < 0,$$ (10.50)

then the trivial solution of system (10.27) is asymptotically mean-square stable.
Rewrite (10.50) in the form

$$\left(B_1 + \frac{\gamma^{-1}}{2}C_1\right)\left(A_1 - \frac{\gamma^{-1}}{2}C_2\right)^{-1} < q < \left(A_2 - \frac{\gamma}{2}C_1\right)\left(B_2 + \frac{\gamma}{2}C_2\right)^{-1}. \quad (10.51)$$

So, if

$$\left(B_1 + \frac{\gamma^{-1}}{2}C_1\right)\left(A_1 - \frac{\gamma^{-1}}{2}C_2\right)^{-1} < \left(A_2 - \frac{\gamma}{2}C_1\right)\left(B_2 + \frac{\gamma}{2}C_2\right)^{-1}, \quad (10.52)$$

then there exists $q > 0$ such that (10.51) holds.
Rewriting (10.52) in the form

$$\frac{\gamma}{2}(A_1C_1 + B_1C_2) + \frac{\gamma^{-1}}{2}(A_2C_2 + B_2C_1) < A_1A_2 - B_1B_2$$

and calculating the infimum of the left-hand part of the obtained inequality with
respect to $\gamma > 0$, we obtain (10.41). So, if (10.41) holds, then there exist positive
numbers q and γ such that (10.50) holds, and therefore the trivial solution of system
(10.27) is asymptotically mean-square stable. The proof is completed.   □

Put now

$$D_1 = \frac{a_1}{b_1} - Ah_1 - \frac{\delta_1}{b}, \qquad D_2 = 1 - bh_2 - \frac{a_1\delta_2}{Ab_1}, \quad (10.53)$$

and note that the first condition (10.34) for system (10.30) is a trivial one and the
second condition takes the form $Aa_2^{-1}b_1h_1 + bh_2 < 1$ or, via the representation
(10.17) for A,

$$b_1h_1a + (a_2h_2 - a_1h_1)b < a_2. \quad (10.54)$$

Corollary 10.1 *If $D_1 > 0$, $D_2 > 0$, and conditions (10.54) and*

$$\sqrt{A(D_1h_1 + h_2)(\delta_2h_1 + D_2bh_2)} + \frac{\delta_2}{b} < D_1D_2 \quad (10.55)$$

*hold, then the trivial solution of system (10.30) is asymptotically mean-square sta-
ble, and the equilibrium point of system (10.25) is stable in probability.*

Proof Calculating for (10.30) the parameters (10.33), (10.36), (10.38), (10.39), (10.40), we obtain

$$a_{11} = -\frac{a_1 b}{b_1}, \qquad a_{12} = -\frac{a_2 b}{b_1}, \qquad a_{21} = \frac{Ab_1}{a_2}, \qquad a_{22} = 0,$$

$$p_{11}^{(0)} = \frac{b_1}{a_1 b}, \qquad p_{11}^{(1)} = \frac{Ab_1^3}{a_1 a_2^2 b^2}, \qquad p_{22}^{(0)} = \frac{a_2^2}{Aa_1 b_1}, \qquad p_{22}^{(1)} = \frac{a_1}{Ab_1} + \frac{b_1}{a_1 b},$$

$$v_{11}^{(0)} = \frac{Ab_1 h_1}{a_1}, \qquad v_{11}^{(1)} = \frac{A^2 b_1^3 h_1}{a_1 a_2^2 b}, \qquad v_{12}^{(0)} = \frac{a_2 b h_2}{a_1}, \qquad v_{12}^{(1)} = \frac{Ab_1^2 h_2}{a_1 a_2},$$

$$v_{21}^{(0)} = 0, \qquad v_{21}^{(1)} = \frac{Ab_1 h_1}{a_2}, \qquad v_{22}^{(0)} = 0, \qquad v_{22}^{(1)} = b h_2,$$

$$A_1 = \frac{b_1}{a_1} D_1, \qquad A_2 = D_2 - \frac{b_1 \delta_2}{a_1 b},$$

$$B_1 = \frac{Ab_1^3}{a_1 a_2^2 b}\left(Ah_1 + \frac{\delta_1}{b}\right), \qquad B_2 = \frac{a_2^2 \delta_2}{Aa_1 b_1},$$

$$C_1 = \frac{Ab_1}{a_1 a_2}(a_1 h_1 + b_1 h_2), \qquad C_2 = \frac{a_2 b h_2}{a_1}.$$

From this it follows that

$$A_1 C_1 + B_1 C_2 = \frac{Ab_1^2}{a_1 a_2}(D_1 h_1 + h_2),$$

$$A_2 C_2 + B_2 C_1 = \frac{a_2}{a_1}(\delta_2 h_1 + D_2 b h_2), \qquad (10.56)$$

$$A_1 A_2 - B_1 B_2 = \frac{b_1}{a_1}\left(D_1 D_2 - \frac{\delta_2}{b}\right).$$

From the representations for a_{ij}, $i, j = 1, 2$, it follows also that conditions (2.62) hold. Substituting (10.56) into (10.41), we obtain (10.55). The proof is completed. □

Remark 10.1 Note that condition (10.55) does not depend on a_2. The dependence on a_2 is included in condition (10.54).

Remark 10.2 By the absence of the delays, i.e., by $h_1 = h_2 = 0$, condition (10.54) is trivial, and condition (10.55) can be written in the form

$$\delta_1 < b\frac{a_1}{b_1}, \qquad \delta_2 < \frac{Ab_1(a_1 b - b_1 \delta_1)}{Ab_1^2 + a_1(a_1 b - b_1 \delta_1)}.$$

The same conditions can be obtained immediately from Corollary 2.3.

10.3.2 Second Way of Constructing a Lyapunov Functional

Let us consider another way of constructing of a Lyapunov functional for system (10.30).

Theorem 10.2 *If $D_1 > 0$, $D_2 > 0$, and conditions (10.54) and*

$$\left(\sqrt{Abh_2^2 + 4\delta_2 b^{-1} D_1} + \sqrt{Abh_2}\right)\left(\sqrt{Abh_1^2 + 4D_2} + \sqrt{Abh_1}\right) < 4D_1 D_2 \quad (10.57)$$

hold, where A and D_1, D_2 are defined by (10.17) and (10.53), respectively, then the trivial solution of system (10.30) is asymptotically mean-square stable, and the equilibrium point of system (10.25) is stable in probability.

Proof Using (10.17), rewrite system (10.30) in the form

$$\dot{z}_1(t) = -\frac{a_1 b}{b_1} z_1(t) - \frac{a_2 b}{b_1} z_2(t) + \sigma_1 z_1(t)\dot{w}_1(t),$$

$$\dot{Z}_2(t) = \frac{Ab_1}{a_2} z_1(t) + \sigma_2 z_2(t)\dot{w}_2(t),$$

(10.58)

where

$$Z_2(t) = z_2(t) + \frac{Ab_1}{a_2} J_1(z_{1t}) + b J_2(z_{2t}),$$

$$J_i(z_{it}) = \int_{t-h_i}^{t} z_i(s)\,ds, \quad i = 1, 2.$$

(10.59)

Consider now the functional

$$V_1(t) = z_1^2(t) + 2\mu z_1(t)Z_2(t) + \gamma Z_2^2(t), \quad (10.60)$$

where the parameters μ and γ will be chosen below. Then by (10.60), (10.58) we have

$$LV_1(t) = -2\frac{b}{b_1}\left(z_1(t) + \mu Z_2(t)\right)\left(a_1 z_1(t) + a_2 z_2(t)\right) + \sigma_1^2 z_1^2(t)$$

$$+ 2\frac{Ab_1}{a_2}\left(\mu z_1(t) + \gamma Z_2(t)\right)z_1(t) + \gamma \sigma_2^2 z_2^2(t)$$

$$= -2\left(\frac{a_1 b}{b_1} - \mu\frac{Ab_1}{a_2} - \delta_1\right)z_1^2(t) - 2\left(\mu\frac{a_2 b}{b_1} - \gamma\delta_2\right)z_2^2(t)$$

$$+ 2\left(\gamma\frac{Ab_1}{a_2} - \mu\frac{a_1 b}{b_1} - \frac{a_2 b}{b_1}\right)z_1(t)z_2(t)$$

$$+ 2\left(\gamma\frac{Ab_1}{a_2} - \mu\frac{a_1 b}{b_1}\right)z_1(t)\left(\frac{Ab_1}{a_2}J_1(z_{1t}) + b J_2(z_{2t})\right)$$

$$- 2\mu \frac{a_2 b}{b_1} z_2(t)\left(\frac{Ab_1}{a_2} J_1(z_{1t}) + b J_2(z_{2t}) \right). \tag{10.61}$$

Defining now γ by the equality

$$\gamma \frac{Ab_1}{a_2} = \mu \frac{a_1 b}{b_1} + \frac{a_2 b}{b_1}, \tag{10.62}$$

from (10.61) we obtain

$$LV_1(t) = -2\left(\frac{a_1 b}{b_1} - \mu \frac{Ab_1}{a_2} - \delta_1 \right) z_1^2(t) - 2\left(\mu \frac{a_2 b}{b_1} - \gamma \delta_2 \right) z_2^2(t)$$

$$+ 2\frac{a_2 b}{b_1} z_1(t)\left(\frac{Ab_1}{a_2} J_1(z_{1t}) + b J_2(z_{2t}) \right)$$

$$- 2\mu \frac{a_2 b}{b_1} z_2(t)\left(\frac{Ab_1}{a_2} J_1(z_{1t}) + b J_2(z_{2t}) \right).$$

By (10.59) from this, for some positive γ_1, γ_2, we have

$$LV_1(t) \le -2\left(\frac{a_1 b}{b_1} - \mu \frac{Ab_1}{a_2} - \delta_1 \right) z_1^2(t) - 2\left(\mu \frac{a_2 b}{b_1} - \gamma \delta_2 \right) z_2^2(t)$$

$$+ \frac{a_2 b}{b_1}\left(\frac{Ab_1}{a_2} \int_{t-h_1}^{t} \left(z_1^2(t) + z_1^2(s) \right) ds + b \int_{t-h_2}^{t} \left(\gamma_1 z_1^2(t) + \gamma_1^{-1} z_2^2(s) \right) ds \right)$$

$$+ \mu \frac{a_2 b}{b_1}\left(\frac{Ab_1}{a_2} \int_{t-h_1}^{t} \left(\gamma_2 z_2^2(t) + \gamma_2^{-1} z_1^2(s) \right) ds \right.$$

$$+ b \int_{t-h_2}^{t} \left(z_2^2(t) + z_2^2(s) \right) ds \bigg). \tag{10.63}$$

By the representations (10.53) for D_1, D_2 and (10.62) for γ inequality (10.63) can be written in the form

$$LV_1(t) \le \left(-2bD_1 - Abh_1 + 2\mu \frac{Ab_1}{a_2} + \gamma_1 \frac{a_2 b^2 h_2}{b_1} \right) z_1^2(t)$$

$$+ \left(-2\mu \frac{a_2 b D_2}{b_1} - \frac{\mu a_2 b^2 h_2}{b_1} + \frac{2a_2^2 b \delta_2}{Ab_1^2} + \gamma_2 \mu Abh_1 \right) z_2^2(t)$$

$$+ Ab\left(1 + \mu \gamma_2^{-1} \right) \int_{t-h_1}^{t} z_1^2(s)\, ds + \frac{b^2 a_2}{b_1}\left(\gamma_1^{-1} + \mu \right) \int_{t-h_2}^{t} z_2^2(s)\, ds.$$

Put now

$$V_2 = Ab\left(1 + \mu \gamma_2^{-1} \right) \int_{t-h_1}^{t} (s - t + h_1) z_1^2(s)\, ds$$

$$+ \frac{b^2 a_2}{b_1} \left(\gamma_1^{-1} + \mu \right) \int_{t-h_2}^{t} (s - t + h_2) z_2^2(s) \, ds.$$

Then

$$LV_2 = Ab\left(1 + \mu \gamma_2^{-1}\right) \left(h_1 z_1^2(t) - \int_{t-h_1}^{t} z_1^2(s) \, ds \right)$$

$$+ \frac{b^2 a_2}{b_1} \left(\gamma_1^{-1} + \mu \right) \left(h_2 z_2^2(t) - \int_{t-h_2}^{t} z_2^2(s) \, ds \right),$$

and as a result, for the functional $V = V_1 + V_2$, we obtain

$$LV \leq \left(-2b D_1 + 2\mu \frac{Ab_1}{a_2} + \gamma_1 \frac{a_2 b^2 h_2}{b_1} + \gamma_2^{-1} \mu A b h_1 \right) z_1^2(t)$$

$$+ \left(-2\mu \frac{a_2 b}{b_1} D_2 + 2 \frac{a_2^2 b}{Ab_1^2} \delta_2 + \gamma_1^{-1} \frac{a_2 b^2 h_2}{b_1} + \gamma_2 \mu A b h_1 \right) z_2^2(t).$$

By Theorem 2.1, if

$$- 2b D_1 + 2\mu \frac{Ab_1}{a_2} + \gamma_1 \frac{a_2 b^2 h_2}{b_1} + \gamma_2^{-1} \mu A b h_1 < 0,$$

$$\tag{10.64}$$

$$- 2\mu \frac{a_2 b}{b_1} D_2 + 2 \frac{a_2^2 b}{Ab_1^2} \delta_2 + \gamma_1^{-1} \frac{a_2 b^2 h_2}{b_1} + \gamma_2 \mu A b h_1 < 0,$$

then the trivial solution of system (10.30) is asymptotically mean-square stable.

Rewrite (10.64) in the form

$$\frac{2 \frac{a_2^2 b}{Ab_1^2} \delta_2 + \gamma_1^{-1} \frac{a_2 b^2 h_2}{b_1}}{2 \frac{a_2 b}{b_1} D_2 - \gamma_2 A b h_1} < \mu < \frac{2b D_1 - \gamma_1 \frac{a_2 b^2 h_2}{b_1}}{2 \frac{Ab_1}{a_2} + \gamma_2^{-1} A b h_1}. \tag{10.65}$$

So, if the inequality

$$\frac{2 \frac{a_2^2 b}{Ab_1^2} \delta_2 + \gamma_1^{-1} \frac{a_2 b^2 h_2}{b_1}}{2 \frac{a_2 b}{b_1} D_2 - \gamma_2 A b h_1} < \frac{2b D_1 - \gamma_1 \frac{a_2 b^2 h_2}{b_1}}{2 \frac{Ab_1}{a_2} + \gamma_2^{-1} A b h_1} \tag{10.66}$$

holds, then there exists μ such that (10.65) holds too.

It is easy to check that from (10.65), (10.62) the condition $\mu^2 < \gamma$ follows, which ensures the positivity of the functional (10.60).

Representing (10.66) in the form

$$\frac{\frac{a_2^2}{Ab_1^2} \delta_2 + \gamma_1^{-1} \frac{a_2 b h_2}{2b_1}}{D_1 - \gamma_1 \frac{a_2 b h_2}{2b_1}} \times \frac{\frac{Ab_1^2}{a_2^2 b} + \gamma_2^{-1} \frac{Ab_1 h_1}{2a_2}}{D_2 - \gamma_2 \frac{Ab_1 h_1}{2a_2}} < 1$$

and using Lemma 2.4 twice, we obtain (10.57) $\square$

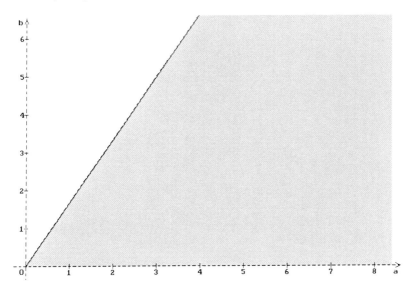

Fig. 10.1 Region of stability in probability for (10.25): $a_1 = 0.6$, $b_1 = 1$, $h_1 = 0$, $h_2 = 0$, $\delta_1 = 0$, $\delta_2 = 0$

Remark 10.3 Note that the representation (10.31)–(10.33) for system (10.30) coincides with (10.58), (10.59). So, conditions (10.55) and (10.57) are equivalent and give the same stability region. For simplicity, let us check this statement by the condition $h_1 = 0$. Indeed, in this case from (10.57) we have

$$Abh_2^2 + 4\delta_2 b^{-1} D_1 < (2D_1\sqrt{D_2} - \sqrt{Ab}h_2)^2 = 4D_1^2 D_2 - 4D_1 h_2\sqrt{AbD_2} + Abh_2^2$$

or $\delta_2 b^{-1} < D_1 D_2 - h_2\sqrt{AbD_2}$, which is equivalent to (10.55) by $h_1 = 0$. Similarly, it is easy to get that (10.55) coincides with (10.57) by the condition $h_2 = 0$ or by the condition $\delta_2 = 0$. In the general case the necessary transformation is bulky enough.

The regions of stability in probability for a positive point of equilibrium of system (10.25), obtained by condition (10.55) (or (10.57)), are shown in the space of the parameters (a, b) for $a_1 = 0.6$, $b_1 = 1$ and different values of the other parameters: in Fig. 10.1 for $h_1 = 0$, $h_2 = 0$, $\delta_1 = 0$, $\delta_2 = 0$, in Fig. 10.2 for $h_1 = 0$, $h_2 = 0$, $\delta_1 = 0.2$, $\delta_2 = 0.3$, in Fig 10.3 for $a_2 = 0.6$, $h_1 = 0.1$, $h_2 = 0.15$, $\delta_1 = 0$, $\delta_2 = 0$, and in Fig. 10.4 for $a_2 = 0.07$, $h_1 = 0.01$, $h_2 = 0.15$, $\delta_1 = 0.05$, $\delta_2 = 0.1$.

The equation of the straight line in Figs. 10.1 and 10.2 is $ab_1 = ba_1$, which corresponds to the condition $A = 0$. In Figs. 10.3 and 10.4 the straight line 1 also corresponds to this equation and the straight line 2 is defined by the equation $b_1 h_1 a + (a_2 h_2 - a_1 h_1)b = a_2$, which follows from condition (10.54).

Note that the stability of the positive equilibrium point of the difference analogue of system (10.25) is investigated in [278].

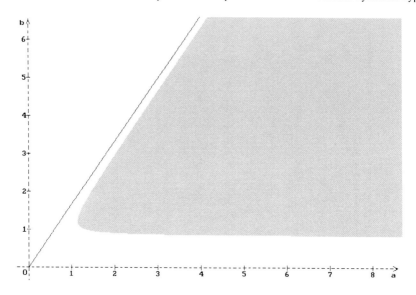

Fig. 10.2 Region of stability in probability for (10.25): $a_1 = 0.6, b_1 = 1, h_1 = 0, h_2 = 0, \delta_1 = 0.2,$ $\delta_2 = 0.3$

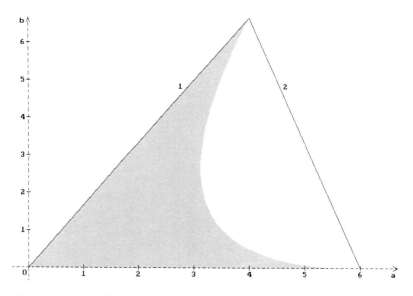

Fig. 10.3 Region of stability in probability for (10.25): $a_1 = 0.6, a_2 = 0.6, b_1 = 1, h_1 = 0.1,$ $h_2 = 0.15, \delta_1 = 0, \delta_2 = 0$

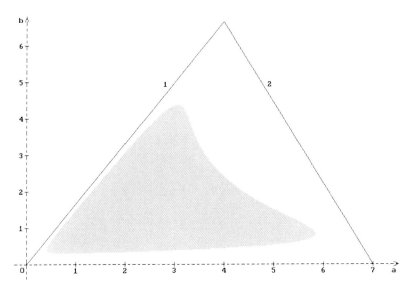

Fig. 10.4 Region of stability in probability for (10.25): $a_1 = 0.6$, $a_2 = 0.07$, $b_1 = 1$, $h_1 = 0.01$, $h_2 = 0.15$, $\delta_1 = 0.05$, $\delta_2 = 0.1$

10.3.3 Stability of the Equilibrium Point of Ratio-Dependent Predator–Prey Model

Consider now system (10.9) with stochastic perturbations, i.e.,

$$\dot{x}_1(t) = x_1(t)\left(a - \int_0^\infty x_1(t-s)\,dK_0(s)\right) - \int_0^\infty \frac{x_1^k(t-s)x_2(t)}{x_1^k(t-s) + a_2 x_2^k(t-s)}\,dK_1(s)$$

$$+ \sigma_1\left(x_1(t) - x_1^*\right)\dot{w}_1(t), \tag{10.67}$$

$$\dot{x}_2(t) = -bx_2(t) + \int_0^\infty \frac{x_1^m(t-s)x_2(t)}{x_1^m(t-s) + b_2 x_2^m(t-s)}\,dR_1(s) + \sigma_2\left(x_1(t) - x_2^*\right)\dot{w}_2(t).$$

System (10.9) was obtained from (10.1) by conditions (10.10). So, by (10.13), (10.14) the positive equilibrium point (x_1^*, x_2^*) of system (10.9) (and also (10.67)) is defined by the conditions

$$x_1^*\left(a - K_0 x_1^*\right) = K_1 f\left(x_1^*, x_2^*\right)x_2^*,$$

$$b = R_1 g\left(x_1^*, x_2^*\right), \qquad a > K_0 x_1^*. \tag{10.68}$$

Suppose that the functions $f(x_1, x_2)$ and $g(x_1, x_2)$ in (10.10) are differentiable and can be represented in the form

$$f\left(y_1 + x_1^*, y_2 + x_2^*\right) = f_0 + f_1 y_1 - f_2 y_2 + o(y_1, y_2),$$

$$g\left(y_1 + x_1^*, y_2 + x_2^*\right) = g_0 + g_1 y_1 - g_2 y_2 + o(y_1, y_2),$$

where $\lim_{|y| \to 0} \frac{o(y_1, y_2)}{|y|} = 0$ for $|y| = \sqrt{y_1^2 + y_2^2}$, and

$$f_0 = f(x_1^*, x_2^*), \qquad f_1 = x_2^* \hat{f}, \qquad f_2 = x_1^* \hat{f}, \qquad \hat{f} = \frac{ka_2(x_1^* x_2^*)^{k-1}}{((x_1^*)^k + a_2(x_2^*)^k)^2},$$

$$g_0 = g(x_1^*, x_2^*), \qquad g_1 = x_2^* \hat{g}, \qquad g_2 = x_1^* \hat{g}, \qquad \hat{g} = \frac{mb_2(x_1^* x_2^*)^{m-1}}{((x_1^*)^m + b_2(x_2^*)^m)^2}.$$

So, the functionals $F_0(x_{1t}, x_{2t})$, $F_1(x_{1t}, x_{2t})$, $G_1(x_{1t}, x_{2t})$ in (10.10) have the representations

$$F_0(y_{1t} + x_1^*, y_{2t} + x_2^*) = K_0 x_1^* + \int_0^\infty y_1(t-s) \, dK_0(s),$$

$$F_1(y_{1t} + x_1^*, y_{2t} + x_2^*) = K_1 f_0 x_2^* + f_1 x_2^* \int_0^\infty y_1(t-s) \, dK_1(s) + K_1 f_0 y_2(t)$$

$$- f_2 x_2^* \int_0^\infty y_2(t-s) \, dK_1(s) + o(y_1, y_2),$$

$$G_1(y_{1t} + x_1^*, y_{2t} + x_2^*) = R_1 g_0 x_2^* + g_1 x_2^* \int_0^\infty y_1(t-s) \, dR_1(s) + R_1 g_0 y_2(t)$$

$$- g_2 x_2^* \int_0^\infty y_2(t-s) \, dR_1(s) + o(y_1, y_2). \tag{10.69}$$

By (10.68), (10.69) the linear part of system (10.67) has the form

$$\dot{z}_1(t) = \left(a - K_0 x_1^*\right) z_1(t) - K_1 f_0 z_2(t) - \int_0^\infty z_1(t-s) \, dK(s)$$

$$+ f_2 x_2^* \int_0^\infty z_2(t-s) \, dK_1(s) + \sigma_1 z_1(t) \dot{w}_1(t), \tag{10.70}$$

$$\dot{z}_2(t) = g_1 x_2^* \int_0^\infty z_1(t-s) \, dR_1(s) - g_2 x_2^* \int_0^\infty z_2(t-s) \, dR_1(s) + \sigma_2 z_2(t) \dot{w}_2(t),$$

where $dK(s) = x_1^* dK_0(s) + f_1 x_2^* dK_1(s)$. Rewrite system (10.70) in the form (10.31) with

$$Z_1(t) = z_1(t) - \int_0^\infty \int_{t-s}^t z_1(\theta) \, d\theta \, dK(s) + f_2 x_2^* \int_0^\infty \int_{t-s}^t z_2(\theta) \, d\theta \, dK_1(s),$$

$$Z_2(t) = z_2(t) + g_1 x_2^* \int_0^\infty \int_{t-s}^t z_1(\theta) \, d\theta \, dR_1(s)$$

$$- g_2 x_2^* \int_0^\infty \int_{t-s}^t z_2(\theta) \, d\theta \, dR_1(s), \tag{10.71}$$

$$a_{11} = K_1 x_2^* \left(\frac{f_0}{x_1^*} - f_1\right) - K_0 x_1^*, \qquad a_{12} = K_1 \left(f_2 x_2^* - f_0\right),$$

$$a_{21} = R_1 g_1 x_2^*, \qquad a_{22} = -R_1 g_2 x_2^*.$$

Further investigation is similar to the previous sections.

For short, consider system (10.67) by conditions (10.11). The point of equilibrium in this case is defined by (10.18). From (10.11), (10.18), (10.70) and (10.71) it follows that system (10.67) and the linear part of this system respectively take the forms

$$
\dot{x}_1(t) = x_1(t)\left(a - a_0 x_1(t) - \frac{a_1 x_2(t)}{x_1(t) + a_2 x_2(t)}\right) + \sigma_1\left(x_1(t) - x_1^*\right)\dot{w}_1(t),
$$
$$
\dot{x}_2(t) = x_2(t)\left(-b + \frac{b_1 x_1(t-h)}{x_1(t-h) + b_2 x_2(t-h)}\right) + \sigma_2\left(x_1(t) - x_2^*\right)\dot{w}_2(t)
\tag{10.72}
$$

and

$$
\dot{z}_1(t) = a_{11} z_1(t) + a_{12} z_2(t) + \sigma_1 z_1(t)\dot{w}_1(t),
$$
$$
\dot{z}_2(t) = a_{21} z_1(t-h) + a_{22} z_2(t-h) + \sigma_2 z_2(t)\dot{w}_2(t),
\tag{10.73}
$$

where

$$
a_{11} = B a_1 \alpha^2 - A, \qquad a_{12} = -B^2 a_1 \alpha^2,
$$
$$
a_{21} = b_1 b_2 \beta^2, \qquad a_{22} = -B b_1 b_2 \beta^2,
$$
$$
A = a - a_1 \alpha, \ B = \frac{b b_2}{b_1 - b}, \ \alpha = \frac{1}{B + a_2}, \ \beta = \frac{1}{B + b_2}.
\tag{10.74}
$$

Let $\hat{A} = \|a_{ij}\|$ be the matrix with the elements defined by (10.74). Suppose that

$$
b \in (0, b_1),
$$
$$
a > \begin{cases} a_1 \alpha & \text{if } a_1 \alpha^2 \le b_1 b_2 \beta^2, \\ a_1 \alpha + B(a_1 \alpha^2 - b_1 b_2 \beta^2) & \text{if } a_1 \alpha^2 > b_1 b_2 \beta^2. \end{cases}
\tag{10.75}
$$

By conditions (10.75) conditions (2.62) for the matrix $\hat{A}$ hold. Indeed,

$$
\mathrm{Tr}(\hat{A}) = B\left(a_1 \alpha^2 - b_1 b_2 \beta^2\right) - A < 0, \qquad \det(\hat{A}) = A B b_1 b_2 \beta^2 > 0.
\tag{10.76}
$$

Let $P = \|p_{ij}\|$ be the matrix with the elements defined by (1.29) for some $q > 0$ and represented in the form (10.35), (10.36). Using (10.35), (10.36), (10.44), (10.76), put

$$
\rho = \rho^{(0)} q + \rho^{(1)}, \qquad \rho^{(0)} = -\frac{a_{12}}{|\mathrm{Tr}(\hat{A})|}, \qquad \rho^{(1)} = \frac{a_{21}}{|\mathrm{Tr}(\hat{A})|},
\tag{10.77}
$$

and

$$
A_1 = 1 - p_{11}^{(0)} \delta_1 - \rho^{(0)} |a_{21}| h, \qquad A_2 = 1 - p_{22}^{(1)} \delta_2 - |a_{22}| h,
$$
$$
B_1 = p_{11}^{(1)} \delta_1 + \rho^{(1)} |a_{21}| h, \qquad B_2 = p_{22}^{(0)} \delta_2, \ \delta_i = \frac{1}{2}\sigma_i^2, \ i = 1, 2,
\tag{10.78}
$$
$$
C_1 = \left(|a_{21}| + \rho^{(1)} |a_{22}|\right) h, \qquad C_2 = \rho^{(0)} |a_{22}| h.
$$

Rewrite system (10.73) in the form

$$\dot{z}_1(t) = a_{11}z_1(t) + a_{12}z_2(t) + \sigma_1 z_1(t)\dot{w}_1(t),$$
$$\dot{Z}_2(t) = a_{21}z_1(t) + a_{22}z_2(t) + \sigma_2 z_2(t)\dot{w}_2(t),$$

(10.79)

where

$$Z_2(t) = z_2(t) + \int_{t-h}^{t} \left(a_{21}z_1(s) + a_{22}z_2(s) \right) ds,$$

(10.80)

and following condition (2.10), suppose that the parameters a_{21} and a_{22} in (10.74) satisfy the condition $h\sqrt{a_{21}^2 + a_{22}^2} < 1$ or, via (10.74), $b_1 b_2 \beta^2 h\sqrt{1 + B^2} < 1$, which is equivalent to

$$(b_1 - b)\sqrt{(b_1 - b)^2 + b^2 b_2^2} < \frac{b_1 b_2}{h}.$$

(10.81)

Theorem 10.3 *Let conditions* (10.75), (10.81) *hold. If* $A_1 > 0$, $A_2 > 0$, *and*

$$\sqrt{(A_1 C_1 + B_1 C_2)(A_2 C_2 + B_2 C_1)} + B_1 B_2 < A_1 A_2,$$

(10.82)

then the trivial solution of system (10.73) *is asymptotically mean-square stable, and the equilibrium point of system* (10.72) *is stable in probability.*

Proof Consider the functional

$$V_1(t) = p_{11}z_1^2(t) + 2p_{12}z_1(t)Z_2(t) + p_{22}Z_2^2(t)$$

with p_{ij}, $i, j = 1, 2$, defined by (1.29). Let L be the generator of system (10.79). Then, using (10.77), similarly to (10.45), for system (10.79), we obtain

$$LV_1(t) = -qz_1^2(t) + \rho Z_2(t)z_1(t) + p_{11}\sigma_1^2 z_1^2(t)$$
$$- \rho z_1(t)z_2(t) - Z_2(t)z_2(t) + p_{22}\sigma_2^2 z_2^2(t).$$

(10.83)

Substituting (10.80) into (10.83) and using some positive γ, we obtain

$$LV_1(t) = -qz_1^2(t) + \rho z_1(t)\left(z_2(t) + \int_{t-h}^{t} \left(a_{21}z_1(s) + a_{22}z_2(s) \right) ds \right) + p_{11}\sigma_1^2 z_1^2(t)$$

$$- \rho z_1(t)z_2(t) - z_2(t)\left(z_2(t) + \int_{t-h}^{t} \left(a_{21}z_1(s) + a_{22}z_2(s) \right) ds \right)$$

$$+ p_{22}\sigma_2^2 z_2^2(t)$$

$$\leq \left(-q + p_{11}\sigma_1^2 \right)z_1^2(t) + \frac{\rho}{2}|a_{21}| \int_{t-h}^{t} \left(z_1^2(t) + z_1^2(s) \right) ds$$

$$+ \frac{\rho}{2}|a_{22}| \int_{t-h}^{t} \left(\gamma^{-1}z_1^2(t) + \gamma z_2^2(s) \right) ds$$

$$+ \left(-1 + p_{22}\sigma_2^2\right) z_2^2(t) + \frac{1}{2}|a_{21}| \int_{t-h}^{t} \left(\gamma z_2^2(t) + \gamma^{-1} z_1^2(s)\right) ds$$

$$+ \frac{1}{2}|a_{22}| \int_{t-h}^{t} \left(z_2^2(t) + z_2^2(s)\right) ds$$

$$= \left(-q + p_{11}\sigma_1^2 + \frac{\rho}{2}|a_{21}|h + \gamma^{-1}\frac{\rho}{2}|a_{22}|h\right) z_1^2(t)$$

$$+ \left(-1 + p_{22}\sigma_2^2 + \frac{1}{2}|a_{22}|h + \frac{\gamma}{2}|a_{21}|h\right) z_2^2(t)$$

$$+ \frac{|a_{21}|}{2}\left(\rho + \gamma^{-1}\right) \int_{t-h}^{t} z_1^2(s)\, ds + \frac{|a_{22}|}{2}(1 + \rho\gamma) \int_{t-h}^{t} z_2^2(s)\, ds.$$

Putting

$$V_2 = \frac{|a_{21}|}{2}\left(\rho + \gamma^{-1}\right) \int_{t-h}^{t} (s - t + h) z_1^2(s)\, ds$$

$$+ \frac{|a_{22}|}{2}(1 + \rho\gamma) \int_{t-h}^{t} (s - t + h) z_2^2(s)\, ds,$$

for the functional $V = V_1 + V_2$, we have

$$LV(t) \le \left[-q + p_{11}\sigma_1^2 + \rho|a_{21}|h + \frac{\gamma^{-1}}{2}h\left(|a_{21}| + \rho|a_{22}|\right) \right] z_1^2(t)$$

$$+ \left[-1 + p_{22}\sigma_2^2 + |a_{22}|h + \frac{\gamma}{2}h\left(|a_{21}| + \rho|a_{22}|\right) \right] z_2^2(t). \quad (10.84)$$

Using the representations (10.35), (10.36), (10.77), (10.78), we can rewrite (10.84) in the form

$$LV(t) \le \left[q\left(-A_1 + \frac{\gamma^{-1}}{2}C_2\right) + B_1 + \frac{\gamma^{-1}}{2}C_1 \right] z_1^2(t)$$

$$+ \left[-A_2 + \frac{\gamma}{2}C_1 + q\left(B_2 + \frac{\gamma}{2}C_2\right) \right] z_2^2(t),$$

which coincides with (10.49). So, from this (10.82) follows, which coincides with (10.41). The proof is completed.   □

The regions of stability in probability for a positive point of equilibrium of system (10.72), obtained by conditions (10.81), (10.82), are shown in the space of the parameters (a, b) for $a_0 = 0.3$, $a_1 = 5$ $a_2 = 0.5$, $b_1 = 6$, $b_2 = 2$, $h = 0.4$ and different values of δ_1, δ_2: in Fig. 10.5 for $\delta_1 = 1.5$, $\delta_2 = 0.05$, in Fig. 10.6 for $\delta_1 = 1$, $\delta_2 = 0.55$.

In the both figures the thick line shows the stability region given by conditions (10.75) that corresponds to the values of the parameters $h = \delta_1 = \delta_2 = 0$.

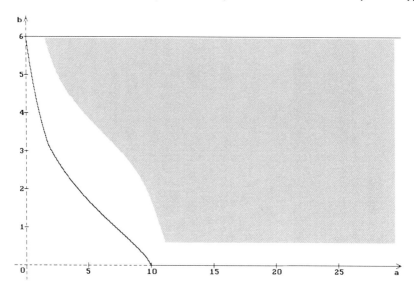

Fig. 10.5 Region of stability in probability for (10.69): $a_0 = 0.3$, $a_1 = 5$, $a_2 = 0.5$, $b_1 = 6$, $b_2 = 2$, $h = 0.4$, $\delta_1 = 1.5$, $\delta_2 = 0.05$

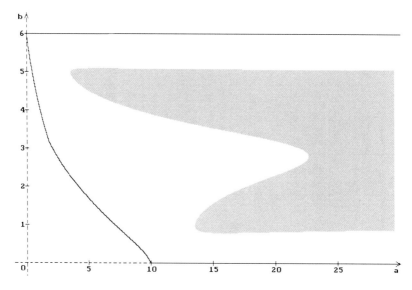

Fig. 10.6 Region of stability in probability for (10.69): $a_0 = 0.3$, $a_1 = 5$, $a_2 = 0.5$, $b_1 = 6$, $b_2 = 2$, $h = 0.4$, $\delta_1 = 1$, $\delta_2 = 0.55$

Chapter 11
Stability of SIR Epidemic Model Equilibrium Points

11.1 Problem Statement

Investigation of different mathematical models for infectious diseases (in particular, so-called SIR epidemic models) has a long story, and until now these models are very popular in researchers (see, for example, [6, 18, 26–28, 41, 45, 58, 71, 110, 117, 118, 123, 124, 152, 154, 168, 179, 187, 188, 193, 202, 203, 212, 213, 218, 223, 289, 294–296, 310, 315, 319, 322–324]). Here sufficient conditions for the stability in probability of two equilibrium points of SIR epidemic model with distributed delays and stochastic perturbations are obtained.

Consider the mathematical model of the spread of infections diseases. Let $S(t)$ be the number of members of a population susceptible to the disease at time t, $I(t)$ be the number of infective members at time t, and $R(t)$ be the number of members that have been removed from the possibility of infection at time t through full immunity, b is the recruitment rate of the population, μ_1, μ_2, and μ_3 are the natural death rates of the susceptible, infective, and recovered individuals, respectively, β is the transmission rate, and λ is the natural recovery rate of the infective individuals. Then the SIR epidemic model can be described by the system of the differential equations

$$\dot{S}(t) = b - \beta S(t) \int_0^\infty I(t-s) \, dF(s) - \mu_1 S(t),$$

$$\dot{I}(t) = \beta S(t) \int_0^\infty I(t-s) \, dF(s) - (\mu_2 + \lambda) I(t), \qquad (11.1)$$

$$\dot{R}(t) = \lambda I(t) - \mu_3 R(t).$$

Here we assume that b, β, λ, μ_1, μ_2, μ_3 are positive constants, the integral is understanding in the Stieltjes sense, and $F(s)$ is a nondecreasing function such that

$$\int_0^\infty dF(s) = 1. \qquad (11.2)$$

L. Shaikhet, *Lyapunov Functionals and Stability of Stochastic Functional Differential Equations*, DOI 10.1007/978-3-319-00101-2_11,
© Springer International Publishing Switzerland 2013

The equilibrium point of (11.1) is defined by the conditions $\dot{S}(t) = 0$, $\dot{I}(t) = 0$, $\dot{R}(t) = 0$ or by the system of algebraic equations

$$\beta S^* I^* + \mu_1 S^* = b, \qquad \beta S^* I^* = (\mu_2 + \lambda) I^*, \qquad \lambda I^* = \mu_3 R^*. \qquad (11.3)$$

From (11.3) it follows that system (11.1) has two points of equilibrium: $E_0 = (b\mu_1^{-1}, 0, 0)$ and

$$E_* = \left(S^*, I^*, R^*\right) \quad \text{with } S^* = \frac{\mu_2 + \lambda}{\beta}, \ I^* = \frac{b(S^*)^{-1} - \mu_1}{\beta}, \ R^* = \frac{\lambda I^*}{\mu_3}. \quad (11.4)$$

Remark 11.1 From (11.4) it follows that by the condition

$$b\beta > \mu_1 (\mu_2 + \lambda) \qquad (11.5)$$

the point E_* is a positive equilibrium point of (11.1).

Below we obtain sufficient conditions for the stability in probability of both equilibrium points E_0 and E_* of (11.1) by stochastic perturbations.

11.2 Stability in Probability of the Equilibrium Point $E_0 = (b\mu_1^{-1}, 0, 0)$

We will assume that (11.1) is influenced by stochastic perturbations of white noise type that are directly proportional to the deviation of the system state $(S(t), I(t), R(t))$ from the equilibrium point E_0 and influence the $\dot{S}(t)$, $\dot{I}(t)$, $\dot{R}(t)$ respectively. So, (11.1) takes the form

$$\dot{S}(t) = b - \beta S(t) J(I_t) - \mu_1 S(t) + \sigma_1 \left(S(t) - b\mu_1^{-1}\right)\dot{w}_1(t),$$
$$\dot{I}(t) = \beta S(t) J(I_t) - (\mu_2 + \lambda) I(t) + \sigma_2 I(t)\dot{w}_2(t), \qquad (11.6)$$
$$\dot{R}(t) = \lambda I(t) - \mu_3 R(t) + \sigma_3 R(t)\dot{w}_3(t),$$

where

$$J(I_t) = \int_0^\infty I(t - s)\, dF(s), \qquad (11.7)$$

σ_1, σ_2, σ_3 are constants, and $w_1(t)$, $w_2(t)$, $w_3(t)$ are mutually independent standard Wiener processes. Note that the equilibrium point E_0 is a solution of (11.6).

Centering (11.6) on the equilibrium point E_0 via the variables $x_1(t) = S(t) - b\mu_1^{-1}$, $x_2(t) = I(t)$, $x_3(t) = R(t)$, we obtain

$$\dot{x}_1(t) = -\mu_1 x_1(t) - \beta b\mu_1^{-1} J(x_{2t}) - \beta x_1(t) J(x_{2t}) + \sigma_1 x_1(t)\dot{w}_1(t),$$
$$\dot{x}_2(t) = -(\mu_2 + \lambda) x_2(t) + \beta b\mu_1^{-1} J(x_{2t}) + \beta x_1(t) J(x_{2t}) + \sigma_2 x_2(t)\dot{w}_2(t), \quad (11.8)$$
$$\dot{x}_3(t) = \lambda x_2(t) - \mu_3 x_3(t) + \sigma_3 x_3(t)\dot{w}_3(t).$$

It is clear that the stability of the equilibrium point E_0 of (11.6) is equivalent to the stability of the trivial solution of (11.8).

Below we will obtain sufficient conditions for the stability in probability of the trivial solution of (11.8). As it was noted in Remark 5.3, for getting sufficient conditions for the stability in probability of the trivial solution of the nonlinear system (11.8) with the order of a nonlinearity higher than one, it is enough to get sufficient conditions for the asymptotic mean-square stability of the trivial solution of the linear part of the considered nonlinear system. Thus, besides (11.8), we will consider the linear part of this system

$$\dot{y}_1(t) = -\mu_1 y_1(t) - \beta b\mu_1^{-1} J(y_{2t}) + \sigma_1 y_1(t)\dot{w}_1(t),$$
$$\dot{y}_2(t) = -(\mu_2 + \lambda)y_2(t) + \beta b\mu_1^{-1} J(y_{2t}) + \sigma_2 y_2(t)\dot{w}_2(t), \qquad (11.9)$$
$$\dot{y}_3(t) = \lambda y_2(t) - \mu_3 y_3(t) + \sigma_3 y_3(t)\dot{w}_3(t),$$

and following the procedure of constructing Lyapunov functionals, we will also consider the auxiliary system without delays

$$\dot{z}_1(t) = -\mu_1 z_1(t) + \sigma_1 z_1(t)\dot{w}_1(t),$$
$$\dot{z}_2(t) = -(\mu_2 + \lambda)z_2(t) + \sigma_2 z_2(t)\dot{w}_2(t), \qquad (11.10)$$
$$\dot{z}_3(t) = \lambda z_2(t) - \mu_3 z_3(t) + \sigma_3 z_3(t)\dot{w}_3(t).$$

Put now

$$\delta_i = \frac{1}{2}\sigma_i^2, \quad i = 1, 2, 3. \qquad (11.11)$$

Lemma 11.1 *The trivial solution of (11.10) is asymptotically mean-square stable if and only if*

$$\delta_1 < \mu_1, \qquad \delta_2 < \mu_2 + \lambda, \qquad \delta_3 < \mu_3. \qquad (11.12)$$

Proof From Remark 2.5 it follows that the first and second inequalities in (11.12) are necessary and sufficient conditions for the asymptotic mean-square stability of the trivial solutions of the first and second equations in (11.10), respectively. To prove the third inequality in (11.12), let us consider separately the system of the second and third equations in (11.10). If $\sigma_2 = \sigma_3 = 0$, then by Corollary 1.1 the trivial solution of this system is asymptotically stable. So, by Theorem 1.3 and (1.29) the matrix equation (1.27) with

$$A = \begin{pmatrix} -(\mu_2 + \lambda) & 0 \\ \lambda & -\mu_3 \end{pmatrix}, \qquad Q = \begin{pmatrix} q & 0 \\ 0 & 1 \end{pmatrix}$$

has a positive definite solution P with the elements

$$p_{11} = \frac{2^{-1}q + \lambda p_{12}}{\mu_2 + \lambda}, \qquad p_{22} = \frac{1}{2\mu_3}, \qquad p_{12} = \frac{\lambda}{2\mu_3(\mu_2 + \mu_3 + \lambda)}. \qquad (11.13)$$

Let L be the generator of the considered system, and

$$V = p_{11}z_2^2(t) + 2p_{12}z_2(t)z_3(t) + p_{22}z_3^2(t), \tag{11.14}$$

where the parameters p_{11}, p_{12}, p_{22} are defined in (11.13). Then by (11.14), (11.10), (11.13), (11.11) we have

$$
\begin{aligned}
LV &= -2\big(p_{11}z_2(t) + p_{12}z_3(t)\big)(\mu_2 + \lambda)z_2(t) + p_{11}\sigma_2^2 z_2^2(t) \\
&\quad + 2\big(p_{12}z_2(t) + p_{22}z_3(t)\big)\big(\lambda z_2(t) - \mu_3 z_3(t)\big) + p_{22}\sigma_3^2 z_3^2(t) \\
&= (2p_{11}\delta_2 - q)z_2^2(t) + 2(-\mu_3 + \delta_3)p_{22}z_3^2(t).
\end{aligned}
$$

If $\delta_2 < \mu_2 + \lambda$, then from (11.13) for big enough q, i.e., $q > 2\lambda p_{12}\delta_2(\mu_2 + \lambda - \delta_2)^{-1}$, we obtain

$$
\begin{aligned}
2p_{11}\delta_2 - q &= \frac{(q + 2\lambda p_{12})\delta_2}{\mu_2 + \lambda} - q \\
&= -\frac{\mu_2 + \lambda - \delta_2}{\mu_2 + \lambda}\left(q - \frac{2\lambda p_{12}\delta_2}{\mu_2 + \lambda - \delta_2}\right) \\
&< 0.
\end{aligned}
$$

So, by Remark 2.1 and (11.12) the trivial solution of (11.10) is asymptotically mean-square stable.

Let us suppose that $\delta_3 \geq \mu_3$. It is easy to see that in this case there exists small enough $q > 0$ such that $LV \geq 0$. So, $EV(z_2(t), z_3(t)) \geq EV(z_2(0), z_3(0)) > 0$, and the trivial solution of the considered system cannot be asymptotically mean-square stable. The proof is completed. □

Theorem 11.1 *If*

$$\delta_1 < \mu_1, \qquad \delta_2 < \mu_2 + \lambda - \beta b\mu_1^{-1}, \qquad \delta_3 < \mu_3, \tag{11.15}$$

then the trivial solution of (11.9) is asymptotically mean-square stable.

Proof Note that in (11.9) the first equation does not depend on $y_3(t)$ and the third equation does not depend on $y_1(t)$. So, (11.9) can be considered as two separate systems, $(y_1(t), y_2(t))$ and $(y_2(t), y_3(t))$.

First, let L be the generator of the system $(y_2(t), y_3(t))$, and

$$V_1 = p_{11}y_2^2(t) + 2p_{12}y_2(t)y_3(t) + p_{22}y_3^2(t), \tag{11.16}$$

where the parameters p_{11}, p_{12}, p_{22} are defined in (11.13). Then by (11.16), (11.9), (11.13), (11.11) we have

$$LV_1 = 2\big(p_{11}y_2(t) + p_{12}y_3(t)\big)\big[-(\mu_2 + \lambda)y_2(t) + \beta b\mu_1^{-1}J(y_{2t})\big] + p_{11}\sigma_2^2 y_2^2(t)$$
$$+ 2\big(p_{12}y_2(t) + p_{22}y_3(t)\big)\big(\lambda y_2(t) - \mu_3 y_3(t)\big) + p_{22}\sigma_3^2 y_3^2(t)$$
$$= (-q + 2\delta_2 p_{11})y_2^2(t) + 2(-\mu_3 + \delta_3)p_{22}y_3^2(t)$$
$$+ 2\beta b\mu_1^{-1}\big(p_{11}y_2(t) + p_{12}y_3(t)\big)J(y_{2t}).$$

Note that, by (11.2) and (11.7), $2y_2(t)J(y_{2t}) \le y_2^2(t) + J(y_{2t}^2)$ and $2y_3(t)J(y_{2t}) \le \alpha^{-1}y_3^2(t) + \alpha J(y_{2t}^2)$ for some $\alpha > 0$. Thus,

$$LV_1 \le (-q + 2\delta_2 p_{11})y_2^2(t) + 2(-\mu_3 + \delta_3)p_{22}y_3^2(t)$$
$$+ \beta b\mu_1^{-1}\big(p_{11}(y_2^2(t) + J(y_{2t}^2)) + p_{12}(\alpha^{-1}y_3^2(t) + \alpha J(y_{2t}^2))\big)$$
$$\le \big[-q + (2\delta_2 + \beta b\mu_1^{-1})p_{11}\big]y_2^2(t)$$
$$+ \big[2(-\mu_3 + \delta_3)p_{22} + \alpha^{-1}\beta b\mu_1^{-1}p_{12}\big]y_3^2(t) + AJ(y_{2t}^2),$$

where $A = \beta b\mu_1^{-1}(p_{11} + \alpha p_{12})$.

Following the procedure of constructing Lyapunov functionals, put

$$V_2 = A \int_0^\infty \int_{t-s}^t y_2^2(\tau)\,d\tau\,dF(s). \tag{11.17}$$

Then, by (11.2) and (11.7), $LV_2 = A(z_2^2(t) - J(z_{2t}^2))$, and, as a result, for the functional $V = V_1 + V_2$, we obtain

$$LV \le \big[-q + (2\delta_2 + \beta b\mu_1^{-1})p_{11} + A\big]y_2^2(t)$$
$$+ \big[2(-\mu_3 + \delta_3)p_{22} + \alpha\beta b\mu_1^{-1}p_{12}\big]y_3^2(t)$$
$$= \big[-q + 2(\delta_2 + \beta b\mu_1^{-1})p_{11} + \alpha\beta b\mu_1^{-1}p_{12}\big]y_2^2(t)$$
$$+ \big[2(-\mu_3 + \delta_3)p_{22} + \alpha^{-1}\beta b\mu_1^{-1}p_{12}\big]y_3^2(t).$$

If

$$\alpha\beta b\mu_1^{-1}p_{12} < q - 2(\delta_2 + \beta b\mu_1^{-1})p_{11},$$
$$\alpha^{-1}\beta b\mu_1^{-1}p_{12} < 2(\mu_3 - \delta_3)p_{22}, \tag{11.18}$$

then by Remark 2.1 the trivial solution of the considered system is asymptotically mean-square stable.

From (11.18) and (11.15) it follows that

$$0 < \frac{\beta b\mu_1^{-1}p_{12}}{2(\mu_3 - \delta_3)p_{22}} < \alpha < \frac{q - 2(\delta_2 + \beta b\mu_1^{-1})p_{11}}{\beta b\mu_1^{-1}p_{12}}. \tag{11.19}$$

So, if

$$\frac{\beta b \mu_1^{-1} p_{12}}{2(\mu_3 - \delta_3) p_{22}} < \frac{q - 2(\delta_2 + \beta b \mu_1^{-1}) p_{11}}{\beta b \mu_1^{-1} p_{12}}, \tag{11.20}$$

then there exists $\alpha > 0$ such that (11.19) holds.

Note that by (11.13) we have

$$q - 2(\delta_2 + \beta b \mu_1^{-1}) p_{11} = q - \frac{(\delta_2 + \beta b \mu_1^{-1})(q + 2\lambda p_{12})}{\mu_2 + \lambda}$$

$$= \frac{q(\mu_2 + \lambda - \beta b \mu_1^{-1} - \delta_2)}{\mu_2 + \lambda} - \frac{2\lambda p_{12}(\delta_2 + \beta b \mu_1^{-1})}{\mu_2 + \lambda}.$$

From this and from (11.15) it follows that for big enough $q > 0$, i.e.,

$$q > \frac{\mu_2 + \lambda}{\mu_2 + \lambda - \beta b \mu_1^{-1} - \delta_2} \left(\frac{(\beta b \mu_1^{-1} p_{12})^2}{2(\mu_3 - \delta_3) p_{22}} + \frac{2\lambda p_{12}(\delta_2 + \beta b \mu_1^{-1})}{\mu_2 + \lambda} \right),$$

equation (11.20) holds. So, the trivial solution of the considered system is asymptotically mean-square stable.

Now let L be the generator of the system $(y_1(t), y_2(t))$ in (11.9), and

$$V_1 = p_{11} y_1^2(t) + p_{22} y_2^2(t), \tag{11.21}$$

where

$$p_{11} = \frac{q}{2\mu_1}, \qquad p_{22} = \frac{1}{2(\mu_2 + \lambda)}. \tag{11.22}$$

Then by (11.21), (11.9), and (11.11) we have

$$LV_1 = 2p_{11} y_1(t)\left(-\mu_1 y_1(t) - \beta b \mu_1^{-1} J(y_{2t})\right) + p_{11} \sigma_1^2 y_1^2(t)$$

$$\quad + 2p_{22} y_2(t)\left(-(\mu_2 + \lambda) y_2(t) + \beta b \mu_1^{-1} J(y_{2t})\right) + p_{22} \sigma_2^2 y_2^2(t)$$

$$= -2(\mu_1 - \delta_1) p_{11} y_1^2(t) - 2(\mu_2 + \lambda - \delta_2) p_{22} y_2^2(t)$$

$$\quad + 2\beta b \mu_1^{-1} J(y_{2t})\left(p_{22} y_2(t) - p_{11} y_1(t)\right).$$

Note that, by (11.2) and (11.7), $2y_2(t) J(y_{2t}) \le y_2^2(t) + J(y_{2t}^2)$ and $2y_1(t) J(y_{2t}) \le \alpha^{-1} y_1^2(t) + \alpha J(y_{2t}^2)$ for some $\alpha > 0$. Thus,

$$LV_1 \le -2(\mu_1 - \delta_1) p_{11} y_1^2(t) - 2(\mu_2 + \lambda - \delta_2) p_{22} y_2^2(t)$$

$$\quad + \beta b \mu_1^{-1} \left[p_{11}\left(\alpha^{-1} y_1^2(t) + \alpha J(y_{2t}^2)\right) + p_{22}\left(y_2^2(t) + J(y_{2t}^2)\right) \right]$$

$$= \left[-2(\mu_1 - \delta_1) + \alpha^{-1} \beta b \mu_1^{-1} \right] p_{11} y_1^2(t)$$

$$\quad + \left[-2(\mu_2 + \lambda - \delta_2) + \beta b \mu_1^{-1} \right] p_{22} y_2^2(t) + AJ(y_{2t}^2),$$

where $A = \beta b \mu_1^{-1}(\alpha p_{11} + p_{22})$. Using (11.17) with this A for the functional $V = V_1 + V_2$, we have

$$LV \leq \left[-2(\mu_1 - \delta_1) + \alpha^{-1}\beta b \mu_1^{-1}\right]p_{11}y_1^2(t)$$
$$+ \left[-2(\mu_2 + \lambda - \delta_2) + \beta b \mu_1^{-1}\right]p_{22}y_2^2(t) + A y_2^2(t)$$
$$= \left[-2(\mu_1 - \delta_1) + \alpha^{-1}\beta b \mu_1^{-1}\right]p_{11}y_1^2(t)$$
$$+ \left[-2\left(\mu_2 + \lambda - \beta b \mu_1^{-1} - \delta_2\right)p_{22} + \alpha\beta b \mu_1^{-1}\alpha p_{11}\right]y_2^2(t).$$

If

$$\alpha^{-1}\beta b \mu_1^{-1} < 2(\mu_1 - \delta_1), \qquad \alpha\beta b \mu_1^{-1}p_{11} < 2\left(\mu_2 + \lambda - \beta b \mu_1^{-1} - \delta_2\right)p_{22},$$
$$(11.23)$$

then by Remark 2.1 the trivial solution of the considered system is asymptotically mean-square stable.

From (11.23), (11.22), and (11.15) it follows that

$$0 < \frac{\beta b \mu_1^{-1}}{2(\mu_1 - \delta_1)} < \alpha < \frac{2(\mu_2 + \lambda - \beta b \mu_1^{-1} - \delta_2)\mu_1}{\beta b \mu_1^{-1}(\mu_2 + \lambda)q}. \qquad (11.24)$$

So, if

$$\frac{\beta b \mu_1^{-1}}{2(\mu_1 - \delta_1)} < \frac{2(\mu_2 + \lambda - \beta b \mu_1^{-1} - \delta_2)\mu_1}{\beta b \mu_1^{-1}(\mu_2 + \lambda)q}, \qquad (11.25)$$

then there exists $\alpha > 0$ such that (11.24) holds. It is easy to see that (11.25) holds by conditions (11.15) for small enough $q > 0$ such that

$$q < \frac{4(\mu_1 - \delta_1)(\mu_2 + \lambda - \beta b \mu_1^{-1} - \delta_2)\mu_1^3}{\beta^2 b^2(\mu_2 + \lambda)}.$$

So, the trivial solution of (11.9) is asymptotically mean-square stable. The proof is completed. ☐

Corollary 11.1 *If conditions (11.15) hold, then the equilibrium point E_0 of (11.6) is stable in probability.*

Remark 11.2 Note that the second condition (11.15) contradicts with (11.5). It means that by conditions (11.15) system (11.1) has no positive equilibrium point (11 4),

11.3 Stability in Probability of the Equilibrium Point $E_* = (S^*, I^*, R^*)$

Here we assume that (11.1) is influenced by stochastic perturbations of white noise type that are directly proportional to the deviation of the system state

$(S(t), I(t), R(t))$ from the equilibrium point $E_* = (S^*, I^*, R^*)$ and influence $\dot{S}(t)$, $\dot{I}(t)$, $\dot{R}(t)$. Then (11.1) takes the form

$$\dot{S}(t) = b - \beta S(t)J(I_t) - \mu_1 S(t) + \sigma_1\big(S(t) - S^*\big)\dot{w}_1(t),$$
$$\dot{I}(t) = \beta S(t)J(I_t) - (\mu_2 + \lambda)I(t) + \sigma_2\big(I(t) - I^*\big)\dot{w}_2(t), \qquad (11.26)$$
$$\dot{R}(t) = \lambda I(t) - \mu_3 R(t) + \sigma_3\big(R(t) - R^*\big)\dot{w}_3(t),$$

where $J(I_t)$, σ_1, σ_2, σ_3, $w_1(t)$, $w_2(t)$, $w_3(t)$ are the same as in (11.6), and the equilibrium point $E_* = (S^*, I^*, R^*)$ is the solution of (11.26).

Centering (11.26) on the positive equilibrium E_* via the variables $x_1(t) = S(t) - S^*$, $x_2(t) = I(t) - I^*$, $x_3(t) = R(t) - R^*$ and using (11.2), (11.3), we obtain

$$\dot{x}_1(t) = -b\big(S^*\big)^{-1}x_1(t) - \beta S^* J(x_{2t}) - \beta x_1(t)J(x_{2t}) + \sigma_1 x_1(t)\dot{w}_1(t),$$
$$\dot{x}_2(t) = \beta I^* x_1(t) - \beta S^* x_2(t) + \beta S^* J(x_{2t})$$
$$\qquad + \beta x_1(t)J(x_{2t}) + \sigma_2 x_2(t)\dot{w}_2(t), \qquad (11.27)$$
$$\dot{x}_3(t) = \lambda x_2(t) - \mu_3 x_3(t) + \sigma_3 x_3(t)\dot{w}_3(t).$$

It is easy to see that the stability of the equilibrium point E_* of (11.26) is equivalent to the stability of the zero solution of (11.27).

Below we will obtain sufficient conditions for the stability in probability of the zero solution of (11.27) using the linear part of this system

$$\dot{y}_1(t) = -b\big(S^*\big)^{-1}y_1(t) - \beta S^* J(y_{2t}) + \sigma_1 y_1(t)\dot{w}_1(t),$$
$$\dot{y}_2(t) = \beta I^* y_1(t) - \beta S^* y_2(t) + \beta S^* J(y_{2t}) + \sigma_2 y_2(t)\dot{w}_2(t), \qquad (11.28)$$
$$\dot{y}_3(t) = \lambda y_2(t) - \mu_3 y_3(t) + \sigma_3 y_3(t)\dot{w}_3(t)$$

and the auxiliary system without delays

$$\dot{z}_1(t) = -b\big(S^*\big)^{-1}z_1(t) + \sigma_1 z_1(t)\dot{w}_1(t),$$
$$\dot{z}_2(t) = \beta I^* z_1(t) - \beta S^* z_2(t) + \sigma_2 z_2(t)\dot{w}_2(t), \qquad (11.29)$$
$$\dot{z}_3(t) = \lambda z_2(t) - \mu_3 z_3(t) + \sigma_3 z_3(t)\dot{w}_3(t).$$

Lemma 11.2 *If*

$$\delta_1 < b\big(S^*\big)^{-1}, \qquad \delta_2 < \beta S^*, \qquad \delta_3 < \mu_3, \qquad (11.30)$$

then the trivial solution of (11.29) *is asymptotically mean-square stable.*

Proof Note that the first inequality in (11.30) is a necessary and sufficient condition for the asymptotic mean-square stability of the trivial solution of the first equation

(11.29). Note also that the system of two first equations in (11.29) has the structure that is the same as the structure of the second and the third equations in (11.10). So, similarly to (11.12), we obtain two first conditions in (11.30).

Finally, note that the third equation in (11.29) with the additive perturbation $z_2(t)$ that satisfies the condition $\lim_{t\to\infty} \mathbf{E}z_2^2(t) = 0$ has the structure that is the same as the structure of the second equation in (11.29) with the additive perturbation $z_1(t)$ that satisfies the same condition $\lim_{t\to\infty} \mathbf{E}z_1^2(t) = 0$. Thus, since the inequality $\delta_2 < \beta S^*$ is a sufficient condition for the asymptotic mean-square stability of the trivial solution of the second equation, the similar inequality $\delta_3 < \mu_3$ is a sufficient condition for asymptotic mean-square stability of the trivial solution of the third equation. The proof is completed.   □

Theorem 11.2 *Let $\delta_3 < \mu_3$ and*

$$\delta_1 < \mu_1, \qquad \delta_2 < \frac{\beta S^* \beta I^*}{b(S^*)^{-1} + \beta S^*} \tag{11.31}$$

or

$$\mu_1 \le \delta_1 < \mu_1 + \beta I^* \frac{\sqrt{4I^*(S^*)^{-1} + 1} - 1}{\sqrt{4I^*(S^*)^{-1} + 1} + 1},$$

$$\delta_2 < \frac{\beta S^* \beta I^*}{b(S^*)^{-1} + \beta S^*} \left(1 - \frac{\beta S^*(\delta_1 - \mu_1)}{(b(S^*)^{-1} - \delta_1)^2}\right). \tag{11.32}$$

Then the trivial solution of (11.28) is asymptotically mean-square stable.

Proof Let L be the generator of the system of two first equations in (11.28). Following the procedure of constructing Lyapunov functionals, we will construct a Lyapunov functional V for this system in the form $V = V_1 + V_2$, where

$$V_1 = p_{11} y_1^2(t) + 2p_{12} y_1(t) y_2(t) + p_{22} y_2^2(t) \tag{11.33}$$

and the parameters p_{11}, p_{12}, p_{22} by (1.29) are

$$p_{11} = \frac{S^*}{2b} q + \frac{\beta I^* S^*}{b} p_{12}, \qquad p_{22} = \frac{1}{2\beta S^*}, \qquad p_{12} = \frac{I^*(S^*)^{-1}}{2(b(S^*)^{-1} + \beta S^*)}. \tag{11.34}$$

Then by (11.33), (11.28) we have

$$LV_1 = 2\big(p_{11} y_1(t) + p_{12} y_2(t)\big)\big(-b(S^*)^{-1} v_1(t) - \beta S^* J(y_{2t})\big) + p_{11}\sigma_1^2 y_1^2(t)$$

$$+ 2\big(p_{12} y_1(t) + p_{22} y_2(t)\big)\big(\beta I^* y_1(t) - \beta S^* y_2(t)$$

$$+ \beta S^* J(y_{2t})\big) + p_{22}\sigma_2^2 y_2^2(t)$$

$$= 2\big((\delta_1 - b(S^*)^{-1})p_{11} + \beta I^* p_{12}\big) y_1^2(t) + 2(\delta_2 - \beta S^*)p_{22} y_2^2(t)$$

$$+ 2\big[\beta I^* p_{22} - \big(b(S^*)\big)^{-1} + \beta S^*\big) p_{12}\big] y_1(t) y_2(t)$$
$$+ 2\beta S^* \big[(p_{12} - p_{11}) y_1(t) + (p_{22} - p_{12}) y_2(t)\big] J(y_{2t}).$$

Note that, by (11.2) and (11.7), $2 y_2(t) J(y_{2t}) \le y_2^2(t) + J(y_{2t}^2)$ and $2 y_1(t) J(y_{2t}) \le \alpha^{-1} y_1^2(t) + \alpha J(y_{2t}^2)$ for some $\alpha > 0$. Besides, by (11.34) and (11.4),

$$\big(\delta_1 - b(S^*)^{-1}\big) p_{11} + \beta I^* p_{12} = (\delta_1 - \mu_1) p_{11} + \beta I^* (p_{12} - p_{11}),$$

$$\beta I^* p_{22} - \big(b(S^*)^{-1} + \beta S^*\big) p_{12} = 0,$$

$$p_{22} - p_{12} = \frac{b(S^*)^{-1} + \beta S^* - \beta I^*}{2\beta S^* (b(S^*)^{-1} + \beta S^*)} = \frac{\mu_1 + \beta S^*}{2\beta S^* (b(S^*)^{-1} + \beta S^*)} > 0.$$

Thus,

$$LV_1 \le \big[2\big((\delta_1 - \mu_1) p_{11} + \beta I^* (p_{12} - p_{11})\big) + \alpha^{-1} \beta S^* |p_{12} - p_{11}|\big] y_1^2(t)$$
$$+ \big[(2\delta_2 - \beta S^*) p_{22} - \beta S^* p_{12}\big] y_2^2(t) + A J(y_{2t}^2),$$

where $A = \beta S^* (\alpha |p_{12} - p_{11}| + p_{22} - p_{12})$.
   Choosing the additional functional V_2 in the form (11.17) with A obtained above, for the functional $V = V_1 + V_2$, we have

$$LV \le \big[2\big((\delta_1 - \mu_1) p_{11} + \beta I^* (p_{12} - p_{11})\big) + \alpha^{-1} \beta S^* |p_{12} - p_{11}|\big] y_1^2(t)$$
$$+ \big[2(\delta_2 p_{22} - \beta S^* p_{12}) + \alpha \beta S^* |p_{12} - p_{11}|\big] y_2^2(t). \tag{11.35}$$

Note that by (11.34) and (11.4)

$$p_{11} - p_{12} = \frac{S^*}{2b} (q - 2\mu_1 p_{12}). \tag{11.36}$$

So, putting $q = 2\mu_1 p_{12} > 0$, we obtain $p_{11} = p_{12}$ and

$$LV \le 2(\delta_1 - \mu_1) p_{11} y_1^2(t) + 2(\delta_2 - \beta S^* p_{12} p_{22}^{-1}) p_{22} y_2^2(t). \tag{11.37}$$

By (11.34) we have $\beta S^* p_{12} p_{22}^{-1} = \beta S^* \beta I^* (b(S^*)^{-1} + \beta S^*)^{-1}$. So, from (11.37) it follows that by conditions (11.31) the trivial solution of (11.28) is asymptotically mean-square stable.
   Suppose now that $\delta_1 \ge \mu_1$ and $q > 2\mu_1 p_{12} > 0$. Then $p_{11} > p_{12}$, and from (11.35) it follows that if

$$0 < \frac{\beta S^* (p_{11} - p_{12})}{2(\beta I^* (p_{11} - p_{12}) - (\delta_1 - \mu_1) p_{11})} < \alpha < \frac{2(\beta S^* p_{12} - \delta_2 p_{22})}{\beta S^* (p_{11} - p_{12})}, \tag{11.38}$$

then the trivial solution of (11.28) is asymptotically mean-square stable.

If

$$0 < \frac{\beta S^*(p_{11} - p_{12})}{2(\beta I^*(p_{11} - p_{12}) - (\delta_1 - \mu_1)p_{11})} < \frac{2(\beta S^* p_{12} - \delta_2 p_{22})}{\beta S^*(p_{11} - p_{12})}, \qquad (11.39)$$

then there exists $\alpha > 0$ such that (11.38) holds. By (11.4), (11.36) from (11.39) it follows that

$$\frac{(\beta S^*)^2}{4(\beta S^* p_{12} - \delta_2 p_{22})} < \frac{\beta I^*(p_{11} - p_{12}) - (\delta_1 - \mu_1)p_{11}}{(p_{11} - p_{12})^2}$$

$$= \frac{(b(S^*)^{-1} - \delta_1)(p_{11} - p_{12}) - (\delta_1 - \mu_1)p_{12}}{(p_{11} - p_{12})^2}$$

$$= \frac{b(S^*)^{-1} - \delta_1}{p_{11} - p_{12}} - \frac{(\delta_1 - \mu_1)p_{12}}{(p_{11} - p_{12})^2}$$

$$= \frac{2b}{(S^*)^2}\left(\frac{S^*(b(S^*)^{-1} - \delta_1)}{q - 2\mu_1 p_{12}} - \frac{2bp_{12}(\delta_1 - \mu_1)}{(q - 2\mu_1 p_{12})^2}\right). \quad (11.40)$$

Note that the right-hand part of the obtained inequality reaches its maximum at

$$q = 2\mu_1 p_{12} + \frac{4(\delta_1 - \mu_1)bp_{12}}{S^*(b(S^*)^{-1} - \delta_1)}. \qquad (11.41)$$

Substituting (11.41) into (11.40), we obtain

$$\frac{(\beta S^*)^2}{\beta S^* p_{12} - \delta_2 p_{22}} < \frac{(b(S^*)^{-1} - \delta_1)^2}{(\delta_1 - \mu_1)p_{12}}$$

or

$$\delta_2 < \frac{\beta S^* p_{12}}{p_{22}}\left(1 - \frac{\beta S^*(\delta_1 - \mu_1)}{(b(S^*)^{-1} - \delta_1)^2}\right),$$

which is equivalent to the second condition in (11.32).

From the positivity of the expression in the brackets, i.e., from the inequality $\beta S^*(\delta_1 - \mu_1) < (b(S^*)^{-1} - \delta_1)^2$, the first condition (11.32) follows. Indeed, put $\Delta = \delta_1 - \mu_1$. Then by (11.3) $b(S^*)^{-1} - \delta_1 = \beta I^* - \Delta$, and we get $\beta S^*\Delta < (\beta I^* - \Delta)^2$ or $\Delta^2 - (2\beta I^* + \beta S^*)\Delta + (\beta I^*)^2 > 0$. From this it follows that

$$\Delta < \frac{1}{2}\left(2\beta I^* + \beta S^* - \sqrt{(2\beta I^* + \beta S^*)^2 - 4(\beta I^*)^2}\right)$$

$$= \beta I^* - \frac{\beta}{2}\left(\sqrt{4I^* S^* + (S^*)^2} - S^*\right)$$

$$= \beta I^* - \frac{2\beta I^* S^*}{\sqrt{4I^* S^* + (S^*)^2} + S^*}$$

$$= \beta I^*\left(1 - \frac{2}{\sqrt{4I^*(S^*)^{-1} + 1} + 1}\right)$$

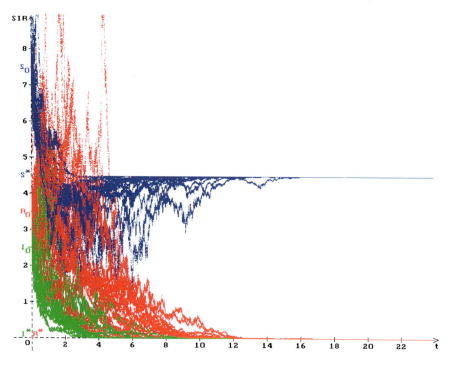

Fig. 11.1 25 trajectories of the processes $S(t)$ (*blue*), $I(t)$ (*green*), $R(t)$ (*red*) with the equilibrium point $S^* = 4.5$, $I^* = 0$, $R^* = 0$, the initial values $S(0) = 7.5$, $I(s) = 2.5$, $s \in [-0.1, 0]$, $R(0) = 3.5$, and the values of the parameters $b = 1.8$, $\beta = 0.2$, $\lambda = 1$, $\mu_1 = 0.4$, $\mu_2 = \mu_3 = 0.5$, $h = 0.1$, $\delta_1 = 0.35$, $\delta_2 = 0.25$, $\delta_3 = 0.15$

$$= \beta I^* \frac{\sqrt{4I^*(S^*)^{-1} + 1} - 1}{\sqrt{4I^*(S^*)^{-1} + 1} + 1},$$

which is equivalent to the first condition in (11.32). The proof is completed.   □

By Theorem 11.2 and Remark 5.3 the following statement holds.

Corollary 11.2 *Let the conditions of Theorem 11.2 hold. Then the equilibrium point E_* of (11.26) is stable in probability.*

Remark 11.3 From (11.4) and (11.31) it follows that

$$b\beta(\mu_2 + \lambda)^{-1} - \mu_1 = \beta I^* > \frac{\beta S^* \beta I^*}{b(S^*)^{-1} + \beta S^*} > \delta_2 > 0.$$

So, (11.5) follows from (11.31). This means that by the conditions of Theorem 11.2 the equilibrium point E_* is a positive equilibrium point of (11.1).

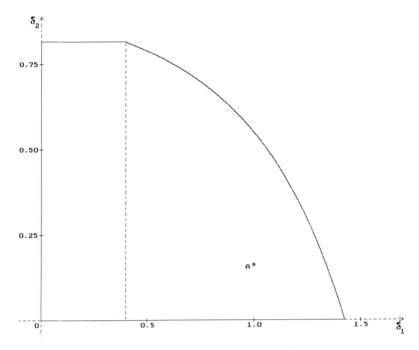

Fig. 11.2 Stability region for E_* given by conditions (11.31), (11.32)

11.4   Numerical Simulation

Let us suppose that in (11.1) $dF(s) = \delta(s - h)ds$, where $\delta(s)$ is Dirac's delta-function, and $h \geq 0$ is the delay. So, by (11.7), $J(I_t) = I(t - h)$.

Put $b = 1.8$, $\beta = 0.2$, $\lambda = 1$, $\mu_1 = 0.4$, $\mu_2 = \mu_3 = 0.5$, $h = 0.1$, and $\delta_1 = 0.35$, $\delta_2 = 0.25$, $\delta_3 = 0.15$. In this case $S^* = b\mu_1^{-1} = 4.5$ and $\delta_1 < \mu_1$, $\delta_2 < \mu_2 + \lambda - \beta b \mu_1^{-1} = 0.6$, $\delta_3 < \mu_3$, i.e., the conditions of Theorem 11.1 hold. So, the equilibrium point $E_0 = (4.5, 0, 0)$ of (11.6) is stable in probability. In Fig. 11.1 for twenty five realizations of each of the processes $S(t)$ (blue), $I(t)$ (green) and $R(t)$ (red) are shown that are the solution of (11.6) with the initial conditions $S(0) = 7.5$, $I(s) = 2.5$, $s \in [-0.1, 0]$, $R(0) = 3.5$. All trajectories converge to the equilibrium point $E_0 = (4.5, 0, 0)$.

Consider now $b = 20$ with the same values of the other parameters. In this case condition (11.5) holds: $b\beta = 4 > \mu_1(\mu_2 + \lambda) = 0.6$. So, (11.6) has a positive equilibrium point (11.4). The stability region given by the conditions (11.31), (11.32) for the given values of the parameters is shown in Fig. 11.2 in the space of the parameters (δ_1, δ_2). The dotted line separates the stability region into two parts given by conditions (11.31) and (11.32), respectively. In Fig. 11.3, for twenty five realizations of each of the processes, $S(t)$ (blue), $I(t)$ (green), and $R(t)$ (red) are shown that are the solution of (11.6) with the initial conditions $S(0) = 3.5$, $I(s) = 16.33$ for $s \in [-0.1, 0]$, and $R(0) = 19.17$ at the point A of the stability region (Fig. 11.2) with the coordinates $\delta_1 = 1$, $\delta_2 = 0.16$ and with $\delta_3 = 0.25$. One can see that all

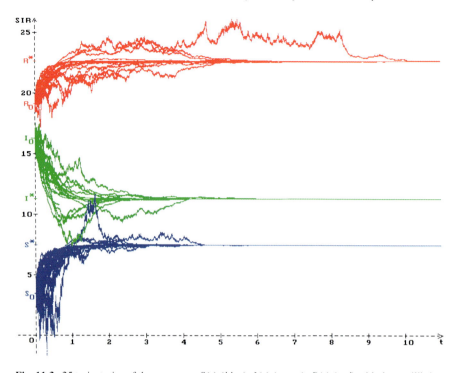

Fig. 11.3 25 trajectories of the processes $S(t)$ (*blue*), $I(t)$ (*green*), $R(t)$ (*red*) with the equilibrium point $S^* = 7.5$, $I^* = 11.33$, $R^* = 22.67$, the initial values $S(0) = 3.5$, $I(s) = 16.33$, $s \in [-0.1, 0]$, $R(0) = 19.17$, and the values of the parameters $b = 20$, $\beta = 0.2$, $\lambda = 1$, $\mu_1 = 0.4$, $\mu_2 = \mu_3 = 0.5$, $h = 0.1$, $\delta_1 = 1$, $\delta_2 = 0.16$, $\delta_3 = 0.25$

trajectories converge to the equilibrium point $E_* = (S^*, I^*, R^*)$, where $S^* = 7.5$, $I^* = 11.33$, $R^* = 22.67$.

Note that numerical simulations of the processes $S(t)$, $I(t)$, and $R(t)$ were obtained via the Euler–Maruyama scheme [200] for stochastic differential equations, and difference analogues of systems (11.6), (11.7) and (11.26), (11.7) with $J(I_t) = I(t - h)$ were used in the form

$$S_{i+1} = S_i + \Delta(b - \beta S_i I_{i-m} - \mu_1 S_i) + \sigma_1 \sqrt{\Delta}(S_i - S^*)\xi_{1,i+1},$$

$$I_{i+1} = I_i + \Delta(\beta S_i I_{i-m} - (\mu_2 + \lambda)I_i) + \sigma_2 \sqrt{\Delta}(I_i - I^*)\xi_{2,i+1},$$

$$R_{i+1} = R_i + \Delta(\lambda I_i - \mu_3 R_i) + \sigma_3 \sqrt{\Delta}(R_i - R^*)\xi_{3,i+1},$$

$$i = 0, 1, 2, \ldots, \quad I_j = I_0, \quad j = -m, \ldots, -1, 0.$$

Here $\xi_{ki} = \Delta^{-\frac{1}{2}}(w_k(t_i) - w_k(t_{i-1}))$ are mutually independent random variables such that $E\xi_{ki} = 0$, $E\xi_{ki}^2 = 1$, $k = 1, 2, 3$, $t_i = i\Delta$, Δ is the step of discretization, $S_i = S(t_i)$, $I_i = I(t_i)$, $R_i = R(t_i)$, $m = h\Delta^{-1}$.

Chapter 12
Stability of Some Social Mathematical Models with Delay Under Stochastic Perturbations

In this chapter we propose a mathematical framework to model some social behavior. To be precise, we propose delayed and stochastic mathematical models to analyze human behaviors related to some addictions: consumption of alcohol and obesity.

12.1 Mathematical Model of Alcohol Consumption

Taking into account the proposal presented in [247], we consider alcohol consumption habit as susceptible to be transmitted by peer pressure or social contact. This fact led us to propose an epidemiologic-type mathematical model to study this social epidemic.

Here we generalize the known nonlinear dynamic model of alcohol consumption [254] by adding distributed delay. We obtain sufficient conditions for the existence of a positive equilibrium point of this system. Similarly to the previous sections, we suppose that this nonlinear system is exposed to additive stochastic perturbations of white noise type that are directly proportional to the deviation of the system state from the equilibrium point. The considered nonlinear system is linearized in the neighborhood of the positive point of equilibrium, and a sufficient condition for asymptotic mean-square stability of the zero solution of the constructed linear system is obtained via the procedure of constructing Lyapunov functionals that is described in Sect. 2.2.2. Since the order of nonlinearity of the considered nonlinear system is higher than one, the obtained condition is also a sufficient one (Sect. 5.3) for stability in probability of the equilibrium point of the initial nonlinear system under stochastic perturbations.

12.1.1 Description of the Model of Alcohol Consumption

Let $A(t)$ be nonconsumers, individuals that have never consumed alcohol or infrequently have alcohol consumption, and $M(t)$ be nonrisk consumers, individuals

L. Shaikhet, *Lyapunov Functionals and Stability of Stochastic Functional Differential Equations*, DOI 10.1007/978-3-319-00101-2_12,
© Springer International Publishing Switzerland 2013

with regular low consumption, to be precise, men consuming less than 50 cc (cubic centimeters) of alcohol every day and women consuming less than 30 cc of alcohol every day. Let $R(t)$ be risk consumers, individuals with regular high consumption, that is, men consuming more than 50 cc of alcohol every day and women who consuming more than 30 cc of alcohol every day.

Considering homogeneous mixing [219], where each individual can contact with any other individual (peer pressure), a dynamic alcohol consumption model is given by the following nonlinear system of ordinary differential equations with distributed delay:

$$\dot{A}(t) = \mu P(t) + \gamma R(t) - d_A A(t) - \beta A(t) \int_0^\infty \frac{M(t-s) + R(t-s)}{P(t-s)} dK(s),$$

$$\dot{M}(t) = \beta A(t) \int_0^\infty \frac{M(t-s) + R(t-s)}{P(t-s)} dK(s) - dM(t) - \alpha M(t), \qquad (12.1)$$

$$\dot{R}(t) = \alpha M(t) - \gamma R(t) - d R(t),$$

$$P(t) = A(t) + M(t) + R(t).$$

Here:

α the rate at which a nonrisk consumer moves to the risk consumption subpopulation (intensity of transition from the group $M(t)$ to the group $R(t)$).

β the transmission rate due to social pressure to increase the alcohol consumption, e.g., family, friends, marketing, TV, etc. (intensity of transition from the group $A(t)$ to the group $M(t)$).

γ the rate at which a risk consumer becomes a nonconsumer (intensity of transition from the group $R(t)$ to the group $A(t)$); so, the scheme of transition from one group to another one is

$$A(t) \xrightarrow{\beta} M(t) \xrightarrow{\alpha} R(t) \xrightarrow{\gamma} A(t).$$

μ the birth rate.
d_A the death rate.
d the augmented death rate due to alcohol consumption (accidents at work, traffic accidents, and diseases derived by alcohol consumption are considered).

We suppose that the parameters $\alpha, \beta, \gamma, \mu, d_A, d$ are nonnegative numbers and $K(s)$ is a nondecreasing function such that

$$\int_0^\infty dK(s) = 1. \qquad (12.2)$$

The integral is understood in the Stieltjes sense.

Remark 12.1 In particular, $dK(s) = \delta(s-h) ds$, where $h > 0$, $\delta(s)$ is Dirac's function, system (12.1) is a system with discrete delay h. The case of a system without delay ($h = 0$) is considered in [254].

12.1.2 Normalization of the Initial Model

Put

$$a(t) = \frac{A(t)}{P(t)}, \qquad m(t) = \frac{M(t)}{P(t)}, \qquad r(t) = \frac{R(t)}{P(t)}. \tag{12.3}$$

From (12.1) and (12.3) it follows that

$$a(t) + m(t) + r(t) = 1. \tag{12.4}$$

Adding the first three equations in (12.1), by (12.4) we obtain

$$\frac{\dot{P}(t)}{P(t)} = \mu - d + (d - d_A)a(t).$$

From this and from (12.3) we have

$$\dot{a}(t) = \frac{\dot{A}(t)P(t) - A(t)\dot{P}(t)}{P^2(t)} = \frac{\dot{A}(t)}{P(t)} - \frac{A(t)}{P(t)} \times \frac{\dot{P}(t)}{P(t)}$$

$$= \frac{\dot{A}(t)}{P(t)} - a(t)\big[\mu - d + (d - d_A)a(t)\big], \tag{12.5}$$

and, similarly,

$$\dot{m}(t) = \frac{\dot{M}(t)}{P(t)} - m(t)\big[\mu - d + (d - d_A)a(t)\big],$$
$$\tag{12.6}$$
$$\dot{r}(t) = \frac{\dot{R}(t)}{P(t)} - r(t)\big[\mu - d + (d - d_A)a(t)\big].$$

Thus, putting

$$I(a_t) = \int_0^\infty a(t - s)\,dK(s), \tag{12.7}$$

by (12.5), (12.6), (12.1), (12.2), and (12.4) we obtain

$$\dot{a}(t) = \mu + \gamma r(t) + \beta a(t)I(a_t) - a(t)\big[\beta + \mu - (d - d_A)(1 - a(t))\big],$$
$$\dot{m}(t) = \beta a(t) - \beta a(t)I(a_t) - m(t)\big[\alpha + \mu + (d - d_A)a(t)\big],$$
$$\dot{r}(t) = \alpha m(t) - r(t)\big[\gamma + \mu + (d - d_A)a(t)\big].$$

In view of (12.4), the last equation can be rejected, and, as a result, we obtain the system of two integro–differential equations

$$\dot{a}(t) = \mu + \gamma - \gamma m(t) + \beta a(t)I(a_t) - a(t)\big[\beta + \mu + \gamma - (d - d_A)(1 - a(t))\big],$$
$$\tag{12.8}$$
$$\dot{m}(t) = \beta a(t) - \beta a(t)I(a_t) - m(t)\big[\alpha + \mu + (d - d_A)a(t)\big].$$

12.1.3 Existence of an Equilibrium Point

By (12.8), (12.2), and (12.4) a point of equilibrium (a^*, m^*, r^*) is defined by the following system of algebraic equations:

$$
\begin{aligned}
(\mu + \gamma)(1 - a^*) &= a^*(\beta - d + d_A)(1 - a^*) + \gamma m^*, \\
\beta a^*(1 - a^*) &= m^*[\alpha + \mu + (d - d_A)a^*], \\
a^* + m^* + r^* &= 1.
\end{aligned}
\tag{12.9}
$$

Lemma 12.1 *If $d \in [d_A, \beta + d_A)$, then system (12.9) has a unique positive solution (a^*, m^*, r^*) if and only if*

$$
\beta > d - d_A + \mu + \frac{\alpha\gamma}{\alpha + \mu + \gamma + d - d_A}.
\tag{12.10}
$$

If $d \geq \beta + d_A$, then system (12.9) has no positive solutions.

Proof Necessity From the first two equations in (12.9) we have

$$
\mu + \gamma = a^*(\beta - d + d_A) + \frac{\gamma\beta a^*}{\alpha + \mu + (d - d_A)a^*}.
\tag{12.11}
$$

Since $a^* \in (0, 1)$, from (12.11) it follows that

$$
\begin{aligned}
\mu + \gamma &= a^*(\beta - d + d_A) + \frac{\gamma\beta}{(\alpha + \mu)(a^*)^{-1} + d - d_A} \\
&< \beta - d + d_A + \frac{\gamma\beta}{\alpha + \mu + d - d_A} \\
&= \frac{\alpha + \mu + \gamma + d - d_A}{\alpha + \mu + d - d_A}\beta - d + d_A,
\end{aligned}
$$

which is equivalent to (12.10) since

$$
\begin{aligned}
\beta &> \frac{(\mu + \gamma + d - d_A)(\alpha + \mu + d - d_A)}{\alpha + \mu + \gamma + d - d_A} \\
&= d - d_A + \mu + \frac{\alpha\gamma}{\alpha + \mu + \gamma + d - d_A}.
\end{aligned}
$$

Sufficiency Rewrite (12.11) in the form

$$
\begin{aligned}
Q(a^*)^2 + Ba^* - C &= 0, \\
B &= (\beta - d + d_A)(\alpha + \gamma + \mu) - \mu(d - d_A), \\
Q &= (\beta - d + d_A)(d - d_A), \quad C = (\mu + \alpha)(\mu + \gamma).
\end{aligned}
\tag{12.12}
$$

Thus, by (12.9) the equilibrium point (a^*, m^*, r^*) is defined by the system of algebraic equations (12.12), and

$$m^* = \frac{\beta a^* (1 - a^*)}{\alpha + \mu + (d - d_A)a^*}, \qquad r^* = 1 - a^* - m^*. \tag{12.13}$$

It is easy to check that by the condition $d \in [d_A, \beta + d_A)$ (or $Q \ge 0$) the existence of a solution a^* of (12.12) in the interval $(0, 1)$ is equivalent to the condition $C < Q + B$, which is equivalent to (12.10).

If $d \ge \beta + d_A$, then $Q \le 0$ and $B < 0$. So, (12.12) cannot have positive roots. The proof is completed.   □

Example 12.1 Following [254], put

$$\alpha = 0.000110247, \qquad \beta = 0.0284534, \qquad \gamma = 0.00144,$$
$$\mu = 0.01, \qquad d = 0.009, \qquad d_A = 0.008. \tag{12.14}$$

Then condition (12.10) holds, and the solution of (12.12)–(12.13) is

$$a^* = 0.364739, \qquad m^* = 0.629383, \qquad r^* = 0.00587794, \tag{12.15}$$

or, in percents, $a^* = 36.47\ \%$, $m^* = 62.94\ \%$, $r^* = 0.59\ \%$.

12.1.4 Stochastic Perturbations, Centralization, and Linearization

Let us suppose that system (12.8) is exposed to stochastic perturbations of white noise type $(\dot{w}_1(t), \dot{w}_2(t))$, which are directly proportional to the deviation of system (12.8) state $(a(t), m(t))$ from the equilibrium point (a^*, m^*), i.e.,

$$
\begin{aligned}
\dot{a}(t) &= \mu + \gamma - \gamma m(t) + \beta a(t) I(a_t) - a(t)\big[\beta + \mu + \gamma - (d - d_A)(1 - a(t))\big] \\
&\quad + \sigma_1 \big(a(t) - a^*\big)\dot{w}_1(t), \\
\dot{m}(t) &= \beta a(t) - \beta a(t) I(a_t) - m(t)\big[\alpha + \mu + (d - d_A)a(t)\big] \\
&\quad + \sigma_2\big(m(t) - m^*\big)\dot{w}_2(t).
\end{aligned}
\tag{12.16}
$$

Here $w_1(t)$, $w_2(t)$ are the mutually independent standard Wiener processes, and the stochastic differential equations (12.16) are understood in the Itô sense (Sect. 2.1.2).

To centralize system (12.16) in the equilibrium point, put now $x_1(t) = a(t) - a^*$, $x_2(t) = m(t) - m^*$. Then from (12.16) it follows that

$$
\begin{aligned}
\dot{x}_1(t) &= \mu + \gamma - \gamma\big(m^* + x_2(t)\big) + \beta\big(a^* + x_1(t)\big)\big(a^* + I(x_{1t})\big) \\
&\quad - \big(a^* + x_1(t)\big)\big[\beta + \mu + \gamma - (d - d_A)(1 \quad a^* - x_1(t))\big] + \sigma_1 x_1(t)\dot{w}_1(t), \\
\dot{x}_2(t) &= \beta\big(a^* + x_1(t)\big) - \beta\big(a^* + x_1(t)\big)\big(a^* + I(x_{1t})\big) \\
&\quad - \big(m^* + x_2(t)\big)\big[\alpha + \mu + (d - d_A)\big(a^* + x_1(t)\big)\big] + \sigma_2 x_2(t)\dot{w}_2(t),
\end{aligned}
$$

or

$$
\begin{aligned}
\dot{x}_1(t) = {}& \mu\bigl(1 - a^*\bigr) + \gamma\bigl(1 - a^* - m^*\bigr) - a^*\bigl(1 - a^*\bigr)(\beta - d + d_A) - \mu x_1(t) \\
& + \gamma\bigl(-x_1(t) - x_2(t)\bigr) + x_1(t)\bigl(1 - 2a^*\bigr)(d - d_A) - \beta x_1(t)\bigl(1 - a^*\bigr) \\
& + \beta a^* I(x_{1t}) - x_1^2(t)(d - d_A) + \beta x_1(t) I(x_{1t}) + \sigma_1 x_1(t)\dot{w}_1(t), \\
& \hspace{9.5cm} (12.17) \\
\dot{x}_2(t) = {}& \beta a^*\bigl(1 - a^*\bigr) - m^*\bigl[\alpha + \mu + (d - d_A)a^*\bigr] + \beta x_1(t)\bigl(1 - a^*\bigr) \\
& - m^* x_1(t)(d - d_A) - \beta a^* I(x_{1t}) - x_2(t)\bigl[\alpha + \mu + (d - d_A)a^*\bigr] \\
& - \beta x_1(t) I_1(x_{1t}) - x_2(t)x_1(t)(d - d_A) + \sigma_2 x_2(t)\dot{w}_2(t).
\end{aligned}
$$

By (12.9) from (12.17) it follows that

$$
\begin{aligned}
\dot{x}_1(t) = {}& a_{11}x_1(t) + a_{12}x_2(t) + \beta a^* I(x_{1t}) + \beta x_1(t) I(x_{1t}) \\
& - (d - d_A)x_1^2(t) + \sigma_1 x_1(t)\dot{w}_1(t), \\
\dot{x}_2(t) = {}& a_{21}x_1(t) + a_{22}x_2(t) - \beta a^* I(x_{1t}) - \beta x_1(t) I(x_{1t}) \\
& - (d - d_A)x_1(t)x_2(t) + \sigma_2 x_2(t)\dot{w}_2(t),
\end{aligned}
\tag{12.18}
$$

where

$$
a_{11} = -\bigl[\mu + \gamma + (\beta - d + d_A)\bigl(1 - a^*\bigr) + (d - d_A)a^*\bigr], \qquad a_{12} = -\gamma,
$$

$$
a_{21} = \frac{\beta(\alpha + \mu)(1 - a^*)}{\alpha + \mu + (d - d_A)a^*}, \qquad a_{22} = -\bigl[\alpha + \mu + (d - d_A)a^*\bigr].
\tag{12.19}
$$

Note that for $d \in [d_A, \beta + d_A)$, the numbers a_{11}, a_{12}, a_{22} are negative, and $a_{21} > 0$. Rejecting the nonlinear terms in (12.18), we obtain the linear part of (12.18):

$$
\begin{aligned}
\dot{y}_1(t) &= a_{11}y_1(t) + a_{12}y_2(t) + \beta a^* I(y_{1t}) + \sigma_1 y_1(t)\dot{w}_1(t), \\
\dot{y}_2(t) &= a_{21}y_1(t) + a_{22}y_2(t) - \beta a^* I(y_{1t}) + \sigma_2 y_2(t)\dot{w}_2(t).
\end{aligned}
\tag{12.20}
$$

12.1.5 Stability of the Equilibrium Point

Note that the nonlinear system (12.18) has the order of nonlinearity higher than one. Thus, as it is shown in Sect. 5.3, sufficient conditions for the asymptotic mean-square stability of the zero solution of the linear part (12.20) of the nonlinear system (12.18) at the same time are sufficient conditions for the stability in probability of the zero solution of the nonlinear system (12.18) and therefore are sufficient conditions for stability in probability of the solution (a^*, m^*) of (12.16).

To get sufficient conditions for the asymptotic mean-square stability of the zero solution of (12.20), rewrite this system in the form

$$
\dot{y}(t) = Ay(t) + B(y_t) + \sigma\bigl(y(t)\bigr)\dot{w}(t),
\tag{12.21}
$$

where

$$y(t) = \big(y_1(t), y_2(t)\big)', \qquad w(t) = \big(w_1(t), w_2(t)\big)',$$

$$B(y_t) = \big(\beta a^* I (y_{1t}), -\beta a^* I (y_{1t})\big)', \qquad (12.22)$$

$$A = \begin{pmatrix} a_{11} & a_{12} \\ a_{21} & a_{22} \end{pmatrix}, \qquad \sigma\big(y(t)\big) = \begin{pmatrix} \sigma_1 y_1(t) & 0 \\ 0 & \sigma_2 y_2(t) \end{pmatrix}.$$

Following the procedure of constructing Lyapunov functionals (Sect. 2.2.2), for stability investigation of (12.21), consider the auxiliary differential equation without memory

$$\dot{z}(t) = A z(t) + \sigma\big(z(t)\big)\dot{w}(t). \qquad (12.23)$$

By Remark 2.6 the zero solution of the differential equation $\dot{z}(t) = A z(t)$ is asymptotically stable if and only if conditions (2.62) hold. By Corollary 2.3 conditions (2.66) are sufficient conditions for the asymptotic mean-square stability of the zero solution of (12.23). Below, we suppose that conditions (2.62) and (2.66) hold.

To get stability conditions for (12.20), consider the matrix equation

$$A'P + PA + P_\sigma = -C, \qquad (12.24)$$

where

$$P = \begin{pmatrix} p_{11} & p_{12} \\ p_{12} & p_{22} \end{pmatrix}, \qquad P_\sigma = \begin{pmatrix} p_{11}\sigma_1^2 & 0 \\ 0 & p_{22}\sigma_2^2 \end{pmatrix}, \qquad C = \begin{pmatrix} c & 0 \\ 0 & 1 \end{pmatrix},$$

$c > 0$, and the matrix A is defined in (12.22), (12.19).

If the matrix equation (12.24) has a positive definite solution P, then the function $v(z) = z'Pz$ is a Lyapunov function for (12.23) since

$$Lv = z'\big(A'P + PA + P_\sigma\big)z = -z'Cz.$$

Note that the matrix equation (12.24) can be represented as the system of the equations

$$2(p_{11}a_{11} + p_{12}a_{21} + p_{11}\delta_1) = -c,$$

$$2(p_{12}a_{12} + p_{22}a_{22} + p_{22}\delta_2) = -1, \qquad (12.25)$$

$$p_{11}a_{12} + p_{12}\operatorname{Tr}(A) + p_{22}a_{21} = 0,$$

with the solution

$$p_{11} = -\frac{c + 2a_{21}p_{12}}{2\hat{a}_{11}}, \qquad p_{22} = -\frac{1 + 2a_{12}p_{12}}{2\hat{a}_{22}}, \qquad p_{12} = \frac{a_{21}\hat{a}_{11} + ca_{12}\hat{a}_{22}}{2Z},$$

$$(12.26)$$

where

$$\hat{a}_{ii} = a_{ii} + \delta_i, \qquad \delta_i = \frac{1}{2}\sigma_i^2, \quad i = 1, 2, \qquad (12.27)$$

$$Z = \mathrm{Tr}(A)\hat{a}_{11}\hat{a}_{22} - a_{12}a_{21}(\hat{a}_{11} + \hat{a}_{22}). \tag{12.28}$$

Lemma 12.2 *Let conditions* (2.52), (2.56) *hold, and let*

$$\hat{a}_{11} < 0, \qquad \hat{a}_{22} < 0. \tag{12.29}$$

Then the zero solution of (12.23) *is asymptotically mean-square stable.*

Proof It is enough to show that the matrix $P = \|p_{ij}\|$ with the elements (12.26), which are a solution of the matrix equation (12.24), is positive definite for an arbitrary $c > 0$, i.e., $p_{11} > 0$, $p_{22} > 0$, $p_{11}p_{22} > p_{12}^2$. To this aim, note that by (2.62), (12.19), (12.29) we have $Z < 0$. Note also that by (12.27), (12.29), Remark 2.8, and (2.72) we obtain

$$\delta_1 < |a_{11}| \le \frac{|\mathrm{Tr}(A)|\det(A)}{A_2} \le \frac{A_1}{|\mathrm{Tr}(A)|},$$
$$\delta_2 < |a_{22}| \le \frac{|\mathrm{Tr}(A)|\det(A)}{A_1} \le \frac{A_2}{|\mathrm{Tr}(A)|}, \tag{12.30}$$

where

$$A_i = \det(A) + a_{ii}^2, \quad i = 1, 2. \tag{12.31}$$

Besides, by (12.28), (12.27), (2.62), and (12.31) we have

$$Z + a_{12}a_{21}\hat{a}_{22} = \mathrm{Tr}(A)\hat{a}_{11}\hat{a}_{22} - a_{12}a_{21}(\hat{a}_{11} + \hat{a}_{22}) + a_{12}a_{21}\hat{a}_{22}$$
$$= \left(\mathrm{Tr}(A)\hat{a}_{22} - a_{12}a_{21}\right)\hat{a}_{11}$$
$$= \left(A_2 - |\mathrm{Tr}(A)|\delta_2\right)\hat{a}_{11} \tag{12.32}$$

and, similarly,

$$Z + a_{12}a_{21}\hat{a}_{11} = \left(A_1 - |\mathrm{Tr}(A)|\delta_1\right)\hat{a}_{22}. \tag{12.33}$$

From this and from (12.26), (12.32), (12.30) it follows that for an arbitrary $c > 0$,

$$p_{11} = -\frac{cZ + a_{21}(ca_{12}\hat{a}_{22} + a_{21}\hat{a}_{11})}{2Z\hat{a}_{11}}$$
$$= -\frac{c(Z + a_{12}a_{21}\hat{a}_{22}) + a_{21}^2\hat{a}_{11}}{2Z\hat{a}_{11}}$$
$$= \frac{c(A_2 - |\mathrm{Tr}(A)|\delta_2) + a_{21}^2}{2|Z|}$$
$$> 0, \tag{12.34}$$

and, similarly, by (12.26), (12.33), (12.30) we obtain

$$p_{22} = \frac{A_1 - |\mathrm{Tr}(A)|\delta_1 + ca_{12}^2}{2|Z|} > 0. \tag{12.35}$$

Finally, let us show that $p_{11}p_{22} > p_{12}^2$. Indeed, the inequality

$$\frac{(c + 2a_{21}p_{12})(1 + 2a_{12}p_{12})}{4\hat{a}_{11}\hat{a}_{22}} > p_{12}^2$$

is equivalent to $4Bp_{12}^2 - 2(a_{21} + ca_{12})p_{12} < c$ by $B = \hat{a}_{11}\hat{a}_{22} - a_{12}a_{21} > 0$. Substituting p_{12} from (12.26) into the obtained inequality, we have

$$B(a_{21}\hat{a}_{11} + ca_{12}\hat{a}_{22})^2 - (a_{21} + ca_{12})(a_{21}\hat{a}_{11} + ca_{12}\hat{a}_{22})Z < cZ^2$$

or

$$c^2 a_{12}^2 \hat{a}_{22}(Z - B\hat{a}_{22}) + c\hat{a}_{11}\hat{a}_{22}\big(Z\,\mathrm{Tr}(A) - 2a_{12}a_{21}B\big) + a_{21}^2 \hat{a}_{11}(Z - B\hat{a}_{11}) > 0.$$

Note also that by (12.28), (12.29), (12.19) we obtain

$$\begin{aligned}
\hat{a}_{11}(Z - B\hat{a}_{11}) &= \hat{a}_{11}\big(\mathrm{Tr}(A)\hat{a}_{11}\hat{a}_{22} - a_{12}a_{21}(\hat{a}_{11} + \hat{a}_{22}) \\
&\quad - (\hat{a}_{11}\hat{a}_{22} - a_{12}a_{21})\hat{a}_{11}\big) \\
&= \hat{a}_{11}\big(\mathrm{Tr}(A)\hat{a}_{11}\hat{a}_{22} - a_{12}a_{21}\hat{a}_{22} - \hat{a}_{11}^2\hat{a}_{22}\big) \\
&= \hat{a}_{11}\hat{a}_{22}\big(\mathrm{Tr}(A)\hat{a}_{11} - a_{12}a_{21} - \hat{a}_{11}^2\big) \\
&= \hat{a}_{11}\hat{a}_{22}\big((\mathrm{Tr}(A) - \hat{a}_{11})\hat{a}_{11} - a_{12}a_{21}\big) \\
&= \hat{a}_{11}\hat{a}_{22}\big((a_{22} - \delta_1)\hat{a}_{11} - a_{12}a_{21}\big) > 0
\end{aligned}$$

and, similarly,

$$\hat{a}_{22}(Z - B\hat{a}_{22}) = \hat{a}_{11}\hat{a}_{22}\big((a_{11} - \delta_2)\hat{a}_{22} - a_{12}a_{21}\big) > 0,$$

$$Z\,\mathrm{Tr}(A) - 2a_{12}a_{21}B > 0.$$

So, for an arbitrary $c > 0$, the matrix P with the elements (12.26) is positive definite. The proof is completed.   □

Theorem 12.1 *If conditions (12.29) hold and, for some $c > 0$, the elements (12.26) of the matrix P satisfy the condition*

$$\big(\beta a^* |p_{12} - p_{22}|\big)^2 + 2\beta a^* |p_{11} - p_{12}| < c, \tag{12.36}$$

then the solution (a^*, m^*) of system (12.16) is stable in probability.

Proof Note that the order of nonlinearity of system (12.16) is higher than one. Therefore, from Sect. 5.3, to get conditions for stability in probability of the equilibrium point (a^*, m^*) of this system, it is enough to get conditions for the asymptotic mean-square stability of the zero solution of the linear part (12.20) of this system. Following the procedure of constructing Lyapunov functionals, we will construct a Lyapunov functional for system (12.20) in the form $V = V_1 + V_2$, where $V_1 = y'Py$,

$y = (y_1, y_2)'$, P is a positive definite solution of system (12.25) with the elements (12.26), and V_2 will be chosen below.

Let L be the generator (Sect. 2.1.2) of system (12.20). Then by (12.20) and (12.25) we have

$$LV_1 = 2\big(p_{11}y_1(t) + p_{12}y_2(t)\big)\big(a_{11}y_1(t) + a_{12}y_2(t) + \beta a^* I(y_{1t})\big) + p_{11}\sigma_1^2 y_1^2(t)$$
$$+ 2\big(p_{12}y_1(t) + p_{22}y_2(t)\big)\big(a_{21}y_1(t) + a_{22}y_2(t) - \beta a^* I(y_{1t})\big) + p_{22}\sigma_2^2 y_2^2(t)$$
$$= -cy_1^2(t) - y_2^2(t) + 2\beta a^*\big[(p_{11} - p_{12})y_1(t) + (p_{12} - p_{22})y_2(t)\big]I(y_{1t}).$$

By (12.7), (12.2) we have $2y_1(t)I(y_{1t}) \le y_1^2(t) + I(y_{1t}^2)$ and $2y_2(t)I(y_{1t}) \le vy_2^2(t) + v^{-1}I(y_{1t}^2)$ for some $v > 0$. Using these inequalities, we obtain

$$LV_1 \le -cy_1^2(t) - y_2^2(t) + \beta a^*|p_{11} - p_{12}|\big(y_1^2(t) + I(y_{1t}^2)\big)$$
$$+ \beta a^*|p_{12} - p_{22}|\big(vy_2^2(t) + v^{-1}I(y_{1t}^2)\big)$$
$$= \big(\beta a^*|p_{11} - p_{12}| - c\big)y_1^2(t) + \big(\beta a^*|p_{12} - p_{22}|v - 1\big)y_2^2(t)$$
$$+ qI(y_{1t}^2), \tag{12.37}$$

where

$$q = \beta a^*\big(|p_{11} - p_{12}| + |p_{12} - p_{22}|v^{-1}\big). \tag{12.38}$$

Putting

$$V_2 = q \int_0^\infty \int_{t-s}^t y_1^2(\theta)\, d\theta\, dK(s),$$

by (12.2), (12.7) we have $LV_2 = q(y_1^2(t) - I(y_{1t}^2))$. Therefore, by (12.37), (12.38) for the functional $V = V_1 + V_2$, we have

$$LV \le \big(2\beta a^*|p_{11} - p_{12}| + \beta a^*|p_{12} - p_{22}|v^{-1} - c\big)y_1^2(t)$$
$$+ \big(\beta a^*|p_{12} - p_{22}|v - 1\big)y_2^2(t).$$

Thus, if

$$2\beta a^*|p_{11} - p_{12}| + \beta a^*|p_{12} - p_{22}|v^{-1} < c, \quad \beta a^*|p_{12} - p_{22}|v < 1, \tag{12.39}$$

then by Remark 2.1 the zero solution of (12.20) is asymptotically mean-square stable.

From (12.39) it follows that

$$\frac{\beta a^*|p_{12} - p_{22}|}{c - 2\beta a^*|p_{11} - p_{12}|} < v < \frac{1}{\beta a^*|p_{12} - p_{22}|}. \tag{12.40}$$

Thus, if for some $c > 0$, condition (12.36) holds, then there exists $v > 0$ such that conditions (12.40) (or (12.39)) hold too, and therefore the zero solution of (12.20)

is asymptotically mean-square stable. From this it follows also that the zero solution of (12.18) and therefore the equilibrium point of system (12.16) are stable in
probability. The proof is completed. □

Example 12.2 Consider system (12.16) with the values of the parameters $\alpha, \beta, \gamma, \mu$,
d, d_A and the equilibrium point (a^*, m^*) given in (12.14), (12.15). As an example,
consider the levels of noises $\sigma_1 = 0.028969$, $\sigma_2 = 0.142252$ or $\delta_1 = 0.000420$, $\delta_2 = 0.010118$. From (12.19) it follows that the values of system (12.20) parameters are
$a_{11} = -0.029245$, $a_{12} = -0.001440$, $a_{21} = 0.017446$, $a_{22} = -0.010475$ and the
conditions (12.29) hold: $\hat{a}_{11} = -0.028825 < 0$, $\hat{a}_{22} = -0.000357 < 0$. Put $c = 20$.
Then by (12.26) $p_{11} = 477.4438$, $p_{12} = 215.6615$, $p_{22} = 530.4124$, and condition
(12.36) holds:

$$\left(\beta a^*|p_{12} - p_{22}|\right)^2 + 2\beta a^*|p_{11} - p_{12}| = 16.1036 < 20.$$

Thus, the solution (a^*, m^*) of system (12.16) is stable in probability.

Example 12.3 Consider system (12.16) with the previous values of the all parameters except for the levels of noises that are $\sigma_1 = 0.0075$, $\sigma_2 = 0.0077$ or
$\delta_1 = 0.000028$, $\delta_2 = 0.000030$. These values of σ_1 and σ_2 are selected taking into
account sample errors of the monitoring of the alcohol consumption in Spain [291].
The parameters a_{11}, a_{21}, a_{22} are the same as in the previous example, and conditions (12.29) hold: $\hat{a}_{11} = -0.029217 < 0$, $\hat{a}_{22} = -0.010445 < 0$. Put $c = 4$. Then
by (12.26) $p_{11} = 78.6856$, $p_{12} = 17.1347$, $p_{22} = 45.5060$, and condition (12.36)
holds:

$$\left(\beta a^*|p_{12} - p_{22}|\right)^2 + 2\beta a^*|p_{11} - p_{12}| = 1.3643 < 4.$$

Thus, the solution (a^*, m^*) of system (12.16) is stable in probability.

Let us now get three corollaries from Theorem 12.1 that simplify a verification
of the stability condition (12.36). By (12.26) and (12.28) we have

$$
\begin{aligned}
p_{12} - p_{11} &= p_{12} + \frac{c + 2a_{21}p_{12}}{2\hat{a}_{11}} \\
&= \left(1 + \frac{a_{21}}{\hat{a}_{11}}\right)\frac{a_{21}\hat{a}_{11} + ca_{12}\hat{a}_{22}}{2Z} + \frac{c}{2\hat{a}_{11}} \\
&= \frac{(a_{21} + \hat{a}_{11})a_{12}\hat{a}_{22} + Z}{2Z\hat{a}_{11}}c + \frac{(a_{21} + \hat{a}_{11})a_{21}}{2Z} \\
&= B_0 c + B_1,
\end{aligned}
\tag{12.41}
$$

where

$$
B_0 = \frac{(\mathrm{Tr}(A) + a_{12})\hat{a}_{22} - a_{12}a_{21}}{2Z}, \qquad B_1 = \frac{(a_{21} + \hat{a}_{11})a_{21}}{2Z}
\tag{12.42}
$$

and, similarly,

$$p_{12} - p_{22} = p_{12} + \frac{1 + 2a_{12}p_{12}}{2\hat{a}_{22}}$$

$$= \left(1 + \frac{a_{12}}{\hat{a}_{22}}\right) \frac{a_{21}\hat{a}_{11} + ca_{12}\hat{a}_{22}}{2Z} + \frac{1}{2\hat{a}_{22}}$$

$$= \frac{(a_{12} + \hat{a}_{22})a_{12}c}{2Z} + \frac{(a_{12} + \hat{a}_{22})a_{21}\hat{a}_{11} + Z}{2Z\hat{a}_{22}}$$

$$= D_0 c + D_1, \tag{12.43}$$

where

$$D_0 = \frac{(a_{12} + \hat{a}_{22})a_{12}}{2Z}, \qquad D_1 = \frac{(\mathrm{Tr}(A) + a_{21})\hat{a}_{11} - a_{12}a_{21}}{2Z}. \tag{12.44}$$

Remark 12.2 Put

$$f(c) = (\beta a^* D_0)^2 \left(c + \frac{D_1}{D_0}\right)^2 + 2\beta a^* |B_0| \left|c + \frac{B_1}{B_0}\right| - c. \tag{12.45}$$

From (12.19), (12.29) it follows that $B_0 < 0$. By (12.41)–(12.45) and $B_0 < 0$ condition (12.36) is equivalent to the condition $f(c) < 0$.

Put now

$$S = (\beta a^* D_0)^2 \left(\frac{D_1}{D_0} - \frac{B_1}{B_0}\right)^2 + \frac{B_1}{B_0},$$

$$R_+ = 2\beta a^* |B_0| \left(\frac{1 - 2\beta a^* |B_0|}{2(\beta a^* D_0)^2} - \frac{D_1}{D_0} + \frac{B_1}{B_0}\right),$$

$$R_- = -2\beta a^* |B_0| \left(\frac{1 + 2\beta a^* |B_0|}{2(\beta a^* D_0)^2} - \frac{D_1}{D_0} + \frac{B_1}{B_0}\right), \tag{12.46}$$

$$Q = \frac{1}{4(\beta a^* D_0)^2} - \frac{D_1}{D_0} - \frac{B_0^2}{D_0^2}.$$

Corollary 12.1 *If conditions (12.29) hold and $S < 0$, then the solution (a^*, m^*) of system (12.16) is stable in probability.*

Proof From $S < 0$ and $B_0 < 0$ it follows that $B_1 > 0$. Putting $c_0 = -B_1 B_0^{-1} > 0$, we obtain $f(c_0) = S < 0$, i.e., condition (12.36) holds. The proof is completed. $\square$

Corollary 12.2 *If conditions (12.29) hold and $0 \le R_+ < Q$, then the solution (a^*, m^*) of system (12.16) is stable in probability.*

Proof Let us suppose that $c + B_1 B_0^{-1} \geq 0$. Then the minimum of the function $f(c)$ is reached by

$$c_0 = \frac{1 - 2\beta a^* |B_0|}{2(\beta a^* D_0)^2} - \frac{D_1}{D_0} \geq -\frac{B_1}{B_0}.$$

Substituting c_0 into the function $f(c)$, we obtain that the condition $f(c_0) < 0$ is equivalent to the condition $0 \leq R_+ < Q$. The proof is completed. □

Corollary 12.3 *If conditions* (12.29) *hold and* $0 < R_- < Q$, *then the solution* (a^*, m^*) *of system* (12.16) *is stable in probability.*

Proof Let us suppose that $c + B_1 B_0^{-1} < 0$. Then the minimum of the function $f(c)$ is reached by

$$c_0 = \frac{1 + 2\beta a^* |B_0|}{2(\beta a^* D_0)^2} - \frac{D_1}{D_0} < -\frac{B_1}{B_0}.$$

Substituting c_0 into the function $f(c)$, we obtain that the condition $f(c_0) < 0$ is equivalent to the condition $0 < R_- < Q$. The proof is completed. □

Example 12.4 Consider system (12.16) with the values of the parameters from Example 12.2. Calculating S, R_+, Q, we obtain: $S = 4.50 > 0$, $R_+ = 736 < Q = 1320$. From Corollary 12.2 it follows that the solution (a^*, m^*) of system (12.16) is stable in probability.

Example 12.5 Consider system (12.16) with the values of the parameters from Example 12.3. Calculating S, R_+, Q, we obtain: $S = -0.39 < 0$, $R_+ = 2462 < Q = 4754$. From both Corollary 12.1 and Corollary 12.2 it follows that the solution (a^*, m^*) of system (12.16) is stable in probability.

12.1.6 Numerical Simulation

Let us suppose that in (12.1) $dK(s) = \delta(s - h)ds$, where $\delta(s)$ is Dirac's delta-function, and $h \geq 0$ is the delay.

In Fig. 12.1 25 trajectories of the solution of (12.16), (12.4) are shown for the values of the parameters from Examples 12.1 and 12.2: $\alpha = 0.000110247$, $\beta = 0.0284534$, $\gamma = 0.00144$, $\mu = 0.01$, $d = 0.009$, $d_A = 0.008$, the initial values $a_0 = 0.43$, $m_0 = 0.53$, $r_0 = 0.04$, the levels of noises $\sigma_1 = 0.028969$, $\sigma_2 = 0.142252$, and delay $h = 0.1$. We can see that all trajectories go to the equilibrium point $a^* = 0.364739$, $m^* = 0.629383$, $r^* = 0.00587794$.

Note that numerical simulations of the processes $a(t)$, $m(t)$, and $r(t)$ were obtained via the difference analogues of (12.16), (12.4) in the form

$$a_{i+1} = a_i + \Delta\big[\mu + \gamma - \gamma\mu_i + \beta a_i a_{i-m} - a_i\big(\beta + \mu + \gamma - (d - d_A)(1 - a_i)\big)\big]$$
$$+ \sigma_1\big(a_i - a^*\big)(w_{1,i+1} - w_{1i}),$$

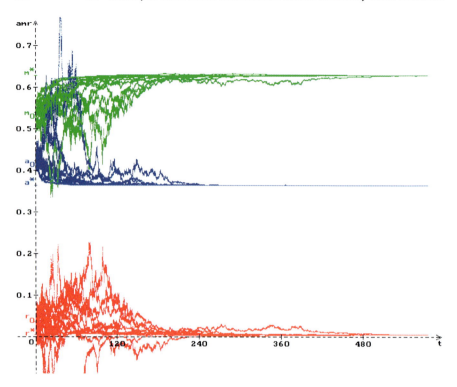

Fig. 12.1 25 trajectories of the processes $a(t)$ (*blue*), $m(t)$ (*green*), $r(t)$ (*red*) with the values of the parameters $\alpha = 0.000110247$, $\beta = 0.0284534$, $\gamma = 0.00144$, $\mu = 0.01$, $d = 0.009$, $d_A = 0.008$, the levels of noises $\sigma_1 = 0.028969$, $\sigma_2 = 0.142252$, the delay $h = 0.1$, the initial values $a(s) = 0.43$, $s \in [-0.1, 0]$, $m(0) = 0.53$, $r(0) = 0.04$, and the equilibrium point $a^* = 0.364739$, $m^* = 0.629383$, $r^* = 0.00587794$

$$m_{i+1} = m_i + \Delta\big[\beta a_i - \beta a_i a_{i-m} - m_i\big(\alpha + \mu + (d - d_A)a_i\big)\big]$$
$$+ \sigma_2\big(m_i - m^*\big)(w_{2,i+1} - w_{2i}),$$
$$r_{i+1} = 1 - a_{i+1} - m_{i+1},$$
$$i = 0, 1, 2, \ldots, \quad a_j = a_0, \quad j = -m, \ldots, -1, 0.$$

Here Δ is the discretization step (which was chosen as $\Delta = 0.01$), $a_i = a(t_i)$, $m_i = m(t_i)$, $r_i = r(t_i)$, $w_{ki} = w_k(t_i)$, $k = 1, 2$, $t_i = i\Delta$, $m = h\Delta^{-1}$, trajectories of the Wiener processes $w_1(t)$ and $w_2(t)$ are simulated by the algorithm described in Sect. 2.1.1.

12.2 Mathematical Model of Social Obesity Epidemic

Social obesity epidemic models are popular enough with researches (see, for instance, [14, 34, 42, 51, 68, 115, 227, 255]). Here the known nonlinear social obesity epidemic model [255] is generalized to the system with distributed delay. It is supposed also that this nonlinear system is exposed to additive stochastic perturbations of white noise type that are directly proportional to the deviation of the system state from the equilibrium point. The research that is similar to the previous one is applied to this model.

12.2.1 Description of the Considered Model

For constructing the mathematical obesity model [255] the 24- to 65-year-old population is divided into three subpopulations based on the so-called body mass index (BMI $=$ Weight/Height^2). The classes or subpopulations are: individuals at a normal weight (BMI $<$ 25 kg/m^2) $N(t)$, people who are overweight (25 kg/m$^2 \leq$ BMI $<$ 30 kg/m^2) $S(t)$, and obese individuals (BMI $\geq$ 30 kg/m^2) $O(t)$.

The transition between the different subpopulations is determined as follows: once an adult starts an unhealthy lifestyle, he/she becomes addicted to the unhealthy lifestyle and starts a progression to being overweight $S(t)$ because of this lifestyle. If this adult continues with his/her unhealthy lifestyle, he/she can become an obese individual $O(t)$. In both these classes individuals can stop his/her unhealthy lifestyle and then move to classes $N(t)$ and $S(t)$, respectively.

The transitions between the subpopulations $N(t)$, $S(t)$, and $O(t)$ are governed by terms proportional to the sizes of these subpopulations. Conversely, the transitions from normal to overweight occur through the transmission of an unhealthy lifestyle from the overweight and obese subpopulations to the normal-interactions weight subpopulation, depending on the meet population, depending on the meetings among them. This transition is modeled using the term

$$\beta N(t) \int_0^\infty \big(S(t-s) + O(t-s)\big) \, dK(s),$$

where $K(s)$ is a nondecreasing function that satisfies condition (12.2), and the integral is understood in the Stieltjes sense. The subpopulations' sizes and their behaviors with time determine the dynamic evolution of adulthood excess weight.

Without loss of generality and for the sake of clarity, the 24- to 65-year-old adult population is normalized to unity, and it is supposed for all $t \geq 0$ that

$$N(t) \geq 0, \qquad S(t) \geq 0, \qquad O(t) \geq 0, \qquad (12.47)$$

$$N(t) + S(t) + O(t) = 1. \qquad (12.48)$$

Thus, under the above assumptions, the following nonlinear system of integro–differential equations is obtained:

$$\dot{N}(t) = \mu N_0 - \mu N(t) - \beta N(t) \int_0^\infty \big(S(t-s) + O(t-s)\big) \, dK(s) + \rho S(t),$$

$$\dot{S}(t) = \mu S_0 + \beta N(t) \int_0^\infty \big(S(t-s) + O(t-s)\big) dK(s)$$

$$\qquad (12.49)$$

$$- (\mu + \gamma + \rho)S(t) + \varepsilon O(t),$$

$$\dot{O}(t) = \mu O_0 + \gamma S(t) - (\mu + \varepsilon)O(t), \qquad t \geq 0,$$

$$N(0) = N_0, \ S(s) = S_0, \ O(s) = O_0, \ s \leq 0.$$

The time-invariant parameters of this system of equations are:

ε the rate at which an obese adult with a healthy lifestyle becomes an overweight individual (intensity of transition from the group $O(t)$ to the group $S(t)$).

ρ the rate at which an overweight individual moves to the normal-weight subpopulation (intensity of transition from the group $S(t)$ to the group $N(t)$).

β the transmission rate because of social pressure to adopt an unhealthy lifestyle, e.g., TV, friends, family, job, and so on (intensity of transition from the group $N(t)$ to the group $S(t)$).

γ the rate at which an overweight 24- to 65-year-old adult becomes an obese individual because of unhealthy lifestyle (intensity of transition from the group $S(t)$ into the group $O(t)$); so, the scheme of transition from one group to another one is

$$O(t) \xrightarrow{\varepsilon} S(t) \xrightarrow{\rho} N(t) \xrightarrow{\beta} S(t) \xrightarrow{\gamma} O(t).$$

μ the average stay time in the system of 24- to 65-year-old adults ($\mu = 1/$ (65 years $-$ 24 years) $\cdot$ 52 weeks/year).

N_0 the proportion of normal weight coming from the 23-year age group.

S_0 the proportion of overweight coming from the 23-year age group.

O_0 the proportion of obese coming from the 23-year age group.

Here the parameters ε, ρ, β, γ, μ are nonnegative numbers, and N_0, S_0, O_0 satisfy the conditions of type (12.47), (12.48).

By condition (12.48) system (12.49) can be simplified to the following system of two equations:

$$\dot{N}(t) = \mu N_0 - \mu N(t) - \beta N(t) \int_0^\infty \big(1 - N(t-s)\big) dK(s) + \rho S(t),$$

$$\dot{S}(t) = \mu S_0 + \beta N(t) \int_0^\infty \big(1 - N(t-s)\big) dK(s) - (\mu + \gamma + \rho)S(t) \qquad (12.50)$$

$$+ \varepsilon \big(1 - N(t) - S(t)\big), \qquad t \geq 0,$$

$$N(s) = N_0, \ s \leq 0, \ S(0) = S_0.$$

12.2.2 Existence of an Equilibrium Point

The equilibrium point (N^*, S^*) of system (12.50) is defined by the conditions $\dot{N}(t) = 0$, $\dot{S}(t) = 0$ and by (12.50), (12.47) is a solution of the following system of algebraic equations:

$$\mu N_0 - \mu N^* - \beta N^*(1 - N^*) + \rho S^* = 0,$$
$$\mu S_0 + \beta N^*(1 - N^*) - (\mu + \gamma + \rho)S^* + \varepsilon(1 - S^* - N^*) = 0. \tag{12.51}$$

From (12.51) it follows that

$$S^* = \rho^{-1}\left[\mu(N^* - N_0) + \beta N^*(1 - N^*)\right],$$
$$S^* = k\rho^{-1}\left[\mu S_0 + (\varepsilon + \beta N^*)(1 - N^*)\right], \tag{12.52}$$

where

$$k = \rho(\mu + \gamma + \rho + \varepsilon)^{-1} < 1. \tag{12.53}$$

By (12.52), (12.53) we obtain that N^* is a root of the quadratic equation

$$\beta(1 - k)(N^*)^2 - (\mu + k\varepsilon + \beta(1 - k))N^* + \mu(N_0 + k S_0) + k\varepsilon = 0. \tag{12.54}$$

Lemma 12.3 *Assume that $N_0 + k S_0 < 1$. If $\beta > 0$, then (12.54) has two real roots, $N_1^* \in (0, 1)$ and $N_2^* > 1$. If $\beta = 0$ and $\mu k\varepsilon > 0$, then (12.54) has one root $N^* \in (N_0 + k S_0, 1)$.*

Proof From $N_0 + k S_0 < 1$ and $\beta > 0$ we have

$$D = \sqrt{(\mu + k\varepsilon + \beta(1 - k))^2 - 4\beta(1 - k)(\mu(N_0 + k S_0) + k\varepsilon)}$$
$$> \sqrt{(\mu + k\varepsilon + \beta(1 - k))^2 - 4\beta(1 - k)(\mu + k\varepsilon)}$$
$$= |\mu + k\varepsilon - \beta(1 - k)|, \tag{12.55}$$

i.e., $D > |\mu + k\varepsilon - \beta(1 - k)| \geq 0$, and therefore the quadratic equation (12.54) has two real roots

$$N_1^* = \frac{\mu + k\varepsilon + \beta(1 - k) - D}{2\beta(1 - k)}, \qquad N_2^* = \frac{\mu + k\varepsilon + \beta(1 - k) + D}{2\beta(1 - k)}. \tag{12.56}$$

If $\mu + k\varepsilon < \beta(1 - k)$, then

$$N_1^* < \frac{\mu + k\varepsilon}{\beta(1 - k)} < 1, \qquad N_2^* > 1.$$

If $\mu + k\varepsilon \geq \beta(1 - k)$, then

$$N_1^* < 1, \qquad N_2^* > \frac{\mu + k\varepsilon}{\beta(1 - k)} \geq 1.$$

If $\beta = 0$, then from (12.54) it follows that

$$1 > N^* = \frac{\mu(N_0 + kS_0) + k\varepsilon}{\mu + k\varepsilon} > N_0 + kS_0.$$

The proof is completed. $\square$

Lemma 12.4 *Assume that* $N_0 = 1$. *If* $\mu + k\varepsilon < \beta(1 - k)$, *then* (12.54) *has two roots on the interval* $(0,1]$: $N_1^* \in (0, 1)$ *and* $N_2^* = 1$. *If* $\mu + k\varepsilon \geq \beta(1 - k)$ *then* (12.54) *has one root only on the interval* $(0,1]$: $N_1^* = 1$.

Proof From $N_0 = 1$ and (12.47) we have $S_0 = 0$. Then, similarly to (12.55), $D = |\mu + k\varepsilon - \beta(1 - k)|$. If $\mu + k\varepsilon < \beta(1 - k)$, then $D = \beta(1 - k) - (\mu + k\varepsilon)$, and by (12.56) we obtain

$$N_1^* = \frac{\mu + k\varepsilon}{\beta(1 - k)} < 1, \qquad N_2^* = 1.$$

If $\mu + k\varepsilon > \beta(1 - k)$, then $D = \mu + k\varepsilon - \beta(1 - k)$, and by (12.56) we have

$$N_1^* = 1, \qquad N_2^* = \frac{\mu + k\varepsilon}{\beta(1 - k)} > 1.$$

If $\mu + k\varepsilon = \beta(1 - k)$, then $D = 0$ and $N_1^* = N_2^* = 1$. The proof is completed. $\square$

Example 12.6 Following [255], put

$$\mu = 0.000469, \qquad \gamma = 0.0003, \qquad \varepsilon = 0.000004, \qquad \rho = 0.000035,$$

$$\beta = 0.00085, \qquad N_0 = 0.704, \qquad S_0 = 0.25, \qquad O_0 = 0.046.$$

Then by (12.56), (12.52), (12.48) we obtain

$$N^* = 0.3311, \qquad S^* = 0.3814, \qquad O^* = 0.2875.$$

Putting $\beta = 0$ with the same values of the other parameters, by Lemma 12.3 we obtain

$$N^* = 0.7149 > N_0 + kS_0 = 0.7148, \qquad S^* = 0.1465, \qquad O^* = 0.1386.$$

Put now $N_0 = 1$, $S_0 = O_0 = 0$. By Lemma 12.4, if $\beta = 0.00085$, i.e., if $\beta > 0$, then $\beta > (\mu + k\varepsilon)(1 - k)^{-1} = 0.00049$ and $N^* = 0.5770$, $S^* = 0.2588$, $O^* = 0.1642$. If $\beta = 0$, then $N^* = 1$, $S^* = O^* = 0$.

12.2.3 *Stochastic Perturbations, Centralization, and Linearization*

Let us suppose that system (12.50) is exposed to stochastic perturbations of white noise type $(\dot{w}_1(t), \dot{w}_2(t))$ that are directly proportional to the deviation of system

(12.50) state $(N(t), S(t))$ from the equilibrium point (N^*, S^*), i.e.,

$$\dot{N}(t) = \mu N_0 - \mu N(t) - \beta N(t) \int_0^\infty \left(1 - N(t-s)\right) dK(s) + \rho S(t)$$

$$+ \sigma_1 \left(N(t) - N^*\right) \dot{w}_1(t), \qquad t \geq 0,$$

$$\dot{S}(t) = \mu S_0 + \beta N(t) \int_0^\infty \left(1 - N(t-s)\right) dK(s) - (\mu + \gamma + \rho) S(t) \qquad (12.57)$$

$$+ \varepsilon \left(1 - N(t) - S(t)\right) + \sigma_2 \left(S(t) - S^*\right) \dot{w}_2(t), \qquad t \geq 0,$$

$$N(s) = N_0, \ s \leq 0, \ S(0) = S_0.$$

Here $w_1(t)$, $w_2(t)$ are mutually independent standard Wiener processes, and the stochastic differential equations of system (12.57) are understood in the Itô sense (Sect. 2.1.2). Note that the equilibrium point (N^*, S^*) of system (12.50) is a solution of (12.57) too.

To centralize system (12.57) at the equilibrium point, put now $x_1 = N - N^*$, $x_2 = S - S^*$. Then by (12.57), (12.53) we have

$$\dot{x}_1(t) = a_{11} x_1(t) + a_{12} x_2(t) + \beta N^* I(x_{1t}) + \beta x_1(t) I(x_{1t})$$

$$+ \sigma_1 x_1(t) \dot{w}_1(t),$$

$$\dot{x}_2(t) = a_{21} x_1(t) + a_{22} x_2(t) - \beta N^* I(x_{1t}) - \beta x_1(t) I(x_{1t}) \qquad (12.58)$$

$$+ \sigma_2 x_2(t) \dot{w}_2(t),$$

where

$$a_{11} = -\mu - \beta \left(1 - N^*\right), \qquad a_{12} = \rho,$$

$$a_{21} = -\varepsilon + \beta \left(1 - N^*\right), \qquad a_{22} = -k^{-1} \rho, \qquad (12.59)$$

$$I(x_{1t}) = \int_0^\infty x_1(t-s) dK(s).$$

Example 12.7 Using the values of the parameters from Example 12.6, by (12.59) we obtain

$$a_{11} = -0.0010376, \qquad a_{12} = 0.000035,$$

$$a_{21} = 0.0005646, \qquad a_{22} = -0.000808.$$

It is clear that the stability of the equilibrium point of system (12.57) is equivalent to the stability of the zero solution of (12.58). Rejecting the nonlinear terms in (12.58), we obtain the linear part of system (12.58)

$$\dot{y}_1(t) = a_{11} y_1(t) + a_{12} y_2(t) + \beta N^* I(y_{1t}) + \sigma_1 y_1(t) \dot{w}_1(t),$$

$$\dot{y}_2(t) = a_{21} y_1(t) + a_{22} y_2(t) - \beta N^* I(y_{1t}) + \sigma_2 y_2(t) \dot{w}_2(t). \qquad (12.60)$$

12.2.4 Stability of an Equilibrium Point

Note that the nonlinear system (12.58) has the order of nonlinearity higher than one. Thus, sufficient conditions for the asymptotic mean-square stability of the zero solution of the linear part (12.60) at the same time are (Sect. 5.3) sufficient conditions for the stability in probability of the zero solution of the nonlinear system (12.58) and therefore are sufficient conditions for the stability in probability of the solution (N^*, S^*) of system (12.57).

To get sufficient conditions for the asymptotic mean-square stability of the zero solution of (12.60), rewrite this system in the form

$$\dot{y}(t) = Ay(t) + B(y_t) + \sigma(y(t))\dot{w}(t), \tag{12.61}$$

where

$$y(t) = (y_1(t), y_2(t))', \qquad w(t) = (w_1(t), w_2(t))',$$

$$B(y_t) = (\beta N^* I(y_{1t}), -\beta N^* I(y_{1t}))', \tag{12.62}$$

$$A = \begin{pmatrix} a_{11} & a_{12} \\ a_{21} & a_{22} \end{pmatrix}, \qquad \sigma(y(t)) = \begin{pmatrix} \sigma_1 y_1(t) & 0 \\ 0 & \sigma_2 y_2(t) \end{pmatrix},$$

and a_{ij}, $i, j = 1, 2$, are defined by (12.59).

Following the procedure of constructing Lyapunov functionals, for stability investigation of (12.61), consider the auxiliary differential equation without memory

$$\dot{z}(t) = Az(t) + \sigma(z(t))\dot{w}(t). \tag{12.63}$$

Remark 12.3 By (12.62), (12.59) for the matrix A, conditions (2.62) hold:

$$\mathrm{Tr}(A) = -\left[\mu + \rho k^{-1} + \beta(1 - N^*)\right] < 0,$$

$$\det(A) = \rho k^{-1}\left[\mu + k\varepsilon + \beta(1 - k)(1 - N^*)\right] > 0. \tag{12.64}$$

Example 12.8 Using the values of the parameters from Example 12.6, we have

$$\mathrm{Tr}(A) = -0.0018456, \qquad \det(A) = 0.0000008.$$

Consider $\hat{a}_{ii}$, $i = 1, 2$, defined in (12.27).

Lemma 12.5 *If*

$$a_{21} \leq 0 \tag{12.65}$$

and

$$\hat{a}_{11} < 0, \qquad \hat{a}_{22} < 0, \tag{12.66}$$

then the zero solution of (12.63) *is asymptotically mean-square stable.*

Proof By (12.59), (12.65) the matrix A from (12.63) satisfies the condition $a_{12}a_{21} \leq 0$. By (12.19) the same condition is satisfied by the matrix A from (12.23). So, further, the proof coincides with that of Lemma 12.2.   □

Lemma 12.6 *If*

$$a_{21} > 0 \tag{12.67}$$

and

$$\max(\delta_1, \delta_2) < \frac{\det(A)}{|\mathrm{Tr}(A)|}, \tag{12.68}$$

then the zero solution of (12.63) *is asymptotically mean-square stable.*

Proof Similarly to Lemma 12.2, it is enough to show that the matrix $P = \|p_{ij}\|$ with the elements (12.26) is positive definite.

Note that by (12.31), (12.59), (12.67) we have

$$A_i = a_{11}a_{22} - a_{12}a_{21} + a_{ii}^2 \leq a_{ii}\,\mathrm{Tr}(A), \quad i = 1, 2.$$

From this and from (12.59), (12.64), (12.68) it follows that

$$\delta_i < \frac{\det(A)}{|\mathrm{Tr}(A)|} \leq \frac{A_i}{|\mathrm{Tr}(A)|} \leq |a_{ii}|, \quad i = 1, 2. \tag{12.69}$$

By (12.27), (12.28), (12.59), (12.64), (12.67), (12.68) we have

$$\begin{aligned}
Z &= \mathrm{Tr}(A)(a_{11} + \delta_1)(a_{22} + \delta_2) - a_{12}a_{21}\big(\mathrm{Tr}(A) + \delta_1 + \delta_2\big) \\
&= \mathrm{Tr}(A)\det(A) + \mathrm{Tr}(A)\delta_1 a_{22} + \mathrm{Tr}(A)\delta_2 a_{11} \\
&\quad + \mathrm{Tr}(A)\delta_1\delta_2 - a_{12}a_{21}(\delta_1 + \delta_2) \\
&= -|\mathrm{Tr}(A)|\det(A) + A_2\delta_1 + A_1\delta_2 - |\mathrm{Tr}(A)|\delta_1\delta_2 \\
&< -|\mathrm{Tr}(A)|\det(A) + (A_1 + A_2)\frac{\det(A)}{|\mathrm{Tr}(A)|} \\
&= -\big(|\mathrm{Tr}(A)|^2 - A_1 - A_2\big)\frac{\det(A)}{|\mathrm{Tr}(A)|} \\
&= -\frac{2a_{12}a_{21}\det(A)}{|\mathrm{Tr}(A)|} \\
&< 0.
\end{aligned} \tag{12.70}$$

From (12.34), (12.35), (12.69), (12.70) we obtain that $p_{11} > 0$, $p_{22} > 0$ for arbitrary $c > 0$.

Let us show that $p_{11}p_{22} > p_{12}^2$. Indeed, by (12.26), (12.34), (12.35) this inequality takes the form

$$\big(c(A_2 - |\mathrm{Tr}(A)|\delta_2) + a_{21}^2\big)\big(A_1 - |\mathrm{Tr}(A)|\delta_1 + ca_{12}^2\big) > (ca_{12}\hat{a}_{22} + a_{21}\hat{a}_{11})^2,$$

which is equivalent to the condition

$$
\begin{aligned}
c^2 a_{12}^2 \big(\det(A) &- |\mathrm{Tr}(A)|\delta_2 + a_{22}^2 - \hat{a}_{22}^2 \big) \\
&+ c\big[\big(A_1 - |\mathrm{Tr}(A)|\delta_1 \big)\big(\det(A) - |\mathrm{Tr}(A)|\delta_2 \big) \\
&+ a_{22}^2 \big(\det(A) - |\mathrm{Tr}(A)|\delta_1 \big) \\
&+ \big(\det(A) \big)^2 + 2 a_{12} a_{21} \big(a_{11} a_{22} - \hat{a}_{11} \hat{a}_{22} \big) \big] \\
&+ a_{21}^2 \big(\det(A) - |\mathrm{Tr}(A)|\delta_1 + a_{11}^2 - \hat{a}_{11}^2 \big) \\
&> 0.
\end{aligned}
\tag{12.71}
$$

By (12.67), (12.68), (12.69) and $|a_{ii}| \geq |\hat{a}_{ii}|$, $i = 1, 2$, condition (12.71) holds for arbitrary $c > 0$. So, for arbitrary $c > 0$, the matrix P with the elements (12.26) is positive definite. The proof is completed. $\square$

Remark 12.4 If condition (12.65) holds, i.e., $a_{21} \leq 0$, then from (12.59) and from the proofs of Lemmas 12.3 and 12.4 it follows that $\beta \in [0, (\mu + \varepsilon)(1 - k)^{-1}]$. On the other hand, if $\beta > (\mu + \varepsilon)(1 - k)^{-1}$, then condition (12.67) holds, i.e., $a_{21} > 0$. For example, by the values of the parameters from Example 12.6 we have $\beta = 0.00085 > (\mu + \varepsilon)(1 - k)^{-1} = 0.0004945$ and $a_{21} = 0.0005646 > 0$.

Theorem 12.2 *If conditions (12.65), (12.66) or (12.67), (12.68) hold and if, for some $c > 0$, the elements (12.26) of the matrix P satisfy the condition*

$$
\big(\beta N^* |p_{12} - p_{22}| \big)^2 + 2\beta N^* |p_{11} - p_{12}| < c,
\tag{12.72}
$$

then the solution (N^*, S^*) of system (12.57) is stable in probability.

Proof Note that the stability in probability of the solution (N^*, S^*) of system (12.57) is equivalent to the stability in probability of the zero solution of system (12.58) and the order of nonlinearity of system (12.58) is higher than one. So, to get for this system conditions for stability in probability, it is enough (Sect. 5.3) to get conditions for the asymptotic mean-square stability of the zero solution of the linear part (12.60) of this system. Following the procedure of constructing Lyapunov functionals, we will construct a Lyapunov functional for system (12.60) in the form $V = V_1 + V_2$, where $V_1 = y'Py$, $y = (y_1, y_2)'$, P is the positive definite solution of system (12.25) with the elements (12.26), and V_2 will be chosen below.

Let L be the generator of system (12.60). Then by (12.60), (12.25) we have

$$
\begin{aligned}
LV_1 &= 2\big(p_{11} y_1(t) + p_{12} y_2(t) \big)\big(a_{11} y_1(t) + a_{12} y_2(t) + \beta N^* I(y_{1t}) \big) + p_{11}\sigma_1^2 y_1^2(t) \\
&\quad + 2\big(p_{12} y_1(t) + p_{22} y_2(t) \big)\big(a_{21} y_1(t) + a_{22} y_2(t) - \beta N^* I(y_{1t}) \big) \\
&\quad + p_{22}\sigma_2^2 y_2^2(t) \\
&= -c y_1^2(t) - y_2^2(t) + 2\beta N^* \big[(p_{11} - p_{12}) y_1(t) + (p_{12} - p_{22}) y_2(t) \big] I(y_{1t}).
\end{aligned}
$$

By (12.2), (12.59) we have $2y_1(t)I(y_{1t}) \leq y_1^2(t) + I(y_{1t}^2)$ and $2y_2(t)I(y_{1t}) \leq \nu y_2^2(t) + \nu^{-1}I(y_{1t}^2)$ for some $\nu > 0$. Using these inequalities, we obtain

$$
\begin{aligned}
LV_1 &\leq -cy_1^2(t) - y_2^2(t) + \beta N^*|p_{11} - p_{12}|\left(y_1^2(t) + I\left(y_{1t}^2\right)\right) \\
&\quad + \beta N^*|p_{12} - p_{22}|\left(\nu y_2^2(t) + \nu^{-1}I\left(y_{1t}^2\right)\right) \\
&= \left(\beta N^*|p_{11} - p_{12}| - c\right)y_1^2(t) + \left(\beta N^*|p_{12} - p_{22}|\nu - 1\right)y_2^2(t) \\
&\quad + qI\left(y_{1t}^2\right),
\end{aligned}
\tag{12.73}
$$

where

$$
q = \beta N^*\left(|p_{11} - p_{12}| + |p_{12} - p_{22}|\nu^{-1}\right).
\tag{12.74}
$$

Putting

$$
V_2 = q \int_0^\infty \int_{t-s}^t y_1^2(\theta)\, d\theta\, dK(s),
$$

by (12.2), (12.59) we get $LV_2 = q(y_1^2(t) - I(y_{1t}^2))$. Therefore, by (12.73), (12.74), for the functional $V = V_1 + V_2$, we have

$$
\begin{aligned}
LV &\leq \left(2\beta N^*|p_{11} - p_{12}| + \beta N^*|p_{12} - p_{22}|\nu^{-1} - c\right)y_1^2(t) \\
&\quad + \left(\beta N^*|p_{12} - p_{22}|\nu - 1\right)y_2^2(t).
\end{aligned}
$$

Thus, if

$$
2\beta N^*|p_{11} - p_{12}| + \beta N^*|p_{12} - p_{22}|\nu^{-1} < c, \quad \beta N^*|p_{12} - p_{22}|\nu < 1, \tag{12.75}
$$

then by Remark 2.1 the zero solution of (12.60) is asymptotically mean-square stable.

From (12.75) it follows that

$$
\frac{\beta N^*|p_{12} - p_{22}|}{c - 2\beta N^*|p_{11} - p_{12}|} < \nu < \frac{1}{\beta N^*|p_{12} - p_{22}|}.
\tag{12.76}
$$

Thus, if for some $c > 0$, condition (12.72) holds, then there exists $\nu > 0$ such that conditions (12.76) (or (12.75)) hold too, and therefore the zero solution of (12.60) is asymptotically mean-square stable. From this it follows that the zero solution of (12.58) and therefore the equilibrium point (N^*, S^*) of system (12.57) is stable in probability. The proof is completed. □

Example 12.9 Consider system (12.50) with the values of the parameters ε, μ, ρ, β, γ and the equilibrium point (N^*, S^*) given in Example 12.6. As an example, consider the levels of noises $\sigma_1 = 0.028256$, $\sigma_2 = 0.029031$. From (12.27) it follows that $\delta_1 = 0.0003992$, $\delta_2 = 0.0004214$ and condition (12.68) holds: $\max(\delta_1, \delta_2) < \det(A)|\mathrm{Tr}(A)|^{-1} = 0.0004436$.

Put $c = 10$. Then by (12.26) $p_{11} = 8335.7$, $p_{12} = 569.4$, $p_{22} = 1344.7$, and condition (12.72) holds:

$$\left(\beta N^* |p_{12} - p_{22}| \right)^2 + 2\beta N^* |p_{11} - p_{12}| = 4.419 < 10.$$

Thus, the solution (N^*, S^*) of system (12.57) is stable in probability.

Using the representations (12.41)–(12.44), we get three corollaries from Theorem 12.2, which simplify a verification of the stability condition (12.72).
Put

$$f(c) = \left(\beta N^* D_0 \right)^2 \left(c + \frac{D_1}{D_0} \right)^2 + 2\beta N^* |B_0| \left| c + \frac{B_1}{B_0} \right| - c, \qquad (12.77)$$

$$S = \left(\beta N^* D_0 \right)^2 \left(\frac{D_1}{D_0} - \frac{B_1}{B_0} \right)^2 + \frac{B_1}{B_0},$$

$$R_+ = 2\beta N^* |B_0| \left(\frac{1 - 2\beta N^* |B_0|}{2(\beta N^* D_0)^2} - \frac{D_1}{D_0} + \frac{B_1}{B_0} \right),$$

$$R_- = -2\beta N^* |B_0| \left(\frac{1 + 2\beta N^* |B_0|}{2(\beta N^* D_0)^2} - \frac{D_1}{D_0} + \frac{B_1}{B_0} \right),$$

$$(12.78)$$

$$Q = \frac{1}{4(\beta N^* D_0)^2} - \frac{D_1}{D_0} - \frac{B_0^2}{D_0^2},$$

where B_0, B_1, D_0, D_1 are defined by (12.41)–(12.44). So, condition (12.72) is equivalent to the condition $f(c) < 0$.

Corollary 12.4 *If conditions (12.65), (12.66) or (12.67), (12.68) hold and $S < 0$, then the solution (N^*, S^*) of system (12.57) is stable in probability.*

Proof By (12.78) from $S < 0$ it follows that $B_1 B_0^{-1} < 0$. Substituting $c_0 = -B_1 B_0^{-1} > 0$ into (12.77), we obtain $f(c_0) = S < 0$, i.e., condition (12.72) holds. The proof is completed. □

Corollary 12.5 *If conditions (12.65), (12.66) or (12.67), (12.68) hold and $0 \le R_+ < Q$, then the solution (N^*, S^*) of system (12.57) is stable in probability.*

Corollary 12.6 *If conditions (12.65), (12.66) or (12.67), (12.68) hold and $0 < R_- < Q$, then the solution (N^*, S^*) of system (12.57) is stable in probability.*

The proofs of Corollaries 12.5 and 12.6 are similar to those of Corollaries 12.2 and 12.3.

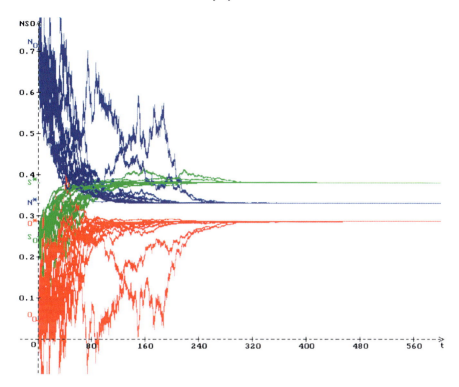

Fig. 12.2   25 trajectories of the processes $N(t)$ (*blue*), $S(t)$ (*green*), $O(t)$ (*red*) with the values of the parameters $\mu = 0.000469$, $\gamma = 0.0003$, $\varepsilon = 0.000004$, $\rho = 0.000035$, $\beta = 0.00085$, $h = 0.1$, $\delta_1 = 0.0003992$, $\delta_2 = 0.0004214$, the initial values $N(s) = 0.704$, $s \in [-0.1, 0]$, $S(0) = 0.25$, $O(0) = 0.046$, and the equilibrium point $N^* = 0.3311$, $S^* = 0.3814$, $O^* = 0.2875$

Example 12.10   Consider system (12.57) with the values of the parameters from Example 12.6 and $\delta_1 = 0.0003992$, $\delta_2 = 0.0002661$. Calculating S, R_+, Q by (12.78), we obtain: $S = -0.0100916 < 0$, $R_+ = 7499 < Q = 18161$. By both Corollaries 12.4 and 12.5 the solution (N^*, S^*) of system (12.57) is stable in probability.

Example 12.11   Consider system (12.57) with the values of the parameters from Example 12.6 and $\delta_1 = 0.0003992$, $\delta_2 = 0.0004214$. Calculating S, R_+, Q by (12.78), we obtain: $S = 0.0051611 > 0$, $R_+ = 7811 < Q = 18914$. The condition of Corollary 12.4 does not hold, but from Corollary 12.5 it follows that the solution (N^*, S^*) of system (12.57) is stable in probability.

12.2.5 Numerical Simulation

Let us suppose that in (12.49) $dK(s) = \delta(s - h)\,ds$, where $\delta(s)$ is Dirac's delta-function, and $h \geq 0$ is a delay.

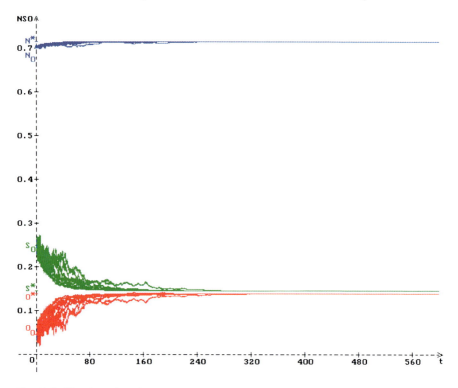

Fig. 12.3 25 trajectories of the processes $N(t)$ (*blue*), $S(t)$ (*green*), $O(t)$ (*red*) with the values of the parameters $\mu = 0.000469$, $\gamma = 0.0003$, $\varepsilon = 0.000004$, $\rho = 0.000035$, $\beta = 0$, $h = 0.1$, $\delta_1 = 0.0003992$, $\delta_2 = 0.0004214$, the initial values $N(s) = 0.704$, $s \in [-0.1, 0]$, $S(0) = 0.25$, $O(0) = 0.046$, and the equilibrium point $N^* = 0.7149$, $S^* = 0.1465$, $O^* = 0.1386$

In Fig. 12.2 25 trajectories of the solution of (12.57), (12.48) are shown for the values of the parameters from Examples 12.6 and 12.9: $\mu = 0.000469$, $\gamma = 0.0003$, $\varepsilon = 0.000004$, $\rho = 0.000035$, $\beta = 0.00085$, the initial values $N_0 = 0.704$, $S_0 = 0.25$, $O_0 = 0.046$, the levels of noises $\sigma_1 = 0.028256$, $\sigma_2 = 0.029031$, and the delay $h = 0.1$. One can see that all trajectories go to the equilibrium point $N^* = 0.3311$, $S^* = 0.3814$, $O^* = 0.2875$.

Putting $\beta = 0$ with the same values of the other parameters, one can see that in accordance with Example 12.6, all trajectories go to another equilibrium point $N^* = 0.7149$, $S^* = 0.1465$, $O^* = 0.1386$ (Fig. 12.3).

Change now the initial values on $N_0 = 1$, $S_0 = O_0 = 0$, and put again $\beta = 0.00085$. In accordance with Example 12.6, corresponding trajectories of the solution go to the equilibrium point $N^* = 0.5770$, $S^* = 0.2588$, $O^* = 0.1642$ (Fig. 12.4).

Note that numerical simulations of the processes $N(t)$, $S(t)$, and $O(t)$ were obtained via the difference analogues of (12.57), (12.48) in the form

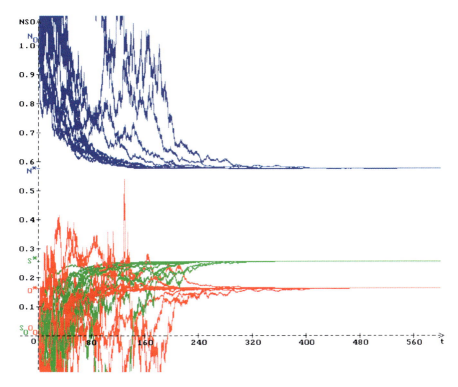

Fig. 12.4 25 trajectories of the processes $N(t)$ (*blue*), $S(t)$ (*green*), $O(t)$ (*red*) with the values of the parameters $\mu = 0.000469$, $\gamma = 0.0003$, $\varepsilon = 0.000004$, $\rho = 0.000035$, $\beta = 0.00085$, $h = 0.1$, $\delta_1 = 0.0003992$, $\delta_2 = 0.0004214$, the initial values $N(s) = 1$, $s \in [-0.1, 0]$, $S(0) = 0$, $O(0) = 0$, and the equilibrium point $N^* = 0.5770$, $S^* = 0.2588$, $O^* = 0.1642$

$$N_{i+1} = N_i + \Delta\left[\mu N_0 - \mu N_i - \beta N_i (1 - N_{i-m}) + \rho S_i\right]$$
$$+ \sigma_1 (N_i - N^*)(w_{1,i+1} - w_{1i}),$$
$$S_{i+1} = S_i + \Delta\left[\mu S_0 + \beta N_i N_{i-m} - (\mu + \gamma + \rho)S_i + \varepsilon(1 - N_i - S_i)\right]$$
$$+ \sigma_2 (S_i - S^*)(w_{2,i+1} - w_{2i}),$$
$$O_{i+1} = 1 - N_{i+1} - S_{i+1},$$
$$i = 0, 1, 2, \ldots, \quad N_j = N_0, \quad j = -m, \ldots, -1, 0.$$

Here Δ is the discretization step (chosen as $\Delta = 0.01$), $N_i = N(t_i)$, $S_i = S(t_i)$, $O_i = O(t_i)$, $w_{ki} = w_k(t_i)$, $k = 1, 2$, $t_i = i\Delta$, $m - h\Delta^{-1}$, and trajectories of the Wiener processes $w_1(t)$ and $w_2(t)$ are simulated by the algorithm described in Sect. 2.1.1.

References

1. Acheson DJ (1993) A pendulum theorem. Proc R Soc Lond Ser A, Math Phys Sci 443(1917):239–245
2. Acheson DJ, Mullin T (1993) Upside-down pendulums. Nature 366(6452):215–216
3. Agarwal RP, Grace SR (2000) Asymptotic stability of certain neutral differential equations. Math Comput Model 31:9–15
4. Agarwal RP, O'Regan D, Wong PJY (1999) Positive solutions of differential, difference and integral equations. Kluwer Academic, Dordrecht
5. Agarwal RP, Grace SR, O'Regan D (2000) Oscillation theory for difference and functional differential equations. Kluwer Academic, Dordrecht
6. Alonso-Quesada S, De la Sen M, Agarwal R, Ibeas A (2012) An observer-based vaccination control law for a SEIR epidemic model based on feedback linearization techniques for nonlinear systems. Adv Differ Equ 2012:161
7. Alzabut JO (2010) Almost periodic solutions for impulsive delay Nicholson's blowflies population model. J Comput Appl Math 234:233–239
8. Alzabut JO, Bolat Y, Abdeljawad T (2012) Almost periodic dynamic of a discrete Nicholson's blowflies model involving a linear harvesting term. Adv Differ Equ 2012:158
9. Andreyeva EA, Kolmanovskii VB, Shaikhet LE (1992) Control of hereditary systems. Nauka, Moscow. pp 336 (in Russian)
10. Appleby JAD, Buckwar E (2010) A constructive comparison technique for determining the asymptotic behaviour of linear functional differential equations with unbounded delay. Differ Equ Dyn Syst 18(3):271–301
11. Appleby JAD, Krol K (2011) Long memory in a linear stochastic Volterra equation. J Math Anal Appl 380(2):814–830
12. Appleby JAD, Reynolds DW (2004) On necessary and sufficient conditions for exponential stability in linear Volterra integro–differential equations. J Integral Equ Appl 16:221–240
13. Appleby JAD, Gyori I, Reynolds DW (2006) On exact rates of decay of solutions of linear systems of Volterra equations with delay. J Math Anal Appl 320:56–77
14. Aranceta J, Serra L, Foz M, Moreno B (2005) Prevalencia de obesidad en Espana. Méd Clin 125:460–466
15. Arino O, Elabdllaoui A, Mikaram J, Chattopadhyay J (2004) Infection on prey population may act as a biological control in a ratio-dependent predator–prey model. Nonlinearity 17:1101–1116
16. Arnold L (1974) Stochastic differential equations. Wiley, New York
17. Arnold L (1998) Random dynamical systems. Springer, Berlin
18. Bailey NTJ (1957) The mathematical theory of epidemics. Griffin, London
19. Bandyopadhyay M, Chattopadhyay J (2005) Ratio dependent predator–prey model: effect of environmental fluctuation and stability. Nonlinearity 18:913–936

20. Bandyopadhyay M, Saha T, Pal R (2008) Deterministic and stochastic analysis of a delayed allelopathic phytoplankton model within fluctuating environment. Nonlinear Anal Hybrid Syst 2:958–970
21. Bedelbaev AK (1958) About construction of Lyapunov functions as a quadratic form and its application to stability of controlled systems. Avtomatika 1:37–43 (in Ukrainian)
22. Bellman R, Cooke KL (1963) Differential difference equations. Academic Press, New York
23. Beretta E, Kuang Y (1998) Global analysis in some delayed ratio-dependent predator–prey systems. Nonlinear Anal 32(4):381–408
24. Beretta E, Takeuchi Y (1994) Qualitative properties of chemostat equations with time delays: boundedness, local and global asymptotic stability. Differ Equ Dyn Syst 2(1):19–40
25. Beretta E, Takeuchi Y (1994) Qualitative properties of chemostat equations with time delays II. Differ Equ Dyn Syst 2(4):263–288
26. Beretta E, Takeuchi Y (1995) Global stability of an SIR epidemic model with time delays. J Math Biol 33:250–260
27. Beretta E, Kolmanovskii V, Shaikhet L (1998) Stability of epidemic model with time delays influenced by stochastic perturbations. Math Comput Simul 45(3–4):269–277 (Special Issue "Delay Systems")
28. Beretta E, Hara T, Ma W, Takeuchi Y (2001) Global asymptotic stability of an SIR epidemic model with distributed time delay. Nonlinear Anal 47:4107–4115
29. Berezansky L, Braverman E, Idels L (2004) Delay differential logistic equations with harvesting. Math Comput Model 40:1509–1525
30. Berezansky L, Braverman E, Idels L (2010) Nicholson's blowflies differential equations revisited: main results and open problems. Appl Math Model 34:1405–1417
31. Berezansky L, Idels L, Troib L (2011) Global dynamics of Nicholson-type delay systems with applications. Nonlinear Anal, Real World Appl 12:436–445
32. Biran Y, Innis B (1979) Optimal control of bilinear systems: time-varying effects of cancer drugs. Automatica 15
33. Blackburn JA, Smith HJT, Gronbech-Jensen N (1992) Stability and Hopf bifurcations in an inverted pendulum. Am J Phys 60(10):903–908
34. Blanchflower DG, Oswald AJ, Landeghem BV (2009) Imitative obesity and relative utility. J Eur Econ Assoc 7:528–538
35. Blizorukov MG (1996) On the construction of solutions of linear difference systems with continuous time. Differ Uravn (Minsk) 32:127–128. Translation in Differential Equations, 133–134
36. Borne P, Kolmanovskii V, Shaikhet L (1999) Steady-state solutions of nonlinear model of inverted pendulum. Theory Stoch Process 5(21)(3–4):203–209. Proceedings of the third Ukrainian–Scandinavian conference in probability theory and mathematical statistics, 8–12 June 1999, Kyiv, Ukraine
37. Borne P, Kolmanovskii V, Shaikhet L (2000) Stabilization of inverted pendulum by control with delay. Dyn Syst Appl 9(4):501–514
38. Bradul N, Shaikhet L (2007) Stability of the positive point of equilibrium of Nicholson's blowflies equation with stochastic perturbations: numerical analysis. Discrete Dyn Nat Soc 2007:92959. doi:10.1155/2007/92959. 25 pages
39. Bradul N, Shaikhet L (2009) Stability of difference analogue of mathematical predator–prey model by stochastic perturbations. Vest Odesskogo Naz Univ Mat Mekh 14(20):7–23 (in Russian)
40. Braverman E, Kinzebulatov D (2006) Nicholson's blowflies equation with a distributed delay. Can Appl Math Q 14(2):107–128
41. Buonomo B, Rionero S (2010) On the Lyapunov stability for SIRS epidemic models with general nonlinear incidence rate. Appl Math Comput 217(8):4010–4016
42. Burkeand MA, Heiland F (2007) Social dynamics of obesity. Econ Inq 45:571–591
43. Burton TA (1983) Volterra integral and differential equations. Academic Press, New York
44. Burton TA (1985) Stability and periodic solutions of ordinary and functional differential equations. Academic Press, Orlando

45. Busenberg S, Cooke KL (1980) The effect of integral conditions in certain equations modelling epidemics and population growth. J Math Biol 10(1):13–32
46. Bush AW, Cook AE (1976) The effect of time delay and growth rate inhibition in the bacterial treatment of wastewater. J Theor Biol 63:385–395
47. Cai L, Li X, Song X, Yu J (2007) Permanence and stability of an age-structured prey–predator system with delays. Discrete Dyn Nat Soc 2007:54861. 15 pages
48. Caraballo T, Real J, Shaikhet L (2007) Method of Lyapunov functionals construction in stability of delay evolution equations. J Math Anal Appl 334(2):1130–1145. doi:10.1016/j.jmaa. 2007.01.038
49. Carletti M (2002) On the stability properties of a stochastic model for phage-bacteria interaction in open marine environment. Math Biosci 175:117–131
50. Chen F (2005) Periodicity in a ratio-dependent predator–prey system with stage structure for predator. J Appl Math 2:153–169
51. Christakis NA, Fowler JH (2007) The spread of obesity in a large social network over 32 years. N Engl J Med 357:370–379
52. Chueshov I, Scheutzow M, Schmalfu B (2005) Continuity properties of inertial manifolds for stochastic retarded semilinear parabolic equations. In: Deuschel J-D, Greven A (eds) Interacting stochastic systems. Springer, Berlin, pp 355–375
53. Colliugs JB (1997) The effects of the functional response on the bifurcation behavior of a mite predator–prey interaction model. J Math Biol 36:149–168
54. Corduneanu C, Lakshmikantham V (1980) Equations with unbounded delay: a survey. Nonlinear Anal 4:831–877
55. Crisci MR, Kolmanovskii VB, Russo E, Vecchio A (1995) Stability of continuous and discrete Volterra integro–differential equations by Lyapunov approach. J Integral Equ 4(7):393–411
56. Crisci MR, Kolmanovskii VB, Russo E, Vecchio A (1998) Stability of difference Volterra equations: direct Lyapunov method and numerical procedure. Comput Math Appl 36(10–12):77–97
57. Cushing JM (1977) Integro–differential equations and delay models in population dynamics. Lectures notes in biomathematics. Springer, Berlin
58. Diekmann O, Heesterbeek JAP, Metz JAJ (1990) On the definition and the computation of the basic reproduction R0 in models for infectious diseases in heterogeneous populations. J Math Biol 28:365–382
59. Diekmann O, Von Gils SA, Verduyn Lunel SM, Walther HO (1995) Delay equations, functional, complex and nonlinear analysis. Applied math sciences. Springer, Berlin
60. Ding X, Jiang J (2008) Positive periodic solutions in delayed Gauss-type predator–prey systems. J Math Anal Appl 339(2):1220–1230. doi:10.1016/j.jmaa.2007.07.079
61. Ding X, Li W (2006) Stability and bifurcation of numerical discretization Nicholson's blowflies equation with delay. Discrete Dyn Nat Soc 2006:1–12. doi:10.1155/DDNS/2006/19413
62. Driver RD (1962) Existence and stability of solutions of a delay differential system. Arch Ration Mech Anal 10:401–426
63. Driver RD (1997) Ordinary and delay differential equations. Applied math sciences. Springer, Berlin
64. Edwards JT, Ford NJ, Roberts JA, Shaikhet LE (2000) Stability of a discrete nonlinear integro–differential equation of convolution type. Stab Control: Theory Appl 3(1):24–37
65. El-Owaidy HM, Ammar A (1988) Stable oscillations in a predator–prey model with time lag. J Math Anal Appl 130:191–199
66. El'sgol'ts LE, Norkin SB (1973) Introduction to the theory and application of differential equations with deviating arguments. Academic Press, New York
67. Erbe LH, Qingkai K, Zhang BG (1994) Oscillation theory for functional differential equations. Marcel Dekker, New York
68. Etile F (2007) Social norms, ideal body weight and food attitudes. Health Econ 16:945–966

69. Fan M, Wang Q, Zhou X (2003) Dynamics of a nonautonomous ratio-dependent predator–prey system. Proc R Soc Edinb A 133:97–118
70. Fan YH, Li WT, Wang LL (2004) Periodic solutions of delayed ratio-dependent predator–prey model with monotonic and no-monotonic functional response. Nonlinear Anal, Real World Appl 5(2):247–263
71. Fan X, Wang Z, Xu X (2012) Global stability of two-group epidemic models with distributed delays and random perturbation. Abstr Appl Anal 2012:132095
72. Farkas M (1984) Stable oscillations in a predator–prey model with time lag. J Math Anal Appl 102:175–188
73. Feng QX, Yan JR (2002) Global attractivity and oscillation in a kind of Nicholson's blowflies. J Biomath 17(1):21–26
74. Ford NJ, Baker CTH (1996) Qualitative behavior and stability of solutions of discretised nonlinear Volterra integral equations of convolution type. J Comput Appl Math 66:213–225
75. Ford NJ, Baker CTH, Roberts JA (1997) Preserving qualitative behavior and transience in numerical solutions of Volterra integro–differential equations of convolution type: Lyapunov functional approaches. In: Proceeding of 15th world congress on scientific computation, modelling and applied mathematics (IMACS97), Berlin, August 1997. Numerical mathematics, vol 2, pp 445–450
76. Ford NJ, Baker CTH, Roberts JA (1998) Nonlinear Volterra integro–differential equations—stability and numerical stability of θ-methods. MCCM numerical analysis report. J Integral Equ Appl 10:397–416
77. Ford NJ, Edwards JT, Roberts JA, Shaikhet LE (1997) Stability of a difference analogue for a nonlinear integro differential equation of convolution type. Numerical Analysis Report, University of Manchester 312
78. Fridman E (2001) New Lyapunov–Krasovskii functionals for stability of linear retarded and neutral type systems. Syst Control Lett 43:309–319
79. Fridman E (2002) Stability of linear descriptor systems with delays: a Lyapunov-based approach. J Math Anal Appl 273:24–44
80. Gantmacher FR (2000) Matrix theory, vol 2. American Mathematical Society, Providence
81. Gard TC (1988) Introduction to stochastic differential equations. Marcel Dekker, Basel
82. Garvie M (2007) Finite-difference schemes for reaction–diffusion equations modelling predator–prey interactions in MATLAB. Bull Math Biol 69(3):931–956
83. Ge Z, He Y (2008) Diffusion effect and stability analysis of a predator–prey system described by a delayed reaction–diffusion equations. J Math Anal Appl 339(2):1432–1450. doi:10.1016/j.jmaa.2007.07.060
84. Gikhman II, Skorokhod AV (1972) Stochastic differential equations. Springer, Berlin
85. Gikhman II, Skorokhod AV (1974) The theory of stochastic processes, vol I. Springer, Berlin
86. Gikhman II, Skorokhod AV (1975) The theory of stochastic processes, vol II. Springer, Berlin
87. Gikhman II, Skorokhod AV (1979) The theory of stochastic processes, vol III. Springer, Berlin
88. Gil' MI (1998) Stability of finite and infinite dimensional systems. Kluwer Academic, Boston
89. Golec J, Sathananthan S (1999) Sample path approximation for stochastic integro–differential equations. Stoch Anal Appl 17(4):579–588
90. Golec J, Sathananthan S (2001) Strong approximations of stochastic integro–differential equations. Dyn Contin Discrete Impuls Syst, Ser B, Appl Algorithms 8(1):139–151
91. Gopalsamy K (1992) Equations of mathematical ecology. Kluwer Academic, Dordrecht
92. Gopalsamy K (1992) Stability and oscillations in delay differential equations of population dynamics. Mathematics and its applications, vol 74. Kluwer Academic, Dordrecht
93. Gorecki H, Fuksa S, Grabowski P, Korytowski A (1989) Analysis and synthesis of time-delay systems. Wiley, New York
94. Gourley SA, Kuang Y (2004) A stage structured predator–prey model and its dependence on maturation delay and death rate. J Math Biol 4:188–200

95. Grove EA, Ladas G, McGrath LC, Teixeira CT (2001) Existence and behavior of solutions of a rational system. Commun Appl Nonlinear Anal 8(1):1–25

96. Gu K (2000) An integral inequality in the stability problem of time-delay system. In: 39th IEEE conference on decision and control, Sydney, Australia, pp 2805–2810

97. Gu K, Niculescu S-I (2000) Additional dynamics in transformed time-delay systems. IEEE Trans Autom Control 45:572–575

98. Guo MY, Li Z (2006) Global dynamics of a staged progression model for infectious diseases. Math Biosci Eng 3(3):513–525

99. Gurney WSC, Blythe SP, Nisbet RM (1980) Nicholson's blowflies revisited. Nature 287:17–21

100. Gyori I, Ladas G (1991) Oscillation theory of delay differential equations with applications. Oxford mathematical monographs. Oxford University Press, New York

101. Gyori I, Trofimchuk S (1999) Global attractivity in $\dot{x}(t) = -\delta x(t) + pf(x(t-\tau))$. Dyn Syst Appl 8:197–210

102. Gyori I, Trofimchuk SI (2002) On the existence of rapidly oscillatory solutions in the Nicholson's blowflies equation. Nonlinear Anal 48:1033–1042

103. Halanay A (1966) Differential equations: stability, oscillations, time lags. Academic Press, New York

104. Hale JK (1972) Oscillations in neutral functional differential equations. Nonlinear Mechanics, CIME

105. Hale JK (1977) Theory of functional differential equations. Springer, New York

106. Hale JK, Verduyn-Lunel SM (1993) Introduction to functional differential equations. Springer, New York

107. Harrison GW (1979) Global stability of predator–prey interactions. J Math Biol 8:159–171

108. Hastings A (1983) Age dependent predation is not a simple process. I. Continuous time models. Theor Popul Biol 23:347–362

109. Hastings A (1984) Delays in recruitment at different trophic levels effects on stability. J Math Biol 21:35–44

110. Hethcote HW (1976) Qualitative analysis of communicable disease models. Math Biol Sci 28:335–356

111. Hino Y, Murakami S, Naito T (1991) Functional differential equations with infinite delay. Lecture notes in mathematics. Springer, New York

112. Hsu SB, Huang TW (1995) Global stability for a class of predator–prey systems. SIAM J Appl Math 55(3):763–783

113. Huo HF, Li WT (2004) Periodic solution of a delayed predator–prey system with Michaelis–Menten type functional response. J Comput Appl Math 166:453–463

114. Imkeller P, Lederer Ch (2001) Some formulas for Lyapunov exponents and rotation numbers in two dimensions and the stability of the harmonic oscillator and the inverted pendulum. Dyn Syst 16:29–61

115. James PT, Leach R, Kalamara E, Shayeghi M (2001) The worldwide obesity epidemic. Obesity 9:228–233

116. Ji C, Jiang D, Li X (2011) Qualitative analysis of a stochastic ratio-dependent predator–prey system. J Comput Appl Math 235:1326–1341

117. Jovanovic M, Krstic M (2012) Stochastically perturbed vector-borne disease models with direct transmission. Appl Math Model 36:5214–5228

118. Jumpen W, Orankitjaroen S, Boonkrong P, Wiwatanapataphee B (2011) SEIQR-SIS epidemic network model and its stability. Int J Math Comput Simul 5:326–333

119. Kac IYa (1998) Method of Lyapunov functions in problems of stability and stabilization of systems of stochastic structure. Ural State Academy of Railway Communication, Yekaterinburg, pp 222 (in Russian)

120. Kac IYa, Krasovskii NN (1960) About stability of systems with stochastic parameters. Prikl Mat Meh 24(5):809–823 (in Russian)

121. Kapitza PL (1965) Dynamical stability of a pendulum when its point of suspension vibrates, and pendulum with a vibrating suspension. In: ter Haar D (ed) Collected papers of

P.L. Kapitza, vol 2. Pergamon Press, London, pp 714–737

122. Karafyllis I, Pepe P, Jiang ZP (2009) Stability results for systems described by coupled retarded functional differential equations and functional difference equations. Nonlinear Anal, Theory Methods Appl 71:3339–3362

123. Keeling MJ, Rohani P (2008) Modeling infectious diseases in humans and animals. Princeton University Press, Princeton

124. Khan H, Mohapatra RN, Vajravelu K, Liao SJ (2009) The explicit series solution of SIR and SIS epidemic models. Appl Math Comput 215:653–669

125. Khasminskii RZ (1980) Stochastic stability of differential equations. Sijthoff & Noordhoff, Alphen aan den Rijn

126. Kim AV (1999) Functional differential equations: application of I-smooth calculus. Mathematics and its applications. Kluwer Academic, Dordrecht

127. Ko W, Ahn I (2013) A diffusive one-prey and two-competing-predator system with a ratio-dependent functional response: I. Long time behavior and stability of equilibria. J Math Anal Appl 397(1):9–28

128. Ko W, Ahn I (2013) A diffusive one-prey and two-competing-predator system with a ratio-dependent functional response: II. Stationary pattern formation. J Math Anal Appl 397(1):29–45

129. Kocic VL, Ladas G (1990) Oscillation and global attractivity in discrete model of Nicholson's blowflies. Appl Anal 38:21–31

130. Kolmanovskii VB (1993) On stability of some hereditary systems. Avtom Telemeh 11:45–59 (in Russian)

131. Kolmanovskii VB (1995) Applications of differential inequalities for stability of some functional differential equations. Nonlinear Anal, Theory Methods Appl 25(9–10):1017–1028

132. Kolmanovskii VB, Myshkis AD (1992) Applied theory of functional differential equations. Kluwer Academic, Dordrecht

133. Kolmanovskii VB, Myshkis AD (1999) Introduction to the theory and applications of functional differential equations. Kluwer Academic, Dordrecht

134. Kolmanovskii VB, Nosov VR (1981) Stability and periodic regimes of regulating hereditary systems. Nauka, Moscow (in Russian)

135. Kolmanovskii VB, Nosov VR (1986) Stability of functional differential equations. Academic Press, London

136. Kolmanovskii VB, Shaikhet LE (1993) Method for constructing Lyapunov functionals for stochastic systems with aftereffect. Differ Uravn (Minsk) 29(11):1909–1920. 2022, (in Russian). Translated in Differential Equations 29(11):1657–1666 (1993)

137. Kolmanovskii VB, Shaikhet LE (1993) Stability of stochastic systems with aftereffect. Avtom Telemeh 7:66–85 (in Russian). Translation in Automat. Remote Control 54(7):1087–1107, part 1

138. Kolmanovskii VB, Shaikhet LE (1994) New results in stability theory for stochastic functional–differential equations (SFDEs) and their applications. In: Proceedings of dynamic systems and applications, Atlanta, 1993, vol 1. Dynamic Publishers, Atlanta, pp 167–171

139. Kolmanovskii VB, Shaikhet LE (1995) Method for constructing Lyapunov functionals for stochastic differential equations of neutral type. Differ Uravn (Minsk) 31(11):1851–1857 (in Russian). Translated in Differential Equations 31(11):1819–1825 (1995)

140. Kolmanovskii VB, Shaikhet LE (1996) Control of systems with aftereffect. Translations of mathematical monographs, vol 157. American Mathematical Society, Providence

141. Kolmanovskii VB, Shaikhet LE (1997) About stability of some stochastic Volterra equations. Differ Uravn 11:1495–1502 (in Russian)

142. Kolmanovskii VB, Shaikhet LE (1997) Matrix Riccati equations and stability of stochastic linear systems with nonincreasing delays. Funct Differ Equ 4(3–4):279–293

143. Kolmanovskii VB, Shaikhet LE (1998) Riccati equations and stability of stochastic linear systems with distributed delay. In: Bajic V (ed) Advances in systems, signals, control and computers. IAAMSAD and SA branch of the Academy of Nonlinear Sciences, Durban, pp 97–100. ISBN 0-620-23136-X

144. Kolmanovskii VB, Shaikhet LE (1998) Riccati equations in stability of stochastic linear systems with delay. Avtom Telemeh 10:35–54 (in Russian)
145. Kolmanovskii VB, Shaikhet LE (2002) Construction of Lyapunov functionals for stochastic hereditary systems: a survey of some recent results. Math Comput Model 36(6):691–716
146. Kolmanovskii VB, Shaikhet LE (2002) Some peculiarities of the general method of Lyapunov functionals construction. Appl Math Lett 15(3):355–360. doi:10.1016/S0893-9659(01)00143-4
147. Kolmanovskii VB, Shaikhet LE (2003) About one application of the general method of Lyapunov functionals construction. Int J Robust Nonlinear Control 13(9):805–818 (Special Issue on Time Delay Systems)
148. Kolmanovskii VB, Shaikhet LE (2004) About some features of general method of Lyapunov functionals construction. Stab Control: Theory Appl 6(1):49–76
149. Kolmanovskii VB, Kosareva NP, Shaikhet LE (1999) A method for constructing Lyapunov functionals. Differ Uravn (Minsk) 35(11):1553–1565 (in Russian). Translated in Differential Equations 35(11):1573–1586 (1999)
150. Kordonis I-GE, Philos ChG (1999) The behavior of solutions of linear integro–differential equations with unbounded delay. Comput Math Appl 38:45–50
151. Kordonis I-GE, Philos ChG, Purnaras IK (2004) On the behavior of solutions of linear neutral integro–differential equations with unbounded delay. Georgian Math J 11:337–348
152. Korobeinikov A (2006) Lyapunov functions and global stability for SIR and SIRS epidemiological models with non-linear transmission. Bull Math Biol 68:615–626
153. Korobeinikov A (2009) Stability of ecosystem: global properties of a general predator–prey model. IMA J Math Appl Med Biol 26(4):309–321
154. Korobeinikov A, Maini PK (2005) Nonlinear incidence and stability of infectious disease models. Math Med Biol 22:113–128
155. Kovalev AA, Kolmanovskii VB, Shaikhet LE (1998) Riccati equations in the stability of retarded stochastic linear systems. Avtom Telemeh 10:35–54 (in Russian). Translated in Automatic Remote Control 59(10):1379–1394, part 1 (1998)
156. Krasovskii NN (1956) On the application of the second Lyapunov method for equation with time-delay. Prikl Mat Meh 20(1):315–327 (in Russian). Translated in J Appl Math Mekh 20 (1956)
157. Krasovskii NN (1956) On the asymptotic stability of systems with aftereffect. Prikl Mat Meh 20(3):513–518 (in Russian). Translated in J Appl Math Mekh 20 (1956)
158. Krasovskii NN (1959) Some problems of dynamic stability theory. Fizmatgiz, Moscow (in Russian)
159. Krasovskii NN (1963) Stability of motion. Standford Univ. Press, Standford
160. Kuang Y (1993) Delay differential equations with application in population dynamics. Academic Press, New York
161. Kulenovic MRS, Ladas G (1987) Linearized oscillations in population dynamics. Bull Math Biol 49(5):615–627
162. Kulenovic MRS, Nurkanovic M (2002) Asymptotic behavior of a two dimensional linear fractional system of difference equations. Rad Mat 11(1):59–78
163. Kulenovic MRS, Nurkanovic M (2005) Asymptotic behavior of a system of linear fractional difference equations. Arch Inequal Appl 2005(2):127–143
164. Kulenovic MRS, Nurkanovic M (2006) Asymptotic behavior of a competitive system of linear fractional difference equations. Adv Differ Equ 2006(5):19756. doi:10.1155/ADE/2006/19756. 13 pages
165. Kulenovic MRS, Ladas G, Sficas YG (1989) Global attractivity in population dynamics. Comput Math Appl 18(10–11):925–928
166. Kulenovic MRS, Ladas G, Sficas YG (1992) Global attractivity in Nicholson's blowflies. Appl Anal 43:109–124
167. Ladde GS, Lakshmikantham V, Zhang BG (1987) Oscillation theory of differential equations with deviating arguments. Marcel Dekker, New York

168. Lahrouz A, Omari L, Kiouach D (2011) Global analysis of a deterministic and stochastic nonlinear SIRS epidemic model. Nonlinear Anal, Model Control 16:59–76

169. Levi M (1988) Stability of the inverted pendulum—a topological explanation. SIAM Rev 30(4):639–644

170. Levi M, Weckesser W (1995) Stabilization of the inverted linearized pendulum by high frequency vibrations. SIAM Rev 37(2):219–223

171. Levin JJ, Nohel JA (1963) Note on a nonlinear Volterra equation. Proc Am Math Soc 14(6):924–929

172. Levin JJ, Nohel JA (1965) Perturbations of a non-linear Volterra equation. Mich Math J 12:431–444

173. Li J (1996) Global attractivity in a discrete model of Nicholson's blowflies. Ann Differ Equ 12(2):173–182

174. Li J (1996) Global attractivity in Nicholson's blowflies. Appl Math J Chin Univ Ser B 11(4):425–434

175. Li J, Du C (2008) Existence of positive periodic solutions for a generalized Nicholson's blowflies model. J Comput Appl Math 221(1):226–233

176. Li W, Fan Y (2007) Existence and global attractivity of positive periodic solutions for the impulsive delay Nicholson's blowflies model. J Comput Appl Math 201(1):55–68

177. Li YK, Kuang Y (2001) Periodic solutions of periodic delay Lotka–Volterra equations and systems. J Math Anal Appl 255:260–280

178. Li M, Yan J (2000) Oscillation and global attractivity of generalized Nicholson's blowfly model. In: Differential equations and computational simulations, Chengdu, 1999. World Scientific, Singapore, pp 196–201

179. Li MY, Graef JR, Wang L, Karsai J (1999) Global dynamics of a SEIR model with varying total population size. Math Biosci 160:191–213

180. Liao X, Zhou Sh, Ouyang Z (2007) On a stoichiometric two predators on one prey discrete model. Appl Math Lett 20:272–278

181. Liu K (2006) Stability of infinite dimensional stochastic differential equations with applications. Chapman and Hall/CRC, Boca Raton

182. Liz E (2007) A sharp global stability result for a discrete population model. J Math Anal Appl 330:740–743

183. Liz E, Röst G (2009) On global attractor of delay differential equations with unimodal feedback. Discrete Contin Dyn Syst 24(4):1215–1224

184. Long F (2012) Positive almost periodic solution for a class of Nicholson's blowflies model with a linear harvesting term. Nonlinear Anal 13(2):686–693

185. Lubich Ch (1983) On the stability of linear multistep methods for Volterra convolution equations. IMA J Numer Anal 3:439–465

186. Luo Q, Mao X (2007) Stochastic population dynamics under regime switching. J Math Anal Appl 334:69–84

187. Ma W, Takeuchi Y, Hara T, Beretta E (2002) Permanence of an SIR epidemic model with distributed time delays. Tohoku Math J 54:581–591

188. Ma W, Song M, Takeuchi Y (2004) Global stability of an SIR epidemic model with time delay. Appl Math Lett 17:1141–1145

189. Macdonald N (1978) Time lags in biological models. Lecture notes in biomath. Springer, Berlin

190. Mackey MC, Glass L (1997) Oscillation and chaos in physiological control system. Science 197:287–289

191. Mackey MC, Nechaeva IG (1994) Noise and stability in differential delay equations. J Differ Equ 6:395–426

192. Maistrenko YuL, Sharkovsky AN (1986) Difference equations with continuous time as mathematical models of the structure emergences. In: Dynamical systems and environmental models. Mathematical ecology, Eisenach. Akademie-Verlag, Berlin, pp 40–49

193. Makinde OD (2007) Adomian decomposition approach to a SIR epidemic model with constant vaccination strategy. Appl Math Comput 184:842–848

194. Malek-Zavarei M, Jamshidi M (1987) Time-delay systems: analysis, optimization and applications. North-Holland, Amsterdam
195. Mao X (1994) Exponential stability of stochastic differential equations. Marcel Dekker, New York
196. Mao X (2000) Stability of stochastic integro–differential equations. Stoch Anal Appl 18(6):1005–1017
197. Mao X, Shaikhet L (2000) Delay-dependent stability criteria for stochastic differential delay equations with Markovian switching. Stab Control: Theory Appl 3(2):88–102
198. Mao X, Yuan C (2006) Stochastic differential equations with Markovian switching. Imperial College Press, London
199. Marotto F (1982) The dynamics of a discrete population model with threshold. Math Biosci 58:123–128
200. Maruyama G (1955) Continuous Markov processes and stochastic equations. Rend Circ Mat Palermo 4:48–90
201. Mata GJ, Pestana E (2004) Effective Hamiltonian and dynamic stability of the inverted pendulum. Eur J Phys 25:717–721
202. McCluskey CC (2010) Complete global stability for an SIR epidemic model with delay—distributed or discrete. Nonlinear Anal, Real World Appl 11:55–59
203. McCluskey CC (2010) Global stability for an SIR epidemic model with delay and nonlinear incidence. Nonlinear Anal, Real World Appl 11(4):3106–3109
204. Melchor-Aguilar D (2010) On stability of integral delay systems. Appl Math Comput 217:3578–3584
205. Melchor-Aguilar D, Kharitonov V, Lozano R (2010) Stability conditions for integral delay systems. Int J Robust Nonlinear Control 20:1–15
206. Michiels W, Niculescu S-I (2007) Stability and stabilization of time-delay systems: an eigenvalue-based approach. SIAM, Philadelphia, PA
207. Miller RK (1971) Asymptotic stability properties of linear Volterra integro–differential equations. J Differ Equ 10:485–506
208. Miller RK (1972) Nonlinear Volterra integral equations. Benjamin, Elmsford
209. Mitchell R (1972) Stability of the inverted pendulum subjected to almost periodic and stochastic base motion—an application of the method of averaging. Int J Non-Linear Mech 7:101–123
210. Mohammed SEA (1984) Stochastic functional differential equations. Pitman, Boston
211. Mohammed SEA, Scheutzow M (1990) Lyapunov exponents and stationary solutions for affine stochastic delay equations. Stoch Stoch Rep 29:259–283
212. Mollison D (2003) Epidemic models: their structure and relation to data. Publications of the Newton Institute. Cambridge University Press, Cambridge
213. Mukhopadhyay B, Bhattacharyya R (2007) Existence of epidemic waves in a disease transmission model with two-habitat population. Int J Syst Sci 38:699–707
214. Mukhopadhyay B, Bhattacharyya R (2009) A nonlinear mathematical model of virus-tumor-immune system interaction: deterministic and stochastic analysis. Stoch Anal Appl 27:409–429
215. Murakami S (1990) Exponential stability for fundamental solutions of some linear functional differential equations. In: Yoshizwa T, Kato J (eds) Proceedings of the international symposium: functional differential equations. World Scientific, Singapore, pp 259–263
216. Murakami S (1991) Exponential asymptotic stability for scalar linear Volterra equations. Differ Integral Equ 4:519–525
217. Muroya Y (2007) Persistence global stability in discrete models of Lotka–Volterra type. J Math Anal Appl 330:24–33
218. Muroya Y, Enatsu Y, Nakata Y (2011) Global stability of a delayed SIRS epidemic model with a non-monotonic incidence rate. J Math Anal Appl 377(1):1–14
219. Murray JD (2002) Mathematical biology, 3rd edn. Springer, New York
220. Myshkis AD (1951) General theory of differential equations with delay. Translations of mathematical monographs, vol 55. American Mathematical Society, Providence

221. Myshkis AD (1972) Linear differential equations with delay argument. Nauka, Moscow (in Russian)
222. Nam PT, Phat VN (2009) An improved stability criterion for a class of neutral differential equations. Appl Math Lett 22:31–35
223. Naresh R, Tripathi A, Tchuenche JM, Sharma D (2009) Stability analysis of a time delayed SIR epidemic model with nonlinear incidence rate. Comput Math Appl 58(2):348–359
224. Nicholson AJ (1954) An outline of the dynamics of animal populations. Aust J Zool 2:9–65
225. Niculescu S-I (2001) Delay effects on stability: a robust control approach. Springer, Berlin
226. Niculescu S-I, Gu K (eds) (2004) Advances in time-delay systems. Springer, Berlin
227. Oswald AJ, Powdthavee N (2007) Obesity, unhappiness and the challenge of affluence: theory and evidence. Econ J 117:F441–454
228. Ovseyevich AI (2006) The stability of an inverted pendulum when there are rapid random oscillations of the suspension point. Int J Appl Math Mech 70:762–768
229. Park JH (2004) Delay-dependent criterion for asymptotic stability of a class of neutral equations. Appl Math Lett 17:1203–1206
230. Park JH (2005) Delay-dependent criterion for guaranteed cost control of neutral delay systems. J Optim Theory Appl 124:491–502
231. Park JH, Kwon OM (2008) Stability analysis of certain nonlinear differential equation. Chaos Solitons Fractals 27:450–453
232. Paternoster B, Shaikhet L (2008) Stability of equilibrium points of fractional difference equations with stochastic perturbations. Adv Differ Equ 2008:718408. doi:10.1155/2008/718408, 21 pages
233. Peics H (2000) Representation of solutions of difference equations with continuous time. Electron J Qual Theory Differ Equ 21:1–8. Proceedings of the 6th colloquium of differential equations
234. Pelyukh GP (1996) A certain representation of solutions to finite difference equations with continuous argument. Differ Uravn (Minsk) 32(2):256–264. Translated in Differential Equations 32(2):260–268 (1996)
235. Peschel M, Mende W (1986) The predator–prey model: do we live in a Volterra world. Akademie-Verlag, Berlin
236. Philos ChG (1998) Asymptotic behavior, nonoscillation and stability in periodic first-order linear delay differential equations. Proc R Soc Edinb A 128:1371–1387
237. Philos ChG, Purnaras IK (2001) Periodic first order linear neutral delay differential equations. Appl Math Comput 117:203–222
238. Philos ChG, Purnaras IK (2004) Asymptotic properties, nonoscillation and stability for scalar first-order linear autonomous neutral delay differential equations. Electron J Differ Equ 2004(3):1–17
239. Razumikhin BS (1956) About stability of systems with delay. Prikl Mat Meh 20(4):500–512 (in Russian)
240. Razumikhin BS (1958) About stability in the first approximation of systems with delay. Prikl Mat Meh 22(2):155–166 (in Russian)
241. Razumikhin BS (1960) Application of Lyapunov method to problems of stability of systems with delay. Avtom Telemeh 21(6):740–748 (in Russian)
242. Razumikhin BS (1966) Method of stability investigation of systems with delay. Dokl Akad Nauk SSSR 167(6):1234–1236 (in Russian)
243. Repin YuM (1965) Square Lyapunov functionals for systems with delay. Prikl Mat Meh 3:564–566 (in Russian)
244. Resnick SI (1992) Adventures in stochastic processes. Birkhauser, Boston
245. Roach GF (ed) (1984) Mathematics in medicine and biomechanics. Shiva, Nautwick
246. Rodkina A, Schurz H, Shaikhet L (2008) Almost sure stability of some stochastic dynamical systems with memory. Discrete Contin Dyn Syst 21(2):571–593
247. Rosenquist JN, Murabito J, Fowler JH, Christakis NA (2010) The spread of alcohol consumption behavior in a large social network. Ann Intern Med 152(7):426–433

248. Röst G, Wu J (2007) Domain decomposition method for the global dynamics of delay differential equations with unimodal feedback. Proc R Soc Lond Ser A, Math Phys Sci 463:2655–2669

249. Ruan S, Xiao D (2001) Global analysis in a predator–prey system with nonmonotonic functional response. SIAM J Appl Math 61:1445–1472

250. Saker SH (2005) Oscillation of continuous and discrete diffusive delay Nicholson's blowflies models. Appl Math Comput 167(1):179–197

251. Saker SH, Agarwal S (2002) Oscillation and global attractivity in a periodic Nicholson's blowflies model. Math Comput Model 35:719–731

252. Saker SH, Zhang BG (2002) Oscillation in a discrete partial delay Nicholson's blowflies model. Math Comput Model 36(9–10):1021–1026

253. Santonja F-J, Shaikhet L (2012) Analysing social epidemics by delayed stochastic models. Discrete Dyn Nat Soc 2012:530472. 13 pages. doi:10.1155/2012/530472

254. Santonja F-J, Sanchez E, Rubio M, Morera J-L (2010) Alcohol consumption in Spain and its economic cost: a mathematical modeling approach. Math Comput Model 52:999–1003

255. Santonja F-J, Villanueva R-J, Jodar L, Gonzalez-Parra G (2010) Mathematical modelling of social obesity epidemic in the region of Valencia, Spain. Math Comput Model Dyn Syst 16(1):23–34

256. Sanz-Serna JM (2008) Stabilizing with a hammer. Stoch Dyn 8:47–57

257. Sarkar RR, Banerjee S (2005) Cancer self remission and tumor stability—a stochastic approach. Math Biosci 196:65–81

258. Scheutzow M (1996) Stability and instability for stochastic systems with time delay. Z Angew Math Mech 76(3):363–365

259. Scheutzow M (2005) Exponential growth rates for stochastic delay differential equations. Stoch Dyn 5(2):163–174

260. Schurz H (1996) Asymptotical mean square stability of an equilibrium point of some linear numerical solutions with multiplicative noise. Stoch Anal Appl 14:313–354

261. Shaikhet L (1975) Stability investigation of stochastic systems with delay by Lyapunov functionals method. Probl Pereda Inf 11(4):70–76 (in Russian)

262. Shaikhet L (1995) On the stability of solutions of stochastic Volterra equations. Avtom Telemeh 56(8):93–102 (in Russian). Translated in Automatic Remote Control 56(8):1129–1137 (1995), part 2

263. Shaikhet L (1995) Stability in probability of nonlinear stochastic hereditary systems. Dyn Syst Appl 4(2):199–204

264. Shaikhet L (1995) Stability in probability of nonlinear stochastic systems with delay. Mat Zametki 57(1):142–146 (in Russian). Translation in Math. Notes 57(1–2):103–106

265. Shaikhet L (1996) Modern state and development perspectives of Lyapunov functionals method in the stability theory of stochastic hereditary systems. Theory Stoch Process 2(18)(1–2):248–259

266. Shaikhet L (1996) Stability of stochastic hereditary systems with Markov switching. Theory Stoch Process 2(18)(3–4):180–185

267. Shaikhet L (1998) Stability of predator–prey model with aftereffect by stochastic perturbations. Stab Control: Theory Appl 1(1):3–13

268. Shaikhet L (2002) Numerical simulation and stability of stochastic systems with Markovian switching. Neural Parallel Sci Comput 10(2):199–208

269. Shaikhet L (2004) About Lyapunov functionals construction for difference equations with continuous time. Appl Math Lett 17(8):985–991

270. Shaikhet L (2004) Construction of Lyapunov functionals for stochastic difference equations with continuous time. Math Comput Simul 66(6):509–521

271. Shaikhet L (2004) Lyapunov functionals construction for stochastic difference second kind Volterra equations with continuous time. Adv Differ Equ 2004(1):67–91. doi:10.1155/S1687183904308022

272. Shaikhet L (2005) General method of Lyapunov functionals construction in stability investigations of nonlinear stochastic difference equations with continuous time. Stoch Dyn

5(2):175–188. Special Issue Stochastic Dynamics with Delay and Memory

273. Shaikhet L (2005) Stability of difference analogue of linear mathematical inverted pendulum. Discrete Dyn Nat Soc 2005(3):215–226

274. Shaikhet L (2006) About stability of a difference analogue of a nonlinear integro–differential equation of convolution type. Appl Math Lett 19(11):1216–1221

275. Shaikhet L (2008) Stability of a positive point of equilibrium of one nonlinear system with aftereffect and stochastic perturbations. Dyn Syst Appl 17:235–253

276. Shaikhet L (2009) Improved condition for stabilization of controlled inverted pendulum under stochastic perturbations. Discrete Contin Dyn Syst 24(4):1335–1343. doi:10.3934/dcds.2009.24.1335

277. Shaikhet L (2011) About an unsolved stability problem for a stochastic difference equation with continuous time. J Differ Equ Appl 17(3):441–444. doi:10.1080/10236190903489973. (iFirst article, 2010, 1–4, 1563–5120, First published on 05 March 2010)

278. Shaikhet L (2011) Lyapunov functionals and stability of stochastic difference equations. Springer, London

279. Shaikhet L (2012) Two unsolved problems in the stability theory of stochastic differential equations with delay. Appl Math Lett 25(3):636–637. doi:10.1016/j.aml.2011.10.002

280. Shaikhet L, Roberts J (2004) Stochastic Volterra integro–differential equations: stability and numerical methods. University of Manchester. MCCM. Numerical Analysis Report 450, 38p

281. Shaikhet L, Roberts J (2011) Asymptotic stability analysis of a stochastic Volterra integro-differential equation with fading memory. Dyn Contin Discrete Impuls Syst, Ser B, Appl Algorithms 18:749–770

282. Sharp R, Tsai Y-H, Engquist B (2005) Multiple time scale numerical methods for the inverted pendulum problem. In: Multiscale methods in science and engineering. Lecture notes computing science and engineering, vol 44. Springer, Berlin, pp 241–261

283. Shi X, Zhou X, Song X (2010) Dynamical properties of a delay prey-predator model with disease in the prey species only. Discrete Dyn Nat Soc. Article ID 196204, 16 pages

284. So JW-H, Yang Y (1998) Dirichlet problem for the diffusive Nicholson's blowflies equation. J Differ Equ 150(2):317–348

285. So JW-H, Yu JS (1994) Global attractivity and uniformly persistence in Nicholson's blowflies. Differ Equ Dyn Syst 2:11–18

286. So JW-H, Yu JS (1995) On the stability and uniform persistence of a discrete model of Nicholson's blowflies. J Math Anal Appl 193(1):233–244

287. So JW-H, Wu J, Yang Y (2000) Numerical steady state and Hopf bifurcation analysis on the diffusive Nicholson's blowflies equation. Appl Math Comput 111(1):53–69

288. Song XY, Chen LS (2002) Optimal harvesting and stability for a predator–prey system with stage structure. Acta Math Appl Sin 18(3):423–430 (English series)

289. Song XY, Jiang Y, Wei HM (2009) Analysis of a saturation incidence SVEIRS epidemic model with pulse and two time delays. Appl Math Comput 214:381–390

290. Sopronyuk FO, Tsarkov EF (1973) About stability of linear differential equations with delay. Dokl Ukr Akad Nauk, Ser A 4:347–350 (in Ukrainian)

291. Spanish Ministry of Health (2008) National drug observatory reports. Retrieved 13 November 2008 from http://www.pnsd.msc.es/Categoria2/observa/estudios/home.htm

292. Sun YG, Wang L (2006) Note on asymptotic stability of a class of neutral differential equations. Appl Math Lett 19:949–953

293. Swords C, Appleby JAD (2010) Stochastic delay difference and differential equations: applications to financial markets. LAP Lambert Academic, Saarbrücken

294. Takeuchi Y, Ma W, Beretta E (2000) Global asymptotic properties of a delay SIR epidemic model with finite incubation times. Nonlinear Anal 42:931–947

295. Tamizhmani KM, Ramani A, Grammaticos B, Carstea AS (2004) Modelling AIDS epidemic and treatment with difference equations. Adv Differ Equ 3:183–193

296. Tapaswi PK, Chattopadhyay J (1996) Global stability results of a "susceptible-infective-immune-susceptible" (SIRS) epidemic model. Ecol Model 87(1–3):223–226

297. Thomas DM, Robbins F (1999) Analysis of a nonautonomous Nicholson blowfly model. Physica A 273(1–2):198–211
298. Tsypkin YaZ (1946) Stability of systems with a delayed feed-back. Avtom Telemeh 7(2–3):107–129 (in Russian)
299. Tsypkin YaZ (1947) Degree of stability of systems with a delayed feed-back. Avtom Telemeh 8(3):145–155 (in Russian)
300. Tsypkin YaZ (1984) An adaptive control of objects with aftereffect. Nauka, Moscow (in Russian)
301. Verriest EI (2001) New qualitative methods for stability of delay systems. Kybernetika 37:229–238
302. Volterra V (1931) Lesons sur la theorie mathematique de la lutte pour la vie. Gauthier-Villars, Paris
303. Volz R (1982) Global asymptotic stability of a periodical solution to an epidemic model. J Math Biol 15:319–338
304. Wang WD, Chen LS (1997) A predator–prey system with stage structure for predator. Comput Math Appl 33(8):83–91
305. Wang LL, Li WT (2003) Existence and global stability of positive periodic solutions of a predator–prey system with delays. Appl Math Comput 146(1):167–185
306. Wang LL, Li WT (2004) Periodic solutions and stability for a delayed discrete ratio-dependent predator–prey system with Holling-type functional response. Discrete Dyn Nat Soc 2004(2):325–343
307. Wang C, Wei J (2008) Bifurcation analysis on a discrete model of Nicholson's blowflies. J Differ Equ Appl 14:737–746
308. Wang QG, Lee TH, Tan KK (1999) Finite-spectrum assignment for time-delay systems. Lecture notes in control and inform sciences. Springer, Berlin
309. Wang Q, Fan M, Wang K (2003) Dynamics of a class of nonautonomous semi-ratio-dependent predator–prey system with functional responses. J Math Anal Appl 278:443–471
310. Wang W, Xin J, Zhang F (2010) Persistence of an SEIR model with immigration dependent on the prevalence of infection. Discrete Dyn Nat Soc. doi:10.1155/2010/727168. Article ID 727168, 7 pages
311. Wangersky PJ, Cunningham WJ (1957) Time lag in prey–predator population models. Ecology 38(1):136–139
312. Wei J, Li MY (2005) Hopf bifurcation analysis in a delayed Nicholson blowflies equation. Nonlinear Anal 60(7):1351–1367
313. Wolkenfelt PHM (1982) The construction of reducible quadrature rules for Volterra integral and integro–differential equations. IMA J Numer Anal 2:131–152
314. Xiao D, Ruan S (2001) Multiple bifurcations in a delayed predator–prey system with non-monotonic functional response. J Differ Equ 176:494–510
315. Yang J, Wang X (2010) Existence of a nonautonomous SIR epidemic model with age structure. Adv Differ Equ. Article ID 212858, 23 pages
316. Yi T, Zou X (2008) Global attractivity of the diffusive Nicholson blowflies equation with Neumann boundary condition: a nonmonotone case. J Differ Equ 245(11):3376–3388
317. Zeng X (2007) Non-constant positive steady states of a prey–predator system with cross-diffusions. J Math Anal Appl 332(2):989–1009
318. Zevin AA (2010) The necessary and sufficient conditions for the stability of linear systems with an arbitrary delay. J Appl Math Mech 74(4):384–388
319. Zhang J, Jin Z (2010) The analysis of epidemic network model with infectious force in latent and infected period. Discrete Dyn Nat Soc. Article ID 604329, 12 pages
320. Zhang BG, Xu HX (1999) A note on the global attractivity of a discrete model of Nicholson's blowflies. Discrete Dyn Nat Soc 3:51–55
321. Zhang X, Chen L, Neumann UA (2000) The stage-structured predator–prey model and optimal harvesting policy. Math Biosci 168:201–210
322. Zhang F, Li Z, Zhang F (2008) Global stability of an SIR epidemic model with constant infectious period. Appl Math Comput 199:285–291

323. Zhang TL, Liu JL, Teng ZD (2009) Dynamic behaviour for a nonautonomous SIRS epidemic model with distributed delays. Appl Math Comput 214:624–631
324. Zhang Z, Wu J, Suo Y, Song X (2011) The domain of attraction for the endemic equilibrium of an SIRS epidemic model. Math Comput Simul 81:1697–1706
325. Zuo W, Wei J (2011) Stability and Hopf bifurcation in a diffusive predator–prey system with delay effect. Nonlinear Anal, Real World Appl 12(4):1998–2011

Index

A

Additive stochastic perturbations, 297
Alcohol consumption, 297, 298, 307
Alcohol consumption model, 298
Algebraic adjuncts, 28
Algebraic equations, 260, 300
Appropriate characteristic equation, 24
Argument deviations, 1, 2
Asymptotic mean-square stability, vii, 43, 44,
47, 51, 53, 59, 62–64, 70, 71, 78, 119,
120, 123, 124, 130, 131, 134, 135, 146,
147, 160, 163, 169, 179, 187, 189, 194,
196, 198, 226, 251–254, 264, 285, 290,
291, 297, 302, 303, 305, 316, 318
Asymptotic mean-square stability conditions,
153, 189, 257
Asymptotic p-stability, 35
Asymptotic stability, 5, 7, 10, 18, 28, 45, 83,
97, 100, 102, 105, 116, 204, 211, 212,
215, 227, 229, 233
Asymptotically mean-square stable, 34, 36,
39–41, 48, 50, 52, 54, 76, 81, 86, 88,
91, 92, 94, 99, 101, 103, 106, 109, 110,
114, 116, 118, 122, 126, 154, 156, 162,
165, 167, 168, 170, 173, 180, 182, 183,
185, 190, 191, 193, 221, 266, 270, 272,
274, 280, 285–292, 304, 306, 307, 317,
319
Asymptotically p-stable, 34
Asymptotically p-trivial, 34, 35
Asymptotically stable, 18, 22, 24, 26, 52, 53,
56, 74, 86, 88, 96, 209, 214–216, 218,
223, 230, 232, 241, 245, 248, 285

B

Birth rate, 298
Brownian motion, 29

C

Characteristic equation, vi, 22, 25, 26, 66, 83,
214, 216, 219
Characteristic function, 227–229
Characteristic quasipolynomial, 5, 9, 13, 18,
215, 216
Constant delay, 52, 97
Constructing a Lyapunov functional, 171
Consumption of alcohol, 297
Controlled inverted pendulum, 209, 222, 226,
229
Controlled pendulum, 240

D

Death rate, 298
Delay, 321
Derivative, 229
Diagonal matrix, 75
Difference analogue, 12, 20, 24, 151, 199, 202,
229, 256, 275, 296, 309, 322
Difference equation, 1, 39
Difference equations with fractional
nonlinearity, 132
Differential equation, 5
Differential equation of neutral type, 2, 86, 88,
91, 124, 214
Differential equation with delay, 153, 210, 251
Differential equation with deviating argument,
2
Differential equation with fractional
nonlinearity, 132
Differential equation without delay, 91
Differential equation without memory, 303
Differential–difference equation, v, 2
Dirac's delta-function, 215, 309, 321
Dirac's function, 298
Discrete delay, 4, 298

L. Shaikhet, *Lyapunov Functionals and Stability of Stochastic Functional
Differential Equations*, DOI 10.1007/978-3-319-00101-2,
© Springer International Publishing Switzerland 2013